AF581636

Advances in Sensors and Sensing for Technical Condition Assessment and NDT

Advances in Sensors and Sensing for Technical Condition Assessment and NDT

Editors

Daria Wotzka
Tomasz Boczar
Michał Kunicki
Łukasz Nagi
Michał Kozioł

MDPI • Basel • Beijing • Wuhan • Barcelona • Belgrade • Manchester • Tokyo • Cluj • Tianjin

Editors

Daria Wotzka
Faculty of Electrical Engineering
Automatic Control and
Informatics
Opole University of Technology
Opole
Poland

Tomasz Boczar
Faculty of Electrical Engineering
Automatic Control and
Informatics
Opole University of Technology
Opole
Poland

Michał Kunicki
Faculty of Electrical Engineering
Automatic Control and
Informatics
Opole University of Technology
Opole
Poland

Łukasz Nagi
Faculty of Electrical Engineering
Automatic Control and
Informatics
Opole University of Technology
Opole
Poland

Michał Kozioł
Faculty of Electrical Engineering
Automatic Control and
Informatics
Opole University of Technology
Opole
Poland

Editorial Office
MDPI
St. Alban-Anlage 66
4052 Basel, Switzerland

This is a reprint of articles from the Special Issue published online in the open access journal *Sensors* (ISSN 1424-8220) (available at: www.mdpi.com/journal/sensors/special_issues/sens_NDT).

For citation purposes, cite each article independently as indicated on the article page online and as indicated below:

LastName, A.A.; LastName, B.B.; LastName, C.C. Article Title. *Journal Name* **Year**, *Volume Number*, Page Range.

ISBN 978-3-0365-2679-9 (Hbk)
ISBN 978-3-0365-2678-2 (PDF)

Contents

Editorial

Latest Trends in the Improvement of Measuring Methods and Equipment in the Area of NDT

Daria Wotzka *, Michał Kozioł, Tomasz Boczar, Michał Kunicki and Łukasz Nagi

Faculty of Electrical Engineering Automatic Control and Informatics, Opole University of Technology, 45-758 Opole, Poland; m.koziol@po.edu.pl (M.K.); t.boczar@po.edu.pl (T.B.); m.kunicki@po.edu.pl (M.K.); l.nagi@po.edu.pl (Ł.N.)

* Correspondence: d.wotzka@po.edu.pl

The adequate assessment of key apparatus conditions is a hot topic in all branches of industry. Various on-line and off-line diagnostic methods are widely applied to provide early detections of any abnormality in exploitation. Furthermore, different sensors may also be applied to capture selected physical quantities that may be used to indicate the type of potential fault. The essential steps of the signal analysis regarding the technical condition assessment process may be listed as: signal measurement (using relevant sensors), processing, modelling, and classification. In the Special Issue entitled "Advances in Sensors and Sensing for Technical Condition Assessment and NDT", we present the latest research in various areas of technology.

First, we present the outcomes of the teams of researchers who described the results of attempts to improve measurement methods and create devices to reduce errors and enable operations in real-time. In [1], authors deal with the area of high-current pulse measurement, particularly in cases where the current shunts that were used exhibited a limited frequency response. Pulse current values of the order of several kA, recorded with the shunt and Rogowski coil, indicate significant differences in the waveforms. Based on the theoretical and experimental research, the authors proposed a method that can be applied in real-time due to its simplicity. Their method, which greatly improves the accuracy of current wave measurement, uses a shunt circuit model created in ANSYS/Q3D to obtain the correct current wave from the solution of the ordinary differential equation (ODE). The article [2] addresses the issues related to linear displacement measurement systems. The authors presented a new method for the real-time compensation of geometric and thermal errors of an optical linear encoder. The algorithm of operation was designed based on the results of theoretical and experimental studies. A regression process was applied using a parametric function model under various ambient temperature dependences and a field programmable gate array (FPGA) for computation. Process quality and machine performance depend on the accuracy of the integrated linear encoder. The authors proposed a method that can reduce positioning error by as much as 98%. They point towards the need for further work related to other encoder designs and mounting methods. In [3], the authors deal with the area of bridge diagnostics and, in particular, the effect of radiation intensity on the stresses that occur in a five-span continuous prestressed-concrete box–girder bridge structure. The authors recorded solar radiation and surface temperatures, wind speeds, displacement, and strain for a period of 12 months and performed a correlation analysis, which showed that the most significant parameter affecting the tensile stress in the material is the lateral temperature gradient. The authors also performed finite-element analyses and developed a simplified numerical model that they plan to expand in the future. In [4], the authors present a Fizeau fiber interferometer to detect internal defects in aluminum plates using ultrasonography. The authors of [5] consider a combination of piezocomposite transducers and pulse-compression-processing methods to study the propagation of ultrasonic signals through steel pipes insulated with

Citation: Wotzka, D.; Kozioł, M.; Boczar, T.; Kunicki, M.; Nagi, Ł. Latest Trends in the Improvement of Measuring Methods and Equipment in the Area of NDT. *Sensors* **2021**, *21*, 7293. https://doi.org/10.3390/s21217293

Received: 18 October 2021
Accepted: 29 October 2021
Published: 2 November 2021

different materials. Using a combination of both methods, the SNR could be increased when measuring the signals reflected from insulated and cladded steel pipes. They prove that the application of ultrasound in the 100–400 kHz frequency range can be successfully applied for diagnostic purposes.

Artificial intelligence methods are being increasingly used in the field of diagnostics. In [6], the authors consider the electromagnetic non-destructive evaluation of conductive materials. They develop a new algorithm for eddy current testing (ECT), which uses wavelet processing and neural networks to study electromagnetic field fluctuations. They examine material samples made of austenitic stainless steel. They also focus on developing a method that is robust to disturbances and can reduce the error evaluation of material defect depth by 10%, even when the signal-to-noise ratio reaches 10 dB. In [7], the authors aim to improve the diagnosis of on-load tap changers (OLTC) using the acoustic emission method. They investigated the impact of the measuring device parameters on the accuracy of the classification of various OLTC defects. The authors applied time, frequency, and wavelet transformation methods to determine the features for the purpose of classification using various machine learning methods. In [8], the authors depict an automatic method to detect the model, type, and power of a distribution transformer based on the low-frequency noise emitted by the device. They applied machine learning and genetic algorithms for the classification of 16 different distribution transformers, achieving high accuracy.

Infrasound was also considered for the study of wind turbine noise. In [9], the authors looked at the issue of infrasound noise emitted by a 2 MW wind turbine. They described a measurement system that allows for the simultaneous measurement of noise at three locations, as well as signal processing methods, including wavelet coherence. The applied methods enabled determination of the differences and similarities in the noise levels that were registered around the turbine under consideration.

In the rest of the book, we include the results of work aiming to improve the measurement methods and advanced processing tools in various diagnostic areas. In [10], a methodology is presented for measuring and analyzing data to identify a tramway line's axis system and support railway infrastructure management processes in planning and maintenance. The authors conducted experimental work using a global navigation satellite system (GNSS). Their algorithm uses multi-device data and repeated position recordings for identification of the main track directions of the tramway. They emphasized the assessment of the signal's dispersion and repeatability using residuals in relation to the estimated track's direction. In [11], the authors considered medical imaging methods that are capable of noninvasively detecting the compositions and variations of muscles. Their aim was to develop a method for the precise diagnosis of a specific muscular disease. They proposed a window-modulated compounding (WMC) Nakagami parameter ratio imaging and a time–Nakagami parameter ratio curve (TNRC) approach for the estimation of perfusion parameters, which can be used for the diagnosis of a specific muscle disease. Their methods provide reproducibility and robustness and a better sensitivity and tolerance of tissue clutters than conventional methods. Article [12] considers the application of classification methods in the field of joint diagnosis. The authors recorded Vibroarthrography signals, assigned to three stages of patella chondromalacia, osteoarthritis, and a healthy knee joint. The authors developed ten new features, calculated in the frequency domain, which were subjected to classification using ten different classification algorithms. The calculated Bhattacharyya coefficient values indicated an average of 9% better accuracy.

The final research area of diagnostics included in this Special Issue is monitoring the hatching process of eggs. The authors of [13] developed a prototype device, whose effectiveness was confirmed in verification studies. The advantage of the device is its ability to be used in industry for the fast and accurate detection of unfertilized eggs and dead-in-shell eggs on the ninth day of hatching. In this study, the authors used an image-processing method and a spectrophotometry technique in the 580 nm-wavelength LED range as the detection light source.

Funding: This research received no external funding.

Acknowledgments: We would like to thank the authors for their contribution to the creation of this Special Issue, the reviewers for their valuable work in improving the quality of the submitted articles and the editors of *Sensors* for their kind help and support.

Conflicts of Interest: The authors declare no conflict of interest.

References

1. Piekielny, P.; Waindok, A. Using a Current Shunt for the Purpose of High-Current Pulse Measurement. *Sensors* **2020**, *21*, 1835. [CrossRef] [PubMed]
2. Gurauskis, D.; Kilikevičius, A.; Kasparaitis, A. Thermal and Geometric Error Compensation Approach for an Optical Linear Encoder. *Sensors* **2020**, *21*, 360. [CrossRef] [PubMed]
3. Lei, X.; Fan, X.; Jiang, H.; Zhu, K.; Zhan, H. Temperature Field Boundary Conditions and Lateral Temperature Gradient Effect on a PC Box-Girder Bridge Based on Real-Time Solar Radiation and Spatial Temperature Monitoring. *Sensors* **2020**, *21*, 5261. [CrossRef] [PubMed]
4. Liu, W.; Yan, W.; Liu, H.; Tong, C.; Fan, Y. Internal Cylinder Identification Based on Different Transmission of Longitudinal and Shear Ultrasonic Waves. *Sensors* **2020**, *21*, 723. [CrossRef] [PubMed]
5. Hutchins, D.A.; Watson, R.L.; Davis, L.A.J.; Akanji, L.; Billson, D.R.; Burrascano, P.; Laureti, S.; Ricci, M. Ultrasonic Propagation in Highly Attenuating Insulation Materials. *Sensors* **2020**, *21*, 2285. [CrossRef] [PubMed]
6. Smetana, M.; Behun, L.; Gombarska, D.; Janousek, L. New Proposal for Inverse Algorithm Enhancing Noise Robust Eddy-Current Non-Destructive Evaluation. *Sensors* **2020**, *21*, 5548. [CrossRef] [PubMed]
7. Wotzka, D.; Cichoń, A. Study on the Influence of Measuring AE Sensor Type on the E ff ectiveness of OLTC Defect Classification. *Sensors* **2020**, *21*, 3095. [CrossRef] [PubMed]
8. Jancarczyk, D.; Bernaś, M.; Boczar, T. Distribution Transformer Parameters Detection Based on Low-Frequency Noise, Machine Learning Methods, and Evolutionary Algorithm. *Sensors* **2020**, *21*, 4332. [CrossRef] [PubMed]
9. Boczar, T.; Zmarzły, D.; Kozioł, M. Application of Correlation Analysis for Assessment of Infrasound Signals Emission by Wind Turbines. *Sensors* **2020**, *21*, 6891. [CrossRef] [PubMed]
10. Wilk, A.; Specht, C.; Koc, W.; Karwowski, K.; Skibicki, J.; Szmagliński, J.; Chrostowski, P.; Dabrowski, P.; Specht, M.; Zienkiewicz, M.; et al. Evaluation of the Possibility of Identifying a Complex Polygonal Tram Track Layout Using Multiple Satellite Measurements. *Sensors* **2020**, *21*, 4408. [CrossRef] [PubMed]
11. Lin, H.-C.; Wang, S.-H. Window-Modulated Compounding Nakagami Parameter Ratio Approach for Assessing Muscle Perfusion with Contrast-Enhanced Ultrasound Imaging. *Sensors* **2020**, *21*, 3584. [CrossRef] [PubMed]
12. Szmajda, M.; Łysiak, A.; Froń, A.; Bączkowski, D. Vibroarthrographic Signal Spectral Features in 5-Class Knee Joint Classification. *Sensors* **2020**, *21*, 5015. [CrossRef]
13. Tsai, S.-Y.; Li, C.-H.; Jeng, C.-C.; Cheng, C.-W. Quality Assessment during Incubation Using Image Processing. *Sensors* **2020**, *21*, 5951. [CrossRef] [PubMed]

Article

Using a Current Shunt for the Purpose of High-Current Pulse Measurement

Pawel Piekielny † and Andrzej Waindok *,†

Department of Electrical Engineering and Mechatronics, Faculty of Electrical Engineering, Automatic Control and Informatics, Opole University of Technology, PL-45758 Opole, Poland; p.piekielny@po.edu.pl
* Correspondence: a.waindok@po.edu.pl
† These authors contributed equally to this work.

Abstract: Measurement of high-current pulses is crucial in some special applications, e.g., electrodynamic accelerators (EA) and converters. In such cases, the current shunts have limitations concerning the frequency bandwidth. To overcome the problem, a method based on the shunt mathematical model is proposed. In the method, the solution of ordinary differential equations for the RL circuit is carried out in order to obtain the real current shape. To check the method, as a referee, a Rogowski coil dedicated to measuring high-current pulses was used. Additionally, the measurement results were compared with the mathematical model of the tested power supply system. Measurements were made for the short power supply circuit, which allows eliminating the nonlinearity. The calculations were carried out using a circuit model. In order to obtain the parameters of the shunt (resistance and inductance), it was modeled using an ANSYS/Q3D Extractor software. Comparison of calculation and measurement results confirms the correctness of our method. In order to compare results, the normalized root mean square error (NRMSE) was used.

Keywords: high-current pulses; current measurement; high-current shunts; circuit modeling; Rogowski coil; electrodynamic accelerators

Citation: Piekielny, P.; Waindok, A. Using a Current Shunt for the Purpose of High-Current Pulse Measurement. *Sensors* **2021**, *21*, 1835. https://doi.org/10.3390/s21051835

Academic Editor: Pavel Mlejnek

Received: 20 December 2020
Accepted: 2 March 2021
Published: 6 March 2021

1. Introduction

Design or optimization of electrical devices requires formulation of the correct mathematical model. Thus, an important step in the research is the measurement verification of the obtained results [1,2]. Measurements of excitation current for electrodynamic accelerators and devices powered by short-term current pulses for some milliseconds (or shorter) with an amplitude of kA to even MA are difficult due to the high dynamics of the waveform [3,4]. For current measurements with a peak value of hundreds of kA, current shunts and Rogowski coils are mainly used [5]. However, using shunts for current measurement involves intrusion in the circuit. In cases of measurements in circuits with very low resistance (power supply systems for EA), the ratio of shunt and circuit resistance is relatively high [1].

Measuring shunts enable precise measurement of the peak value and shape of the waveform of relatively fast current pulses. Current shunts are characterized by a wide frequency response, short rise time, and high accuracy of resistance values—they are made in the accuracy class from 0.05 to 0.2. Below the upper frequency limit (−3dB), the measuring shunts behave like typical DC resistors with a frequency-independent resistance value. Depending on the design used, they may be insensitive to electromagnetic interference. The design of shunts allows the dissipation of relatively high power. The voltage drop across the shunt is proportional to the flowing current. The shunt has a low value of passive parasitic components, i.e., a low value of capacitance and inductance. These parameters are particularly reduced in shunts adapted to alternating currents (ACs). As a result, in practice, capacity and inductance do not affect the quality of measurements in the range from constant signals to signals with a frequency of several kHz [5–9]. For these reasons, manufacturers provide only the resistance value of the shunt. The coaxial shunt resistors are characterized by a relatively large parasitic inductance compared to the wire constructions (which are used in the presented research) [10]. Various mathematical models of current shunts were analyzed and some tests were conducted to determine their full parameters. In [11], the equivalent inductance of the cagelike shunts was measured against a set of four-terminal resistors. In [12], a simulated annealing algorithm was used in order to obtain the LCR parameters, which minimize the mean squared error across different model schemes and real measurements. In both publications, the measurements are the base for the inductance determination. None of the mentioned publications deal with the problem of eliminating the measurement error, arising due to the parasitic inductance of the shunt in the case of high-current pulses. The publications are fixed on the determination of shunt parameters.

There are many different works about Rogowski coils. Current measurements of them can be influenced by many factors ranging from their construction, the shape of the coil loop, and the location of the measured circuit in the coil loop [13–15]. Accurate calibration methods have been developed for Rogowski coils dedicated to pulsed power systems [16,17]. Precise calibration allows for obtaining more accurate measurement results. Traditional electromagnetic current sensors cannot accurately measure high-amplitude currents, which is due to the problem of saturation of the magnetic core. In the case of Rogowski coils, this problem does not occur because the flexible core of the coil is made of nonferromagnetic materials, and its turns are evenly wound on this core. Rogowski coils are used for noninvasive current measurement when other measuring methods, such as current shunts or current transformers, are impracticable. Due to their noninvasiveness, Rogowski coils are widely used in diagnostics [18–22]. They are devoted for measuring alternating currents in the range from several hundred mA to several hundred kA and frequencies from tenths of Hz to several MHz. The output voltage from the coil is proportional to the rate of current change [23]. Thus, in our investigations, the Rogowski coil was used as a referee for the calculation results and shunt measurements.

In the presented paper, a comparison of both measuring methods for high-current pulses occurring in electrodynamic accelerators (EA) was carried out. A problem was observed with the shape of a current wave measured by the shunt—there was a relatively high step of current value, which is not typical for an RL load. This step in the current wave is not visible in the model of the system and in measurements using the Rogowski coil. Including the small inductance of the shunt in the mathematical model, the same sharp increase was observed. Based on theoretical and experimental investigations, a method for solving this problem was proposed. It uses a circuit model of the shunt in order to obtain the proper current wave based on the solution of an ordinary differential equation (ODE). The presented method corrects the aberration of the shunt reading caused by its inductance. The correction significantly improves the accuracy of the current wave measurement, which was shown in Sections 7 and 8.

2. Mathematical Model of the Current Shunt

The mathematical model of the investigated current shunt is based on its simplified equivalent circuit presented in Figure 1. The voltage drop on the shunt is determined based on the following ordinary differential equation (ODE):

$$L\frac{di}{dt} + Ri = LC\frac{d^2u}{dt^2} + RC\frac{du}{dt} + u. \tag{1}$$

The circuit parameters of the model could be determined either by calculation or measurement methods. To determine parasitic parameters by calculations, it is necessary to use electromagnetic field simulation tools, such as ANSYS Maxwell 3D, FEMM, or Opera 3D, which are based on finite element analysis (FEA) techniques to solve the Maxwell differential field equations. However, when the geometry of the test object becomes more complex, full-field simulation is computationally intensive, the models themselves often show poor convergence, and results can be affected by significant errors. The Partial Equivalent Circuit Method (PEEC), which uses Maxwell's integral equations instead of differential equations and analytically calculates inductance and capacitance based on geometry and material information, is a very common solution [24–26]. Such a reduced model can significantly reduce the simulation cost. Therefore, the PEEC method is suitable for simulating high-level objects or in cases where large numbers of design iterations are involved [24–26].

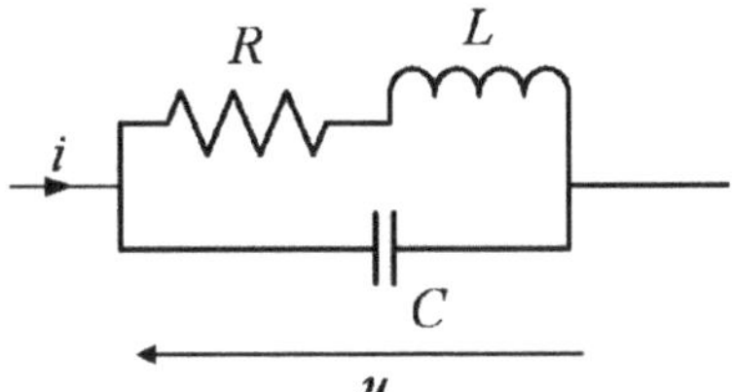

Figure 1. Simplified circuit model of the shunt.

Due to the very small value of the shunt passive inductance and capacitance, and due to complex geometry of the current shunt, they were determined using a mathematical model formulated in the ANSYS Q3D Extractor program. It is a tool dedicated for calculating parasitic parameters in any 2D or 3D objects. The software allows calculating RLCG parameters (resistance, inductance, capacity, and conductivity) depending on the given frequency [27]. Q3D Extractor uses the method of moments (integral equations), partial element equivalent circuits (PEEC) method, and finite element method (FEM) to compute capacitive, conductance, inductance, and resistance matrices. It uses the fast multipole method (FMM) to accelerate the solution of integral equations efficiently. The PEEC method requires only geometric and material information. Q3D Extractor computes the full electromagnetic field pattern using the specified mesh and the electrical parameters from the computed field quantities. To include the skin and proximity effects, a nonuniform current distribution is realized by creating a series of layers and gradings based on the skin depth [28].

For numerical calculations of the power supply system dynamics, the circuit method was used [29–32]. The circuit model of the shunt was implemented in Matlab/Simulink software. Additionally, in order to compare the calculation results with measurement results, the power supply system was also modeled. The completed circuit model in Simulink is presented in Figure 2.

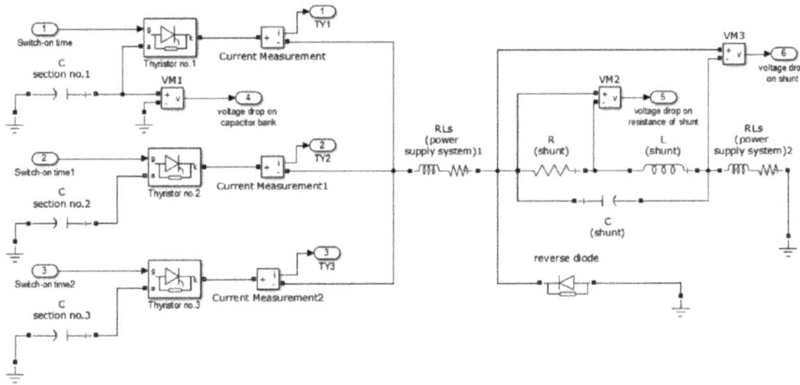

Figure 2. Circuit model of the power supply system and current shunt in Simulink .

Due to the relatively small pulse spectrum (less than 1 kHz), the capacitance value was neglected. Thus, the ODE could be written in a more simple form,

$$L\frac{di}{dt} + Ri = u. \tag{2}$$

The knowledge of the shunt parameters allows for recovering the real current waveform based on the Equation (2). Knowing the voltage wave, a current wave could be obtained by integrating the above equation. In our investigations, we implemented a simple Euler method:

$$i_{k+1} = i_k + h \cdot (u_k - \frac{R}{L} i_k), \tag{3}$$

where

- i_k—current value in k-th step;
- u_k—voltage value on the shunt at k-th step;
- h—integration step equal to the sampling time.

Due to its simplicity, the method allows us to determine the real current wave in real-time.

The Rogowski coil was chosen as a referee for current wave measurement. It is characterized by high bandwidth and precision. The measurements were also compared with the calculation results of the shunt model and supply system. In order to compare the measurement results quantitatively, the normalized root mean square error (NRMSE) was used. It is described by the following equation [33]:

$$NRMSE = \frac{\sqrt{\frac{1}{N}\sum_{i=0}^{N}(y_i^{meas} - y_i^{calc})^2}}{max(y_{meas}) - min(y_{meas})} \cdot 100\%, \tag{4}$$

where

- N—number of measurement points;
- y_i^{meas}—measured value in ith point;
- y_i^{calc}—calculated value in ith point.

3. Experimental Setup

Comparative analysis of the measurement methods was carried out using the real impulse power supply system of EA. As an energy source, a capacitor bank divided into three symmetric sections with a total capacity of 363 mF was used (Figures 3 and 4).

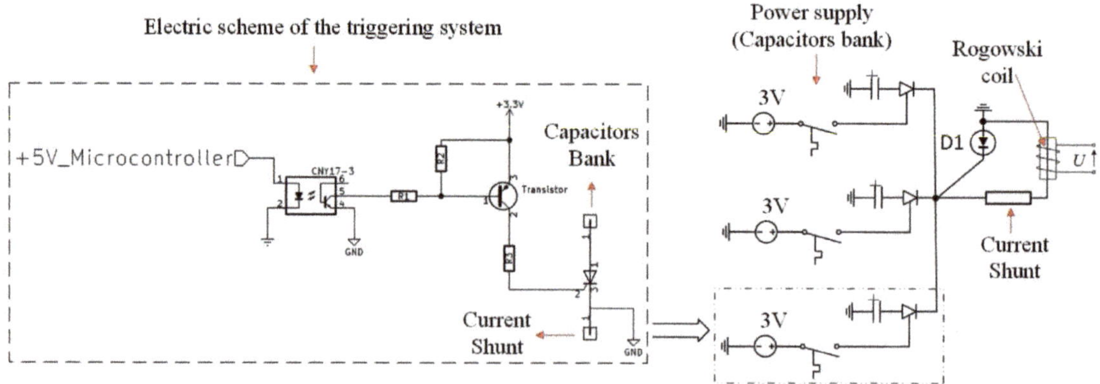

Figure 3. Electrical diagram of the power supply and measurement system.

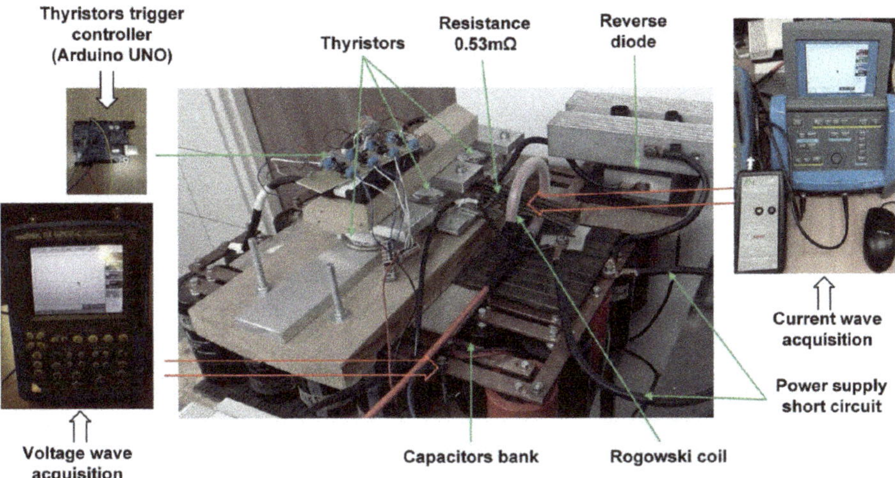

Figure 4. Photographs of the power supply and measurement system.

For switching-on of each section, high-power thyristors were implemented (model T95-1900 from Kubara LAMINA Company) [34]. Both voltage waves on the capacitor bank and current wave were measured using oscilloscopes. The current was measured simultaneously using a Rogowski coil and current shunt. Thus, the comparison of measurement results was possible.

The electric diagram of the test stand is presented in Figure 3. The reverse diode (D1) between the thyristor and the ground of the capacitors' bank is used to prevent capacitors charging in the negative direction. Due to the high current value, three thyristors were used for switching on. The signals from both measuring systems were recorded on the same oscilloscope. The tests started with switching on the thyristors. The capacitor discharged through the short circuit. The thyristors switched off after discharging of capacitors, and the current flowed through the reverse diode D1.

4. Description of the Measuring Current Sensors

4.1. Rogowski Coil—Type CWT1500

In Figure 5, the construction and principle of operation of the used Rogowski coil with a sensitivity of 0.02 mV/A is presented. The voltage induced in the coil is proportional to the rate of current change within the closed coil circuit. By integrating the voltage from the coil, a voltage value proportional to the current flowing in the circuit inside the coil loop can be obtained. The measurement error of the coil depends on the location of the measured circuit in the coil loop. The declared relative error for the central location of the measured circuit is in the order of 0.5%. Along with the change of the position of the perimeter regarding the edge of the coil loop, the measurement error increases. The loop does not need to be circular [23].

The induced voltage in a closed measuring coil is described by the following relationship:

$$e = \mu_0 NA\frac{di}{dt} = H\frac{di}{dt}, \tag{5}$$

where

- $H(Vs/A)$—coil sensitivity;
- $i(A)$—measured current in the circuit passing through the loop of coil;
- N—number of turns of the coil;
- $A(m^2)$—cross-sectional area of the coil core.

The output voltage U_{out} is described by the expression

$$U_{out} = \frac{1}{T_i} \int e dt = R_{SH} I, \quad (6)$$

where

- $T_i = R_0 C_1$—integrator time constant;
- $R_{SH} = \frac{H}{T_i}$—Rogowski coil sensitivity.

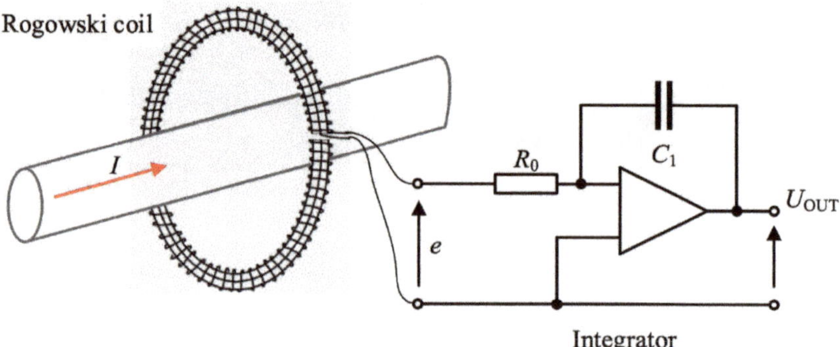

Figure 5. Construction and operation of the Rogowski coil—type CWT1500.

A dedicated data acquisition system was used in order to measure the voltage in the coil. The oscilloscope acquired the output voltage signal. Table 1 contains the most important parameters of the Rogowski coil used. The length of the coil loop is 300 mm. The converter has an output voltage of 6 V for the rated peak current of 300 kA. If the peak current exceeds this value, the integrator saturates and the measured waveform is incorrect (unlike an amplifier for which the output waveform is cut off). The Rogowski coil is connected to the integrator by a coaxial cable, being an inseparable product of the company Power Electronic Measurements Ltd. [35].

Table 1. Parameters of the used Rogowski coil—type CWT1500.

Sensitivity (mV/A)	Peak Current (kA)	Peak *di*/*dt* (kA/µs)	Droop Typ. (%ms)	LF (3 dB) Bandwidth (Hz)	HF (3 dB) Bandwidth (MHz)
0.02	300	40	0.035	0.03	16

4.2. Current Shunt

In Figure 6, the outline of the current shunt of accuracy 0.1 was presented. The shunt used in conjunction with the oscilloscope enables recording of impulses and shock current waveforms with values of the order of several hundred kA.

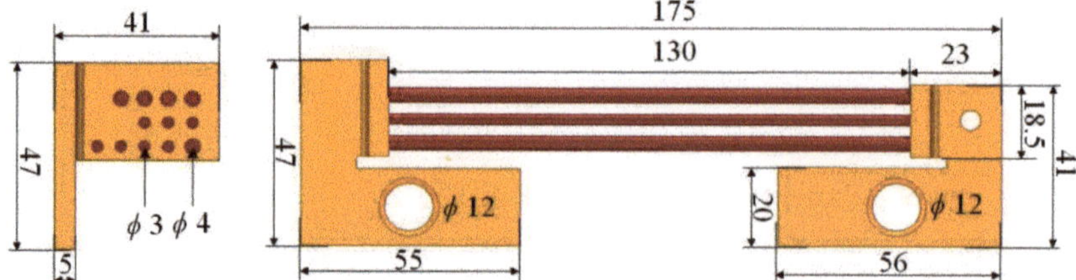

Figure 6. Outline of the investigated current shunt (dimensions in mm).

The current shunt is made of copper mounting bases and thrust rods. The following parameters were adopted for the calculations:

- copper conductivity $\sigma = 58 \cdot 10^6$ (S/m);
- rods' conductivity $\sigma = 23.2 \cdot 10^4$ (S/m).

The conductivity of rods is approximately 200 times lower compared with copper.

The numerical model in the Q3D software is presented in Figure 7. The boundary conditions, i.e., the source and sink were assumed on the bottom part of the terminals. In the investigated case, the pulse spectrum occupation is up to $f = 1$ (kHz). Thus, the skin depth of shunt rods is equal to

$$\delta = \frac{1}{\sqrt{\pi \cdot f \cdot \mu \cdot \sigma}} = \frac{1}{\sqrt{\pi \cdot 1000 \cdot 4\pi \cdot 10^{-7} \cdot 23.2 \cdot 10^4}} = 33 \text{ (mm)} \tag{7}$$

and does not influence the calculation results significantly (the rods' diameters do not exceed 4 (mm)). Although, the skin depth of copper is lower (2.1 (mm)), as the numerical tests show, thus influencing the current density distribution only weakly.

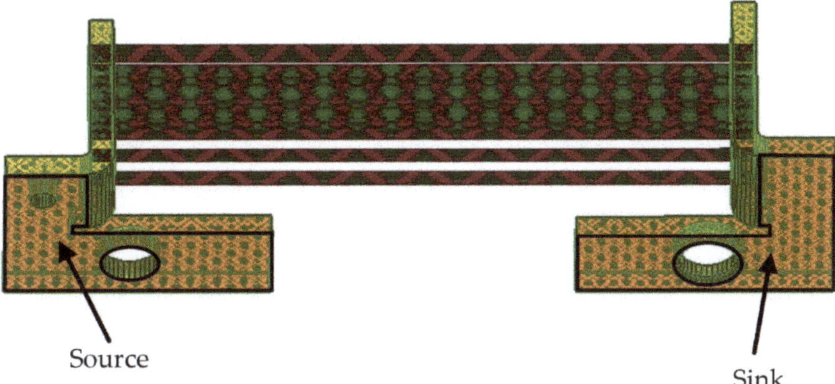

Figure 7. A Q3D model.

In rods of the investigated current shunt, the current density distribution is homogenous (Figure 8b). In the terminals, the highest value is observed in the curved part (Figure 8a). The current density value decreases along with the distance from curve. It is due to the fact that the total current flowing in the terminal is divided into the rod currents. The electric potential decreases approximately linearly along the shunt (Figure 9), which is due to linear conductors. Results of the shunt circuit parameters calculation, made for the DC case, are presented in Table 2.

The calculated shunt resistance was verified experimentally. A value of 0.53 mΩ was obtained, which is very close to the theoretical one. Measurements were carried out using a 24-bit measurement card NI9219 with a measurement range from −125 mV to 125 mV (the measurement resolution of the card was 7.45 nV). The current was measured using a current transducer LEM with a rated measuring current of $I_n = 50$ A. In addition, the shunt inductance value was determined experimentally and a 91.59 nH value was obtained. The value is 7.12 nH higher than the result of numerical calculations. However, the discrepancy is less than 10%. Both experimental and numerical results confirm that the value of the current shunt inductance is an important factor influencing the measurement of high-current pulses.

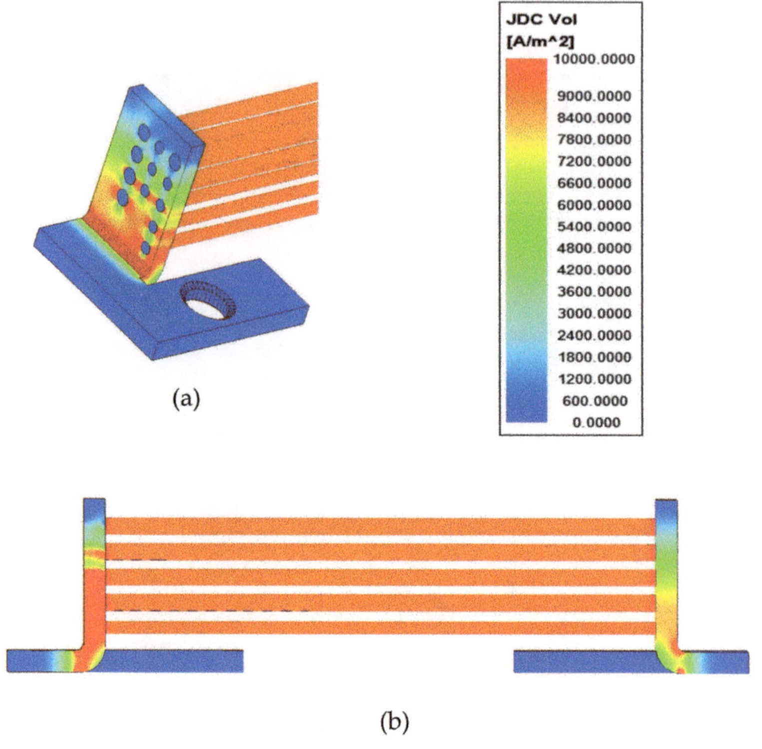

Figure 8. Current density distribution in the shunt: (**a**) front part of one terminal, (**b**) side view.

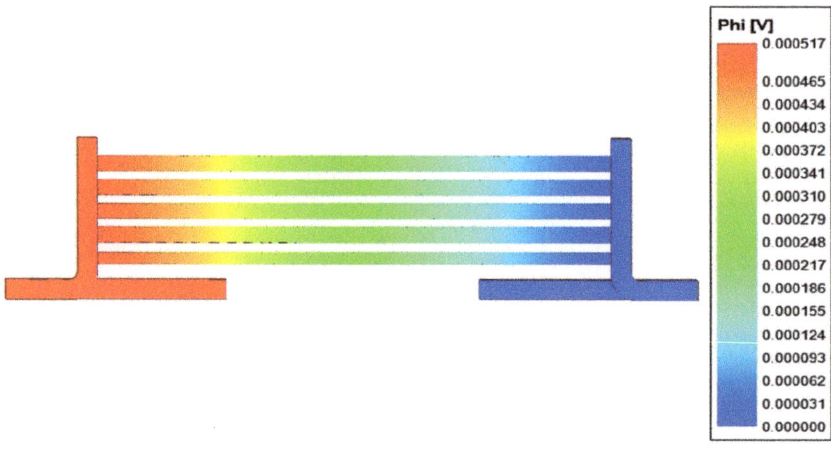

Figure 9. Potential distribution in the investigated shunt.

Table 2. Shunt circuit parameters—results of numerical calculations.

R (mΩ)	C (pF)	L (nH)
0.527	4.73	84.47

5. Comparison of Measurement Results Obtained from Rogowski Coil and Current Shunt

In the first step to compare the waveforms from the Rogowski coil and the current shunt, measurement repeatability was tested. In Figure 10, the waveforms of the currents from the Rogowski coil (Figure 10b) and current shunt (Figure 10a) are presented. The triggering of the impulse supply system was set to be the same for all waves. There are only very small differences in the recorded waveforms observed, which result directly from the fluctuation of the pulse trigger voltage. The RMS error is equal to 0.255 kA in case of the shunt and 0.315 kA in case of the Rogowski coil measurements, which is less than 1% of the maximum current value. This means that the waveforms obtained from both measurement methods are repeatable. Thus, a comparison of them could be made.

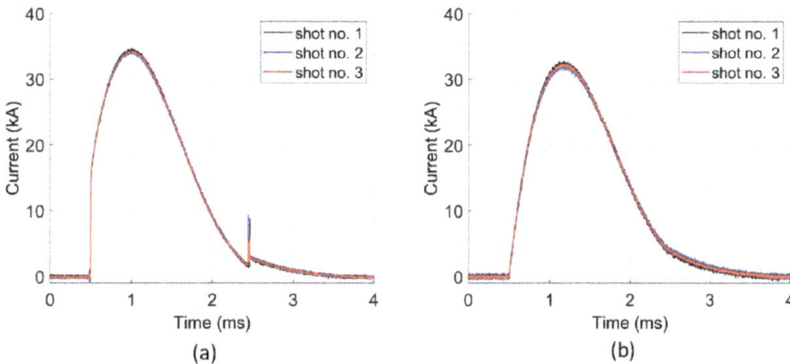

Figure 10. Measurement repeatability of current waves for $U = 98$ V: (**a**) current shunt; (**b**) Rogowski coil.

In Figure 11a–c, a comparison of current waveforms obtained from the Rogowski coil and the current shunt measurements for different values of the initial supply voltage is presented. Independent of the voltage value, the current waveforms for a given measuring method have the same shape—they differ only in the maximum value, which results directly from the initial supply voltage.

The measured current waveforms differ depending on the measurement method used. In the case of a current shunt, there is a more rapid slope of current increase, and hence higher amplitudes in relation to the Rogowski coil were observed. The short pulse, observed in the shunt measurement at 2.5 ms, is arising due the voltage pulse generated on its inductance due to the rapid current change. It does not cause any current flow, thus, it is not visible in the Rogowski coil measurement. The measurement using a current shunt is more sensitive to overvoltages associated with the commutation of semiconductor elements of the power supply system. Due to the finite inductance value of the circuit, the current wave measured by the current shunt is not proper. The current cannot increase abruptly. The problem arises due to the nonzero parasitic inductance of the shunt (Table 2).

In Figure 11d, the waveforms of voltage drop on the capacitor battery for the subsequent series of measurements are presented. The capacitor bank is negatively charged, despite the reverse diode. The negative voltage value is higher for higher current pulses. It is due to the finite reverse time of the diode. The voltage waves cross the zero value at the same time: 1.55 ms. This means that the discharge time does not depend on the supply voltage.

The values of NRMSE for the waveforms presented in Figure 11 are given in Table 3. There is a significant error value observed (approximately 11%). The value increases along with the initial voltage value. In the case of a railgun, the driving force depends directly on the current value. Thus, in order to investigate such a device, it is important to determine the real current wave. In case of the shunt, the measurement error is too large to be credible.

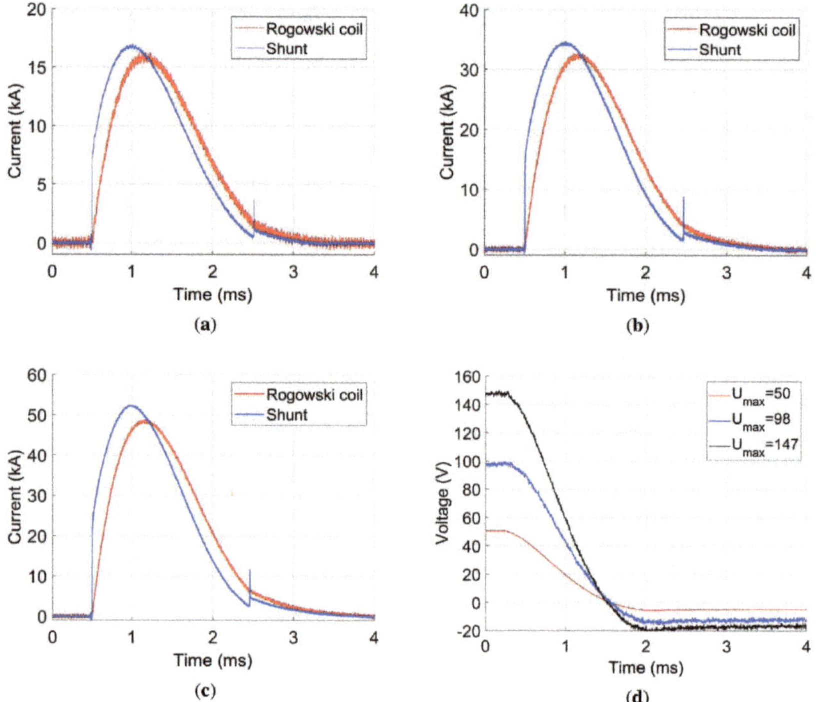

Figure 11. Comparison of power supply waves: (**a**) excitation current vs. time for $U = 50$ V; (**b**) excitation current vs. time for $U = 98$ V; (**c**) excitation current vs. time for $U = 147$ V; (**d**) voltage waves for all initial voltage values.

Table 3. Normalized root mean square error (NRMSE) values for transients given in Figure 11.

Performance Graph	Analyzed Wave	NRMSE (%)
Figure 11a	$i(t)$—for $U = 50$ V	10.67
Figure 11b	$i(t)$—for $U = 98$ V	11.17
Figure 11c	$i(t)$—for $U = 147$ V	11.22

6. Measurement Verification of the Current Shunt Mathematical Model

The mathematical model presented in chapter 2 was verified experimentally. In Tables 4 and 5, measured parameters of the power supply system and thyristors are presented.

Table 4. Power supply parameters (excluding shunt).

Element	Rs (mΩ)	Cs_{total} (mF)	Ls (μH)
Basic elements	1.08	363.0	78.0

Table 5. Thyristor parameters adopted.

Element	Rs (mΩ)	Lon (nH)	Vf (V)	Il (A)	Tq (μs)
Thyristors	0.16	50	0.8	0.1	100

In Figure 12, the comparison between Rogowski coil measurement and the calculation model for the current wave is presented. Different values of initial capacitor voltage were assumed: 50 V, 98 V, and 147 V. Remaining initial conditions were assumed to be zero. A good conformance is observed. Some differences occur, mainly for the amplitude of the measured signal. The NRMSE error between measurements and calculations is between 1.16% and 1.68% (Table 6), which is a low value. This proves the quality of the calculation model and confirms that the Rogowski coil measures the real current wave in the system. The differences are mainly due to some simplification of the mathematical model, e.g., the eddy currents in the wires were neglected. Additionally, the measured system parameters could slightly differ from the real ones, which influences the calculation results.

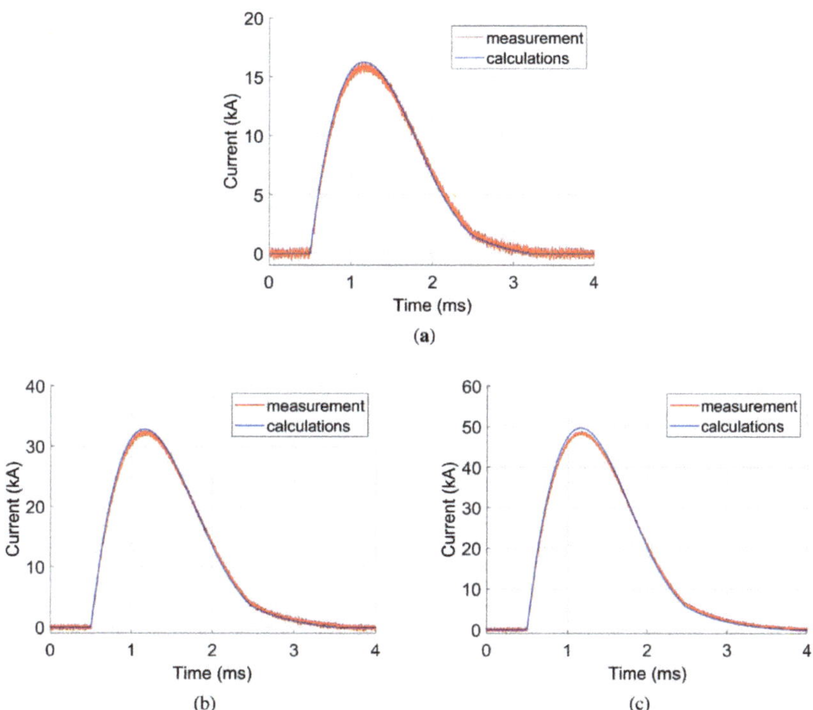

Figure 12. Measurement verification of the model using the Rogowski Coil for different initial voltage values: (**a**) $U = 50$ V; (**b**) $U = 98$ V; (**c**) $U = 147$ V.

Table 6. NRMSE values for transients given in Figure 12.

Case Number	Performance Graph	Wave	NRMSE (%)
1	Figure 12a	$i(t)$—for $U = 50$ V	1.68
2	Figure 12b	$i(t)$—for $U = 98$ V	1.16
3	Figure 12c	$i(t)$—for $U = 147$ V	1.43

A similar comparison between the model and measurement results is presented for the current shunt (Figure 13). In this case, the voltage waves (voltage drop on the shunt) obtained both from calculations and measurements were converted to current waves. In this test, the current shunt resistance value R was determined experimentally, while the inductance value L was chosen to best fit the current shape in calculation and measurement results (Table 7). The resistance and inductance values are very close to that obtained by the ANSYS Q3D Extractor model (Table 2), which confirms its correctness. The capacitance was assumed equal to zero, since it does not influence the results significantly.

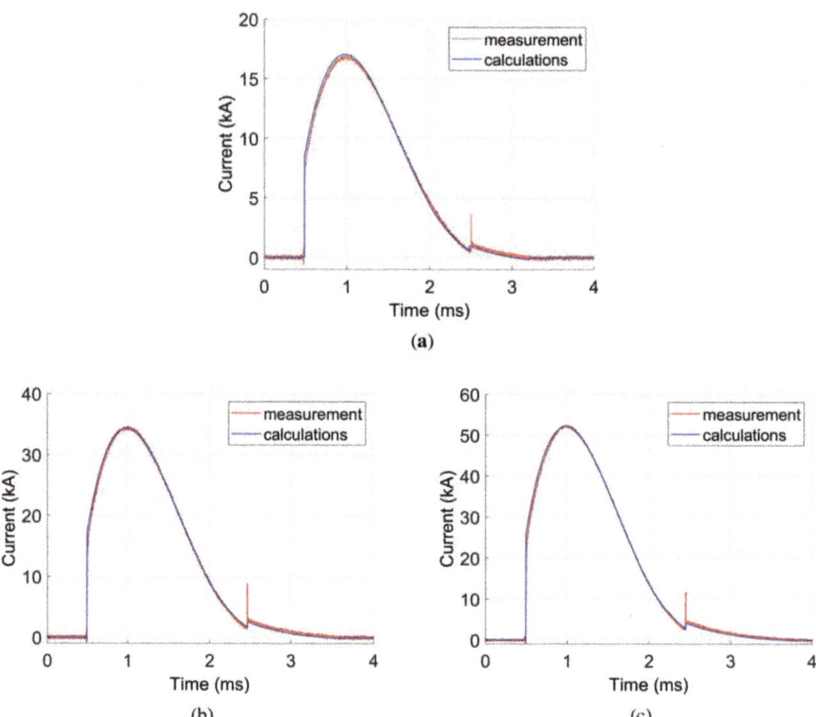

Figure 13. Measurement verification of the shunt circuit model for different initial voltage values: (**a**) $U = 50$ V; (**b**) $U = 98$ V; (**c**) $U = 147$ V.

Table 7. Assumed current shunt parameters.

R (mΩ)	C (pF)	L (nH)
0.53	0	80.0

Taking into account the shunt inductance, a very good agreement between calculations and measurement results was obtained (Figure 13). The NRMSE error between current waves obtained from the mathematical and physical model of the shunt is in between 1.39% and 1.71% (Table 8). The error value is slightly higher than for the Rogowski coil model, which may be due to the sensitivity of the method to overvoltages. The results confirm correctness of the transient mathematical model. Slight differences between measurements and calculations are, similarly as in the case of Rogowski coil, due to some simplification of the mathematical model.

Table 8. NRMSE values for transients given in Figure 13.

Case Number	Performance Graph	Wave	NRMSE (%)
1	Figure 13a	$i(t)$—for $U = 50$ V	1.71
2	Figure 13b	$i(t)$—for $U = 98$ V	1.39
3	Figure 13c	$i(t)$—for $U = 147$ V	1.44

Comparing current waves obtained from the Rogowski coil and current shunt, a significant difference is visible (Figures 12 and 13). Based on the mathematical model, it is assumed that the Rogowski coil measures the real current waveform in the system. In the case of high-current peaks, the current shunt does not give the proper current wave shape. This is due to its parasitic inductance. Although it is relatively small (80 nH), in the case of high-current pulses, it influences results considerably, which is visible both in measurements and numerical models. Two current steps observed in the current wave (at 0.5 ms and 2.5 ms) are due to this inductance. The first step at 0.5 ms occurs after switching on the circuit (a voltage is induced according to the electromotive force $e = Ldi/dt$). The second step is due to diode commutation. The electromotive force induced on the shunt inductance practically does not cause any current flow in the circuit. Thus, these peaks are not visible in the Rogowski coil measurement.

7. Recovering a Real Current Waveform from Shunt Voltage Measurement

The expression (3) for recovering the current waveform is very simple and could be implemented in real-time applications. In the presented investigations, it was implemented in Simulink software. As the input, the voltage wave measured on the current shunt was used. The results obtained using the expression (3) are shown in Figure 14 (black line, Shunt recalculated). For lower voltage values (below 100 V, Figure 14a,b), there is almost no visible difference between the current waves obtained from Rogowski coil and from shunt after applying Equation (3). For higher voltages, only a slight difference is observed (Figure 14c), which could be due to errors in shunt parameters determination. In Table 9, current peak values measured by investigated methods are given.

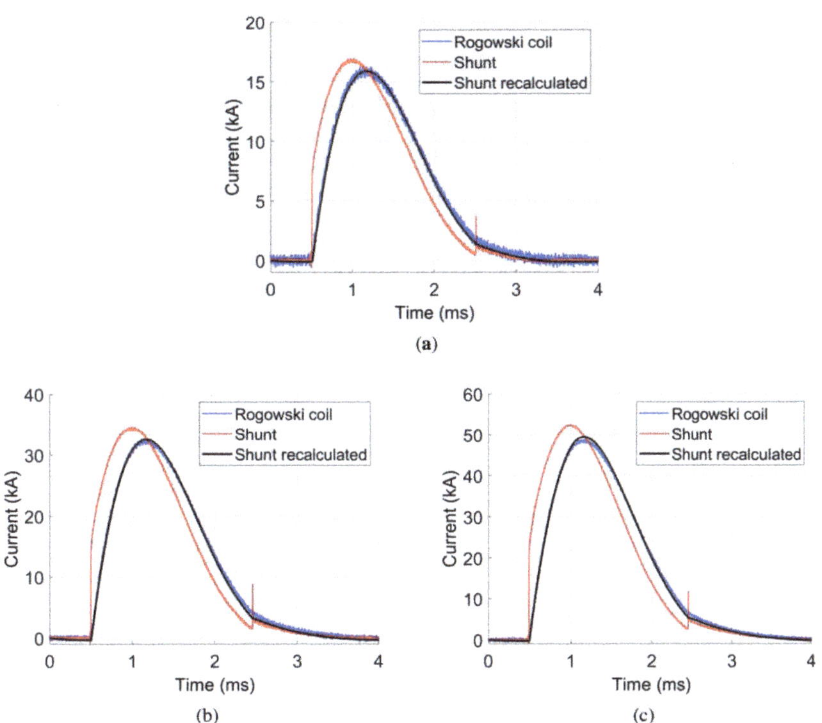

Figure 14. Comparison of different current waveform measurements: (**a**) $U = 50$ V; (**b**) $U = 98$ V; (**c**) $U = 147$ V.

Table 9. Measured current peak values.

U(V)	Rogowski Coil (kA)	Shunt (kA)	Error (%) vs. Rogowski Coil	Shunt Recalculated (kA)	Error (%) vs. Rogowski Coil
50	16.50	16.95	2.73	16.40	0.61
98	32.90	34.72	5.53	33.07	0.52
147	49.05	52.64	7.32	50.08	2.10

The Rogowski coil measurement was assumed as a referee. The values obtained from the shunt after implementing our method are characterized by a significantly lower error than those obtained directly from the voltage drop measurement (red line, Shunt). The NRMSE error for waveforms obtained after applying Equation (3) is between 1.15% and 1.35% (Table 10). The error is relatively low, which satisfies the presented method. Thus, knowing all shunt parameters allows for obtaining the correct wave of the current, even for high-value current pulses.

Table 10. NRMSE value for transients given in Figure 14 (shunt recalculated)

Case Number	Performance Graph	Wave	NRMSE [%]
1	Figure 14a	$i(t)$—for $U = 50$ V	1.35
2	Figure 14b	$i(t)$—for $U = 98$ V	1.24
3	Figure 14c	$i(t)$—for $U = 147$ V	1.15

The most important advantage of the presented method is its simplicity. The only problem is determining the inductance of the current shunt (the resistance is known). This could be done either by measurements or by magnetostatic field calculations using the finite element method (FEM). In the case of the field modeling, all dimensions of the shunt and all material properties must be known.

8. Tests for a Second Current Shunt

In order to test the proposed method more deeply, calculations and measurements for a shunt, which differs from the previous one, were carried out. The outline and dimensions of it are presented in Figure 15. Compared to first one, the shunt is characterized by a lower number of rods and smaller length. In Table 11, the parameters of the shunt obtained from Q3D model are given. These parameters were used to recover the original current waveform. The results are presented in Figure 16. Due to the much lower nominal current of the shunt, the tests were made for lower voltage values.

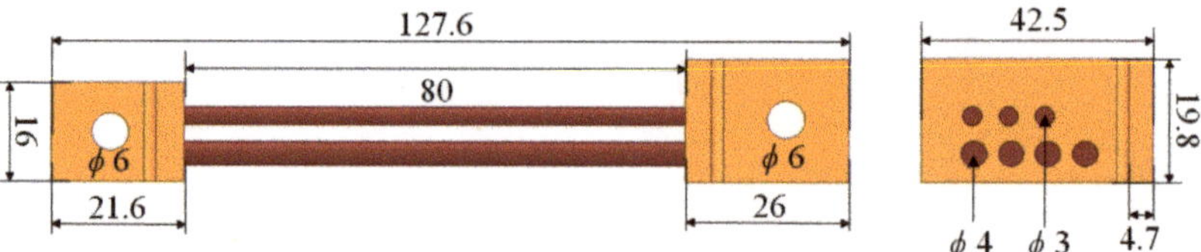

Figure 15. Outline of the second current shunt (dimensions in mm).

Table 11. Second shunt circuit parameters—numerical calculations results (Q3D).

R (mΩ)	L (nH)
0.516	54.74

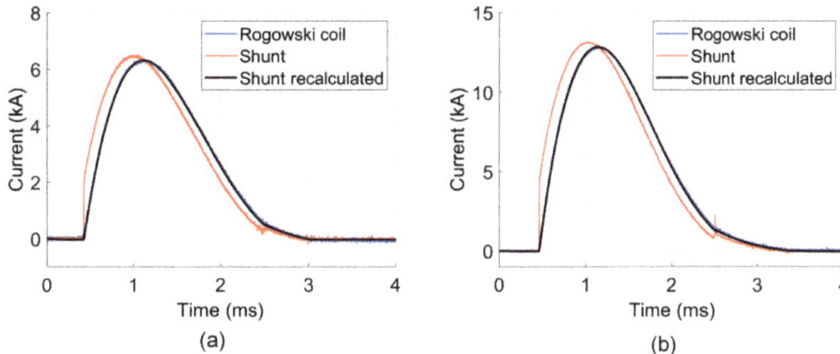

Figure 16. Comparison of different current waveform measurements: (**a**) $U = 20$ V; (**b**) $U = 40$ V.

Similar to the previous measurement, the Rogowski coil results were assumed as a referee. The errors values are given in Tables 12 and 13. The current peak values differ more than 2.5% between Rogowski coil and shunt measurement results. After recalculation, the differences are less than 1%. The NRMSE value decreases after correction from more than 7.5% to below 1%, which is even lower than in the tests shown in chapter 7. Thus, the presented method significantly reduces the measurement error for the current shunt.

Table 12. Measured current peak values.

U (V)	Rogowski Coil (kA)	Shunt (kA)	Error (%) vs. Rogowski Coil	Shunt Recalculated (kA)	Error (%) vs. Rogowski Coil
20	6.35	6.51	2.52	6.30	0.79
40	12.89	13.15	2.02	12.82	0.54

Table 13. NRMSE value for transients given in Figure 16.

Case Number	Performance Graph	Wave	NRMSE (%) Shunt	NRMSE (%) Shunt Recalculated
1	Figure 16a	$i(t)$—for $U = 20$ V	7.56	0.72
2	Figure 16b	$i(t)$—for $U = 40$ V	8.17	0.78

9. Conclusions

For a high-current pulse measurement (some kA) there are very significant differences between current waveforms obtained from a shunt and visible Rogowski coil. The proper current wave is measured by the Rogowski coil, while the shunt does not give the correct result. Both mathematical models and measurements show that this is mainly due to the inductance of the shunt. The value obtained from measurements (91.59 nH) is slightly higher than that obtained from the magnetostatic field model (7.8%) and from synthesis using a circuit model (12.6%). Although the value is very small, it needs to be taken into account in the current peak measurements. The capacitance of the shunt does not influence the calculation results significantly.

From the knowledge of the shunt parameters, it is possible to recover the real current waveform. A simple Euler method used for integration of an ODE describing the RL element is sufficient to significantly improve the accuracy, which was presented for two different current shunts. The simplicity of the method allows using it in real-time calculations. Thus, the shunt could be used in the measurements of high dynamic current waves (e.g., in railguns, magnetizing devices) after applying the correction arising from Equation (3). Taking into account the higher price of Rogowski coil, in the case of some measurements, it is more convenient to use a current shunt including the presented correction method.

In order to explain problems with the measuring of high-current pulses using a current shunt, we used both numerical models and experimental investigations. The most difficult task was choosing an appropriate model of the shunt. In the presented case, the Q3D software was successfully implemented. The presented approach is very useful and not only allows us to explain problems with current pulses measurement, but also indicates the way to solve them.

Author Contributions: Conceptualization, P.P. and A.W.; methodology, P.P. and A.W.; software, P.P. and A.W.; validation, P.P.; formal analysis, P.P. and A.W.; investigation, P.P. and A.W.; resources, P.P. and A.W.; writing–original draft preparation, P.P.; writing–review and editing, A.W.; visualization, P.P.; supervision, A.W. All authors have read and agreed to the published version of the manuscript.

Funding: This research received no external funding.

Institutional Review Board Statement: Not applicable.

Informed Consent Statement: Not applicable.

Conflicts of Interest: The authors declare no conflict of interest.

References

1. Waindok, A.; Piekielny, P. Influence of the power supply parameters on the projectile energy in the permanent magnet electrodynamic accelerator. In Proceedings of the E3S Web of Conferences, Szczyrk, Poland, 25–27 October 2017; Volume 19, pp. 1–6.
2. Tomczuk, B.; Wajnert, D. Field–circuit model of the radial active magnetic bearing system. *Electr. Eng.* **2018**, *100*, 2319–2328.
3. Waindok, A.; Piekielny, P. Transient analysis of a railgun with permanent magnets support. *Acta Mech. Autom.* **2017**, *11*, 302–307.
4. Ding, H.; Jiang, C.; Ding, T.; Xu, Y.; Li, L.; Duan, X.; Pan, Y.; Herlach, F. Prototype test and manufacture of a modular 12.5 MJ capacitive pulsed power supply. *IEEE Trans. Appl. Supercond.* **2010**, *20*, 1676–1680.
5. Lehr, J.; Ron, P. Pulsed Voltage and Current Measurements. In *Foundations of Pulsed Power Technology*; John Wiley & Sons, Wiley-IEEE Press: Hoboken, NJ, USA, **2018**, 493–546.
6. Pacha, M.; Varecha, P.; Sumega, M. HW issues of current sensing by DC-link shunt resistor. In Proceedings of the 2018 ELEKTRO, Mikulov, Czech Republic, 21–23 May 2018; pp. 1–5.
7. Filipski, P.S.; Boecker, M. AC-DC current shunts and system for extended current and frequency ranges. In Proceedings of the IEEE Instrumentation and Measurement Technology Conference Proceedings, Ottawa, ON, Canada, 16–19 May 2006; Volume 55, pp. 1222–1227.
8. Voljč, B.; Lindič, M.; Pinter, B.; Kokalj, M.; Svetik, Z.; Lapuh, R. Evaluation of a 100 A current shunt for the direct measurement of AC current. *IEEE Trans. Instrum. Meas.* **2013**, *62*, 1675–1680.
9. Shalamov, S. Current Measuring Shunts for Impulse Current Test According to IEC 62305–2010. In Proceedings of the 2018 9th International Conference on Ultrawideband and Ultrashort Impulse Signals (UWBUSIS), Odessa, Ukraine, 4–7 September 2018; pp. 127–130.
10. Zhang, W.; Zhang, Z.; Wang, F. Review and Bandwidth Measurement of Coaxial Shunt Resistors for Wide-Bandgap Devices Dynamic Characterization. In Proceedings of the 2019 IEEE Energy Conversion Congress and Exposition (ECCE), Baltimore, MD, USA, 29 September–3 October 2019; pp. 3259–3264.
11. Zhang, J.; Pan, X.; Liu, W.; Gu, Y.; Wang, B.; Zhang, D. Determination of equivalent inductance of current shunts at frequency up to 200 kHz. *IEEE Trans. Instrum. Meas.* **2013**, *62*, 1664–1668.
12. Miličević, A.; Mostarac, P. Simulated Annealing Characterization of Equivalent Circuit Models for Precision Coaxial Current Shunts. In Proceedings of the 2019 2nd International Colloquium on Smart Grid Metrology (SMAGRIMET), Split, Croatia, 9–12 April 2019; pp. 1–5.
13. Xu, M.; Yan, J.; Geng, Y.; Zhang, K.; Sun, C. Research on the factors influencing the measurement errors of the discrete Rogowski coil. *Sensors* **2018**, *18*, 847.
14. Metwally, I.A. Self-integrating Rogowski coil for high-impulse current measurement. *IEEE Trans. Instrum. Meas.* **2009**, *59*, 353–360.
15. Liu, X.; Huang, H.; Jiao, C. Modeling and Analyzing the Mutual Inductance of Rogowski Coils of Arbitrary Skeleton. *Sensors* **2019**, *19*, 3397.
16. Saroj, P.; Agrawal, R.; Roy, A.; Menon, R.; Singh, S.; Raul, S.; Kulkarni, M.; Sharma, A.; Nagesh, K.; Chakravarthy, D. Development and Calibration of Rogowski coils for pulsed power systems. In Proceedings of the 2011 IEEE International Vacuum Electronics Conference (IVEC), Bangalore, India, 21–24 February 2011; pp. 471–472.
17. Djokić, B.V. Improvements in the performance of a calibration system for Rogowski coils at high pulsed currents. *IEEE Trans. Instrum. Meas.* **2016**, *66*, 1636–1641.
18. Habrych, M.; Wisniewski, G.; Miedziński, B.; Lisowiec, A.; Fjałkowski, Z. HDI PCB Rogowski coils for automated electrical power system applications. *IEEE Trans. Power Deliv.* **2017**, *33*, 1536–1544.

19. Suomalainen, E.P.; Hallstrom, J.K. Onsite calibration of a current transformer using a Rogowski coil. *IEEE Trans. Instrum. Meas.* **2008**, *58*, 1054–1058.
20. Ghanbari, T.; Farjah, A. Application of Rogowski search coil for stator fault diagnosis in electrical machines. *IEEE Sensors J.* **2013**, *14*, 311–312.
21. Liu, Y.; Xie, X.; Hu, Y.; Qian, Y.; Sheng, G.; Jiang, X. A novel transient fault current sensor based on the PCB Rogowski coil for overhead transmission lines. *Sensors* **2016**, *16*, 742.
22. Jiao, C.; Zhang, J.; Zhao, Z.; Zhang, Z.; Fan, Y. Research on small square pcb rogowski coil measuring transient current in the power electronics devices. *Sensors* **2019**, *19*, 4176.
23. Robles, G.; Shafiq, M.; Martínez-Tarifa, J.M. Designing a Rogowski Coil with Particle Swarm Optimization. *Proceedings* **2018**, *4*, 10.
24. Ansoft Corporation. *Maxwell FS 3D*, version 6.0; Ansoft Corporation: Pittsburgh, PA, USA, 2018.
25. Ansoft Corporation. *Maxwell SI*, version 4.0; Ansoft Corporation: Pittsburgh, PA, USA, 2018.
26. Ruehli, A.E.; Cangellaris, A.C. Progress in the methodologies for the electrical modeling of interconnects and electronic packages. *Proc. IEEE* **2002**, *89*, 740–770.
27. Che, C.; Zhao, H.; Guo, Y.; Hu, J.; Kim, H. Investigation of Segmentation Method for Enhancing High Frequency Simulation Accuracy of Q3D Extractor. In Proceedings of the 2019 IEEE International Conference on Computational Electromagnetics (ICCEM), Shanghai, China, 20–22 March 2019; pp. 1–3.
28. Yi, R.; Zhao, Z.; Zhong, Y. Modeling of busbars in high power neutral point clamped three-level inverters. *Tsinghua Sci. Technol.* **2008**, *13*, 91–97.
29. Wajnert, D.; Tomczuk, B. Nonlinear magnetic equivalent circuit of the hybrid magnetic bearing. *COMPEL-Int. J. Comput. Math. Electr. Electron. Eng.* **2019**, *38*, 1190–1203.
30. Wajnert, D. A field-circuit model of the hybrid magnetic bearing. *Arch. Mech. Eng.* **2019**, *66*, 191–208.
31. Qian, J.; Chen, X.; Chen, H.; Zeng, L.; Li, X. Magnetic field analysis of Lorentz motors using a novel segmented magnetic equivalent circuit method. *Sensors* **2013**, *13*, 1664–1678.
32. Wajnert, D.; Sykulski, J.K.; Tomczuk, B. An enhanced dynamic simulation model of a hybrid magnetic bearing taking account of the sensor noise. *Sensors* **2020**, *20*, 1116.
33. Waindok, A.; Tomczuk, B.; Koteras, D. Modeling of Magnetic Field and Transients in a Novel Permanent Magnet Valve Actuator. *Sensors* **2020**, *20*, 2709.
34. Waindok, A.; Piekielny, P. Analysis of an iron-core and ironless railguns powered sequentially. *COMPEL-Int. J. Comput. Math. Electr. Electron. Eng.* **2018**, *37*, 1707–1721.
35. CWT Specification, Power Electronic Measurement Ltd., Nottingham, UK, 2018, Available online: http://www.pemuk.com/Userfiles/CWT/cwt_0318.pdf (accessed on 5 March 2021).

sensors

MDPI

Article

Thermal and Geometric Error Compensation Approach for an Optical Linear Encoder

Donatas Gurauskis [1,*], Artūras Kilikevičius [2] and Albinas Kasparaitis [1]

[1] Department of Mechanical and Material Engineering, Vilnius Gediminas Technical University, J. Basanavičiaus g. 28, 03224 Vilnius, Lithuania; albinas.kasparaitis@vgtu.lt
[2] Institute of Mechanical Science, Vilnius Gediminas Technical University, J. Basanavičiaus g. 28, 03224 Vilnius, Lithuania; arturas.kilikevicius@vgtu.lt
* Correspondence: donatas.gurauskis@vgtu.lt

Abstract: Linear displacement measuring systems, like optical encoders, are widely used in various precise positioning applications to form a full closed-loop control system. Thus, the performance of the machine and the quality of its technological process are highly dependent on the accuracy of the linear encoder used. Thermoelastic deformation caused by a various thermal sources and the changing ambient temperature are important factors that introduce errors in an encoder reading. This work presents an experimental realization of the real-time geometric and thermal error compensation of the optical linear encoder. The implemented compensation model is based on the approximation of the tested encoder error by a simple parametric function and calculation of a linear nature error component according to an ambient temperature variation. The calculation of a two-dimensional compensation function and the real-time correction of the investigated linear encoder position readings are realized by using a field programmable gate array (FPGA) computing platform. The results of the performed experimental research verified that the final positioning error could be reduced up to 98%.

Keywords: measuring scale; thermoelastic deformation; coefficient of thermal expansion

check for updates

Citation: Gurauskis, D.; Kilikevičius, A.; Kasparaitis, A. Thermal and Geometric Error Compensation Approach for an Optical Linear Encoder. *Sensors* **2021**, *21*, 360. https://doi.org/10.3390/s21020360

Received: 3 December 2020
Accepted: 5 January 2021
Published: 7 January 2021

1. Introduction

The vast majority of industrial and scientific applications use optical encoders for a position measurement and closed-loop position control, for example, machine tools [1–5], tracking systems [6–8], industrial robots [9–13], positioning stages [14–17], and so on. All these technological machines work under various environment conditions such as temperature, humidity, mechanical vibration, etc. In turn, these effects inevitably generate a corresponding error. According to Ramesh et al. [18,19], the thermal factors account for 40–70% of the total dimensional and shape errors in machine tools. Much scientific research has been performed to analyze, model, and compensate the influence of thermal positioning error in manually or computer-numerical-control (CNC) machines [20–28]. The majority of this research analyzes only the machine tool structure, considering that the used measurement system is not the source of the error by itself [29].

Working environment adversely affects the accuracy of the integrated encoder. Lopez et al. investigate optical encoder errors under vibration at different mounting conditions [30] and optical scanning principles [31]. They also present a methodology for a vibration error compensation [32]. Performance of the encoder could be improved by correcting its output signals in real-time. This could be done by using look-up tables, digital filtering, or other techniques [33–37]. These methods help to reduce a high frequency sub-divisional encoder error that repeats at each period of a scale grating. Temperature changes introduce strains that change the width of a grating period. Therefore, the thermal errors could not be compensated by using these techniques.

The thermal behavior of the encoder is also a very important factor. One way to deal with a changing temperature impact is by using low thermal coefficient (CTE) materials. Glass ceramics like ZERODUR (CTE $= 0 \pm 0.1$ µm/m°C) or ROBAX (CTE $= \sim 0$ µm/m°C) are used for a measuring scale manufacturing. Another way is trying to match the thermal coefficients between used linear encoder, machine tool support, and a workpiece material. In this case, the change in a workpiece size has the same value as the expanded or contracted encoder, so the thermoelastic error is practically eliminated. A well-defined and reproducible thermal behavior of the encoder must be ensured. Unfortunately, it is quite challenging task to do. For example, an enclosed type linear encoder consists of an aluminum extrusion and a measuring scale, usually made from glass or stainless steel. The scale is attached to extrusion by adhesive or a double-sided adhesive tape. Such an assembly demonstrates a complex thermal behavior because of the combination of different CTE materials. Alejandre et al. [38] present the method to determine the real thermal coefficient of a linear encoder. In this research, they investigate an enclosed-type linear encoder and established that the real CTE is influenced by the bonding material between the aluminum extrusion and the glass scale. Moreover, during another study [39], a non-linear thermal behavior in optical linear encoders was noticed. This could be explained as a consequence of varying stresses transmitted from the extrusion to measuring scale.

The thermal error compensation could be an effective and economic method to improve optical encoder accuracy. This procedure is based on encoder error correction by introducing correction coefficients derived by using various mathematical ways and experimental research. Yu et al. [40] improve the rotary encoder accuracy by using Fourier expansion-polynomial fitting technique. In 2020, Jia et al. [41] proposed the compensation approach based on Fourier expansion-back propagation neural network technique optimized by a genetic algorithm. This group of scientists minimized rotary encoder error from 110.2 arc sec to 2.7 arc sec. Hu et al. [42] used the empirical mode decomposition and the linear least square fitting methods for a linear encoder error compensation at different temperatures. In general, there is not much information about the real thermal behavior of optical encoders and their accuracy under a real ambient condition. Even less information is published about a practical realization capability of the embedded error compensation solution.

In a previous work [43], the theoretical investigation of the linear encoder thermal behavior was done using the finite element method. The performed computer simulation analyzed the occurring thermal processes and introduced thermoelastic deformations, when a linear encoder is influenced by various heat sources and the changing temperature of the working environment. The results showed that the analyzed encoder demonstrated systematic behavior, which could be approximated by a simple parametric function. This could be used to compensate the final encoder position value, in order to improve its accuracy.

In this work, the real-time geometric and thermal error compensation approach is proposed. The article is based on theoretical and experimental research of the tested optical linear encoder and practical realization of the composed compensation algorithm. The presented method is optimized by experimentally estimating actual CTE of the linear encoder under test and could reduce the thermoelastic error to the accuracy range specified by the manufacturer. In Section 2, the error compensation background is discussed. Section 3 presents the setup used for the experimental investigation. Equipment used for the tested encoder accuracy measurement at different ambient temperatures and the composed subsequent electronics which realize the error compensation in real-time are specified. Obtained results and performed compensation algorithm optimization, based on a real CTE calculation, are described in Section 4. The discussion about the collected data and a short summary are written in Section 5. In Section 6, the main findings and conclusions are listed.

2. The Error Compensation Method for Linear Encoder

The accuracy of the encoder is one of the most significant parameters. This term describes the difference between the target position (real position value) and actual position—

the encoder position reading. Mathematically, the position reading of the encoder Q could be expressed as:

$$Q = Q_{real} + \delta, \tag{1}$$

where Q_{real}—is a real linear position value, δ—is an error component. Therefore, the accuracy of the encoder is directly related to the size of a measurement error and is sometimes called the position error. In practice, the encoder accuracy measurement is a specific procedure that requires a well-calibrated equipment and a certain environmental condition. The calibrating encoder readings are compared with a reference device position indication, which are accepted as a real position value Q_{real}. Usually, second, a highly accurate encoder or a laser interferometer is used as a reference.

According to ISO 5725-1, when the accuracy term is applied to sets of measurements of the same measurand, it involves a component of systematic error and a component of random error [44]. In this case, the term "trueness" is used to describe the closeness of the mean of a set of measurement results to the actual value and term "precision" is used to describe the closeness of agreement among a set of results. In practice, the trueness is accepted as the accuracy of the encoder and the precision is used to describe the repeatability or reproducibility of the device. In general, linear encoder error δ depends on the position q and the temperature T. Then, the mathematical model of the error is written as:

$$\Delta(q, T) = F(q, T) + \varepsilon, \tag{2}$$

where $\Delta F(q, T)$—parametric function approximating a systematic error component, ε is the residual random component. In optical linear encoders, glass or stainless steel scales with a precise grating patterns are used for a linear position measurement. The manufacturing inaccuracies, like a varying duty cycle of the grating; any kind of scale deformations during the encoder assembly or mounting procedure; and a thermal expansion or contraction of the encoder due to ambient temperature or other temperature sources cause the systematic error component that could be measured, approximated by a parametric function and compensated. The error compensation K is equated to the systematic error component approximating function value with an opposite sign, i.e.,

$$K = -F(q, T). \tag{3}$$

Other effects, like accidental mechanical vibrations or a shock, dust, metal chips or any other contaminants on the measuring scale surface, etc. represent the random error component and cannot be easily compensated.

Theoretically, the simplified mathematical error model for compensation could be expressed as:

$$\Delta(q, T) = F_g(q) + F_{gr}(q) + F_\alpha(\Delta T, q), \tag{4}$$

where the first member is $F_g(q)$—geometric error approximation function. The calibration process of the encoder is done at nominal temperature $T_n = 20\ ^\circ C$, in a special thermostable laboratory room, where ambient temperature varies only about $\pm 0.2\ ^\circ C$. Tested encoder error values are plotted in graph according to the linear position values. Such a plot is called an encoder accuracy graph and is added to each manufactured encoder as a document to ensure the accuracy of the calibrated device. In this case, the compensation model assumes, that the ambient temperature stays constant and is equal to a nominal $T = T_n$. The calibrated encoder error plot is approximated by a single parametric function whose argument is the position value q.

$$\Delta(q) = F_g(q) + \varepsilon_g, \tag{5}$$

The second member $F_{gr}(q)$—thermoelastic error approximation, when thermal gradient is steady and ambient temperature is stable. In real applications, there are various temperature sources around the measuring system. Those sources generate a relative temperature gradient along the encoder. That causes an unwanted deformation of the measuring scale. Estimation of the approximating thermal error function consists of the temperature gradient measurement by means of temperature sensor values in multiple points along the encoder, and calculation of the thermoelastic deformation of the measuring scale by the finite element method. The simulated total displacement values of the scale are approximated by a single parametric function, which is later used to estimate the thermal gradient error size in compensation process.

The third member $F_a(\Delta T, q)$—thermal error component—expresses a linear deformation of the measuring scale due to changing ambient temperature.

$$F_a(\Delta T, q) = \alpha_{corrected} \cdot \Delta T \cdot q, \tag{6}$$

where $\alpha_{corrected}$—corrected coefficient of linear thermal expansion (CTE) of the measuring scale and ΔT—ambient temperature difference from the nominal temperature $(\Delta T = T_n - T)$.

The total error value calculated according the Equation (4) could be considered as a size for the real-time compensation. The determined compensation value at specific position should be relatively added or subtracted from the encoder position readings, in order to get the compensated position value.

3. Experimental Setup

The prototype absolute linear optical encoder LK50 of the company JSC "Precizika Metrology" was chosen for the experimental research. It is a reflective type optical encoder with a measuring scale pattern engraved onto the stainless steel tape surface by a laser. The tape is fitted into the encoder's aluminum extrusion, stretched by using a special rigid spring based mechanism and tightened at both ends. Another stainless steel tape is used as a guideway for a precise positioning and motion of the scanning carriage. The cross-sectional view of the tested optical linear encoder is shown in Figure 1.

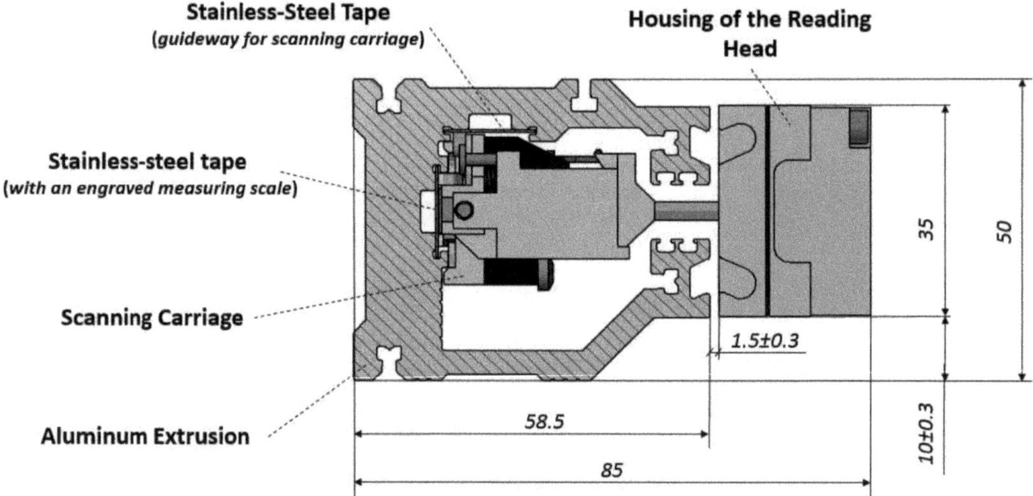

Figure 1. Cross-sectional view of the tested optical linear encoder LK50.

The main parameters of the tested encoder, such as dimensions, measuring length and so on, are specified in Table 1.

In order to investigate the thermal behavior of the encoder, all experiments were carried out in a laboratory room, where the stable ambient temperature could be maintained. The specially customized technological stand was used for encoder mounting and imitation of an appropriate reading head motion along the measuring scale. The aluminum extrusion of the tested encoder was mounted onto the stainless steel support fixed on the granite base. The extrusion was attached with only one fixing screw in the middle of its length. In this way, the ends of the encoder could freely move during the thermal expansion and contraction. The reading head is attached to a moving carriage with an aerostatic bearing.

Table 1. Parameters of tested optical linear encoder.

Concept	Value	Units
Measuring length (ML)	1200	mm
Accuracy (to any meter within the ML)	±5	μm/m
Resolution	0.1	μm
Interface	BiSS-C	-
Aluminum extrusion	Dimensions: 50 × 58.5 × 1485	mm × mm × mm
	Thermal coefficient (CTE): 23×10^{-6}	m/(m °C)
Stainless steel tape	Dimensions: 12 × 0.5 × 1440	mm × mm × mm
	Thermal coefficient (CTE): 10.5×10^{-6}	m/(m °C)

During the tests, readings of the linear encoder were compared to a linear position indication of the laser calibration system "Keysight 5530". The interferometer assembly was placed at the end of the technological stand. The retroreflector assembly is located at the moving carriage. To avoid uncertainties and compensate laser measuring system errors due to changing temperature, the "E1736A USB Sensor Hub" and relatively mounted temperature sensors "E1737A" were used. The composed experimental setup is shown in Figure 2.

The content of the used experimental setup is listed in Table 2 according the position numbers marked in Figure 2.

Table 2. Content of the experimental setup.

Position	Object
1	Granite base
2	Stainless steel support (for encoder mounting)
3	Moving carriage (with aerostatic bearings)
4	Optical linear encoder (device under test)
5	Fixing screws (for encoder reading head)
6	Fixing screw (for encoder aluminum extrusion)
7	Subsequent electronics (for error compensation)
8	Ambient temperature sensor (E1738A)
9	Laser (5519A/B)
10	Interferometer assembly (linear interferometer, linear retroreflector, base, height adjuster, and post)
11	Retroreflector assembly (linear retroreflector, post and height adjuster, base)
12	Temperature sensors (E1737A)
13	USB sensor hub (E1736A)
14	USB axis module (E1735A)
15	PC (with an appropriate software)

Considering the linear position compensation implementation into a real application, the response of the encoder becomes an important factor. For incremental encoders, the response is limited to a specific input signal frequency. The latency depends on the analog amplifier bandwidth, interpolation process, and the resolution. In practical applications, the incremental interface encoder latency is usually ignored, given that the edges of digital output signals have the real-time nature [45]. Unfortunately, the thermal error compensation process realization in incremental encoder is a hard task, because the output signals did not contain any information about the absolute position. They indicate the size of the reading head linear displacement. The absolute linear encoders usually consist of low resolution absolute position track and high resolution incremental track. The combination of the two tracks determines the absolute position value with a high resolution. These data are given on the demand of an application controller by a serial interface. The data transmission time depends on the bit length and overall speed.

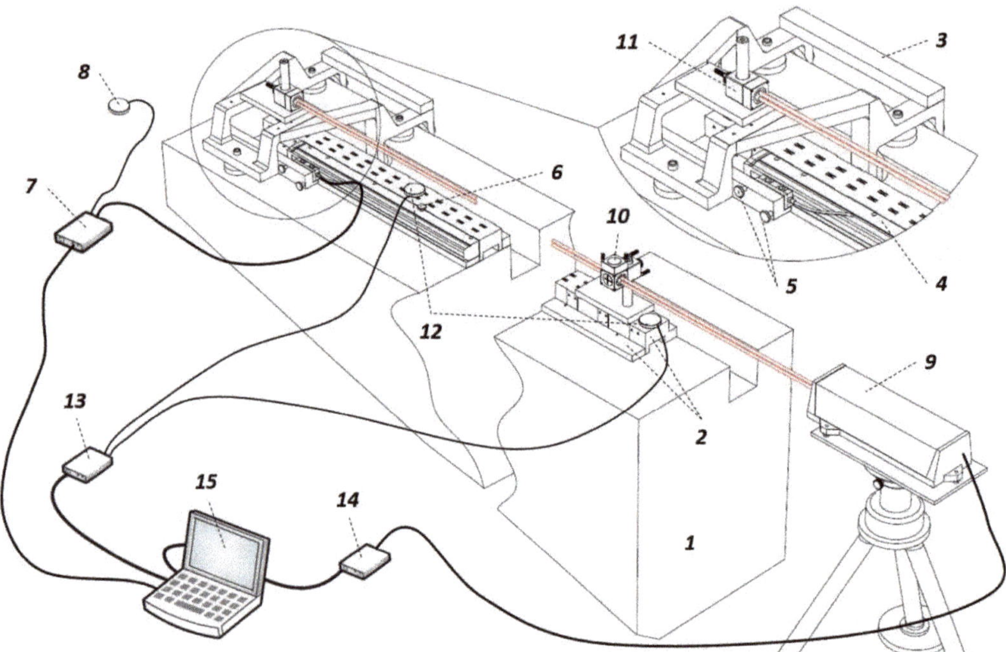

Figure 2. Experimental setup based on "Keysight 5530" system configured for a linear measurement.

The selected absolute optical linear encoder transmits its position by using a bidirectional synchronous serial interface BiSS. Usually it is used in industrial applications, where high transfer rates are required [46]. Depending on cable length, the encoder could handle clock frequencies up to 4 MHz and the calculation time is ≤ 5 µs. The maximum traversing speed is limited up to 2 m/s. The chosen resolution position is outputted with a 30-bit format. Taking into account the calculation time, the absolute position is transmitted upon ≤ 13 µs. Thermal and geometric encoder error compensation is realized by using the composed subsequent electronics. The programmable gate array (FPGA) platform "S7 Mini" with "Xilinx Spartan-7 7S25" is used as a master to request and get the linear encoder position readings, calculate the compensation value according to an integrated mathematical algorithm and external ambient temperature sensor data, and output the compensated position at the real-time. The vanishingly small calculation time of the FPGA could perform the compensation process almost instantly. If the compensation is processed by subsequent electronics, the FPGA has to receive, recognize, compensate, and generate the absolute position value. The whole process takes approximately double the time of the encoder transmission time. In this case, a compensated position is outputted ≤ 26 µs. If the proposed mathematical algorithm could be installed into the integrated FPGA or other controller, the calculation time might be drastically reduced.

The measuring length of the tested linear encoder is 1200 mm. The rectilinear velocity of the moving carriage is 0.2 m/s. To reduce the uncertainty and maintain the reproducibility of the successive measurements, the digital incremental encoder signals are also recorded. According the counted edges of these signals, the absolute position request is sent to the encoder at every 1000 counts i.e., at each 0.1 mm. In such a way, there are 12,000 equally spaced measured positions along the linear encoder.

Additionally, the "Texas Instruments" THVD 1451 RS-485 transceivers are used to deal with differential encoder CLOCK and DATA signals. The simplified block diagram of the composed compensation electronics is shown in Figure 3.

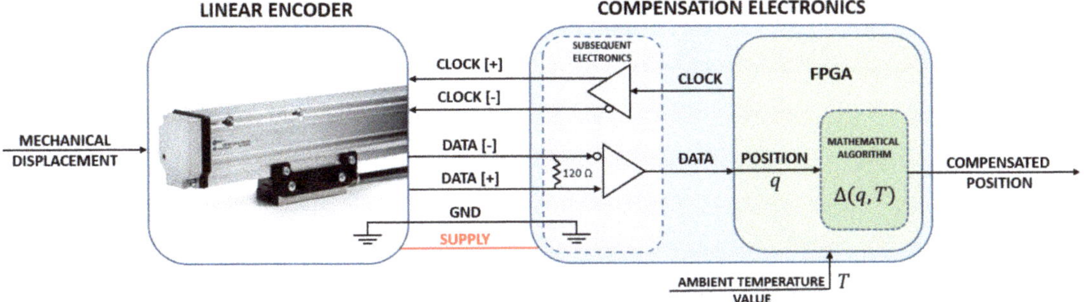

Figure 3. Block diagram of the compensation electronics with a field-programmable gate array (FPGA).

The encoder readings, compensated encoder position values, and respective indications of the laser interferometer measuring system are recorded simultaneously during all experiments. Collected data are processed by using a numerical computing software environment "MATLAB" for the estimation of approximating function, further data analyzation, and graphical representation.

4. Results

Firstly, the whole experimental setup was left in the laboratory room at fixed nominal ambient temperature $T_n = 20 \pm 0.2$ °C for 5 h to stabilize. During the tests, the ambient temperature was changed, so this stabilization process was repeated four times, at each settled temperature (i.e., 20 °C, 17.8 °C, 22.6 °C , and 25.3 °C). Because in the laboratory room there were a number of electronic components which generate approximately the same amount of heat all the time, it is stated that along the linear encoder existed a steady thermal gradient. Temperature differences in various part of the encoder induced thermoelastic deformations of the measuring scale. Encoder mounting could also be the source of the linear position measurement error, because of misalignment or deformations during fixation, lack of support stiffness, inaccurate guideway of the carriage, and so on. All these factors introduced the geometric error component. These conditions are relatively close to some of a real application, where such an encoder could be used.

Five separate unidirectional measurements are taken at each temperature. The average value of these five measurements is calculated. Based on standard ISO 230-2, the half peak-to-peak value of the resulting average position error curve is accepted as the unidirectional systematic positioning error of the encoder. The compensation and minimization of this systematic error is the main goal of this work.

The first five measurements at 20 °C ambient temperature were recorded. The error values at corresponding positions and the average meaning curve are presented in Figure 4. The unidirectional systematic positioning error of the encoder was ±2.2 µm. This value is accepted as the accuracy of non-compensated tested encoder.

The parametric function approximating the average position error curve could be accepted as the combination of the geometric $F_g(q)$ and the thermoelastic $F_{gr}(q)$ error components, i.e., the sum of the first two members of the Equation (4). The fitted approximating function is shown in Figure 5.

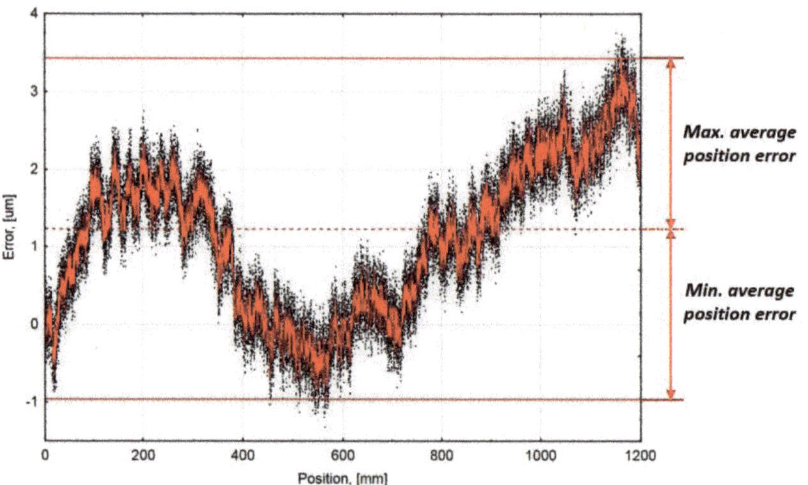

Figure 4. Accuracy graph at 20 °C ambient temperature. (Black dots—the measured error values at corresponding positions; Red curved line—the resulting average position error curve).

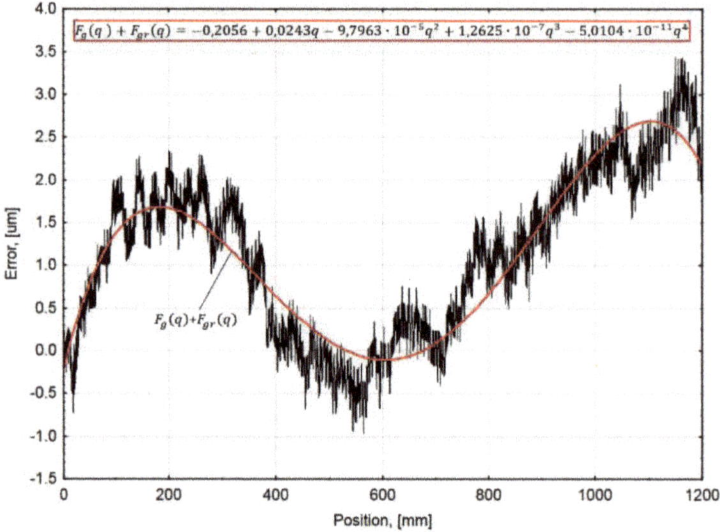

Figure 5. Average accuracy graph at 20 °C ambient temperature approximated by a parametric function. (Red line—approximating function and its equation).

The graph data are quite accurately approximated by a 4th order polynomial function that could be described by the following equation:

$$F_g(q) + F_{gr}(q) = F_G(q) = -0.2056 + 0.0243q - 9.7963 \times 10^{-5}q^2 + 1.2625 \times 10^{-7}q^3 - 5.0104 \times 10^{-11}q^4, \quad (7)$$

This determined function was used as a base for the further thermal and geometric error compensation value calculation. The approximating function was integrated into compensation electronics (FPGA), and the five measurements were repeated. The accuracy graph of an average position error of the compensated linear encoder is shown in Figure 6.

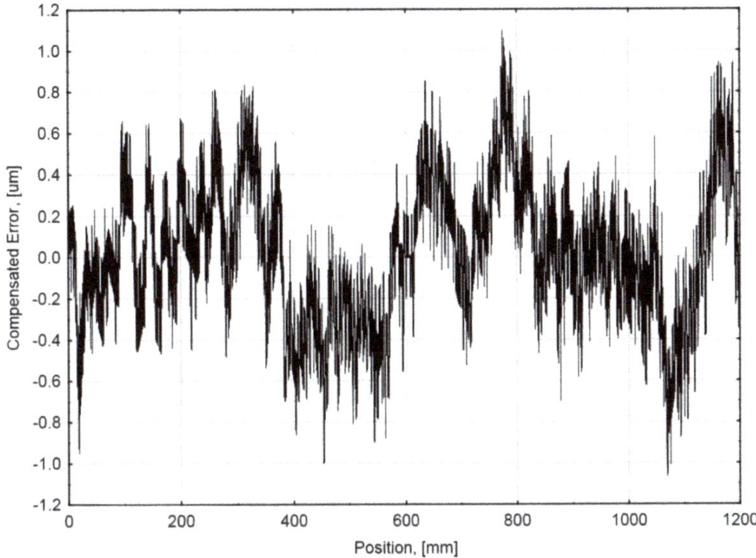

Figure 6. Average accuracy graph of compensated tested encoder at 20 °C ambient temperature.

In order to minimize encoder error at different ambient temperatures, the third component of the general error Equation (4) must be found. The mechanical construction and the thermal behavior of the linear encoder were investigated in a previous work [43] by using the finite element method (FEM). The computer simulation showed that the linear thermal expansion coefficient (CTE) of the measuring scale was greatly changed because of its fixing type, mass, and geometry differences between the scale and the extrusion, etc. Due to these reasons, the CTE of the scale was increased up to $\alpha_{corrected} = 22.9\ \mu m/(m\ °C)$. This value was used for the third member calculation, according the Equation (6). The ambient temperature difference from the nominal temperature was calculated according to the mean value of several external temperature sensor readings. The mathematical algorithm was supplemented and integrated into compensation electronics for other experiments at different ambient temperatures. The average uncompensated linear encoder accuracy graphs and the compensation functions, calculated according the mathematical algorithm, are combined in Figure 7.

How accurately the derived function describes the average uncompensated encoder error was evaluated by a standard deviation of error meanings with respect to the determined function.

$$\sigma_G = \sqrt{\frac{\sum_{i=0}^{N}(\overline{x_i} - F_G(q_i))^2}{N}}, \quad (8)$$

where σ_G—a standard deviation, $(\overline{x_1}, \overline{x_2}, \ldots, \overline{x_N})$—values of calibration process realizations, q_i—value of the argument q (linear position), which help to estimate the compensation function value F_G, and N—number of realizations. To evaluate ~96% of measurements, the standard deviation was multiplied by 2.1. Both values for each ambient temperature are listed in Table 3.

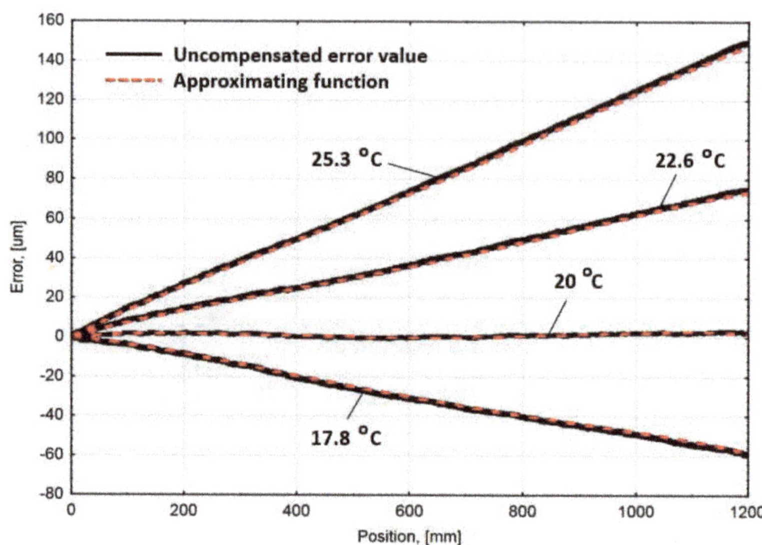

Figure 7. Combined average accuracy graph of uncompensated linear encoder average error at different ambient temperatures, and their compensation functions calculated according the presented mathematical algorithm.

Table 3. Parameters of approximating function and compensated tested encoder accuracy graphs.

Parameter	Ambient Temperature			
	17.8 °C	20 °C	22.6 °C	25.3 °C
Accuracy of approximating function (by mean of standard deviation)	±0.98 µm	±0.34 µm	±1.02 µm	±1.40 µm
Std. dev. of ~96% measurements (Std. dev. multiplied by 2.1)	±2.06 µm	±0.72 µm	±2.14 µm	±2.94 µm
Maximal error value (Non-compensated encoder)	0.07 µm	3.43 µm	75.47 µm	150.03 µm
Minimal error value (Non-compensated encoder)	−60.09 µm	−0.96 µm	0.01 µm	0.15 µm
Average accuracy of non-compensated encoder	±30.08 µm	±2.20 µm	±37.74 µm	±75.09 µm
Maximal error value (Compensated encoder)	0.57 µm	1.10 µm	2.51 µm	3.26 µm
Minimal error value (Compensated encoder)	−2.48 µm	−1.07 µm	−0.73 µm	0.64 µm
Average accuracy of compensated encoder	±1.52 µm	±1.08 µm	±1.62 µm	±1.95 µm

After applying the specified functions, the compensation electronics gave corrected position values. The average compensated linear encoder accuracy graphs and corresponding average uncompensated error values are shown in Figure 8.

The main indicators, such as maximum and minimum values and unidirectional systematic encoder error (average encoder accuracy), etc., are listed in Table 3.

4.1. Estimation of the Real Thermal Coefficient

The accuracy of the compensating value calculation highly depends on how precisely the approximating function is fitted to measured data. Unfortunately, the higher order polynomials or even more complex interpolation functions could cause a practical problem with their integration and calculation time. More efficient and more expensive calculation platforms could be needed. Another way to improve the precision of the presented approach is to estimate the real coefficient of the linear thermal expansion (CTE). Because the mounting of the tested encoder during the performed experiments allows the free axial movement (the encoder could freely expand and contract), the real CTE estimation could be done by analyzing the experimental data of average uncompensated encoder error values.

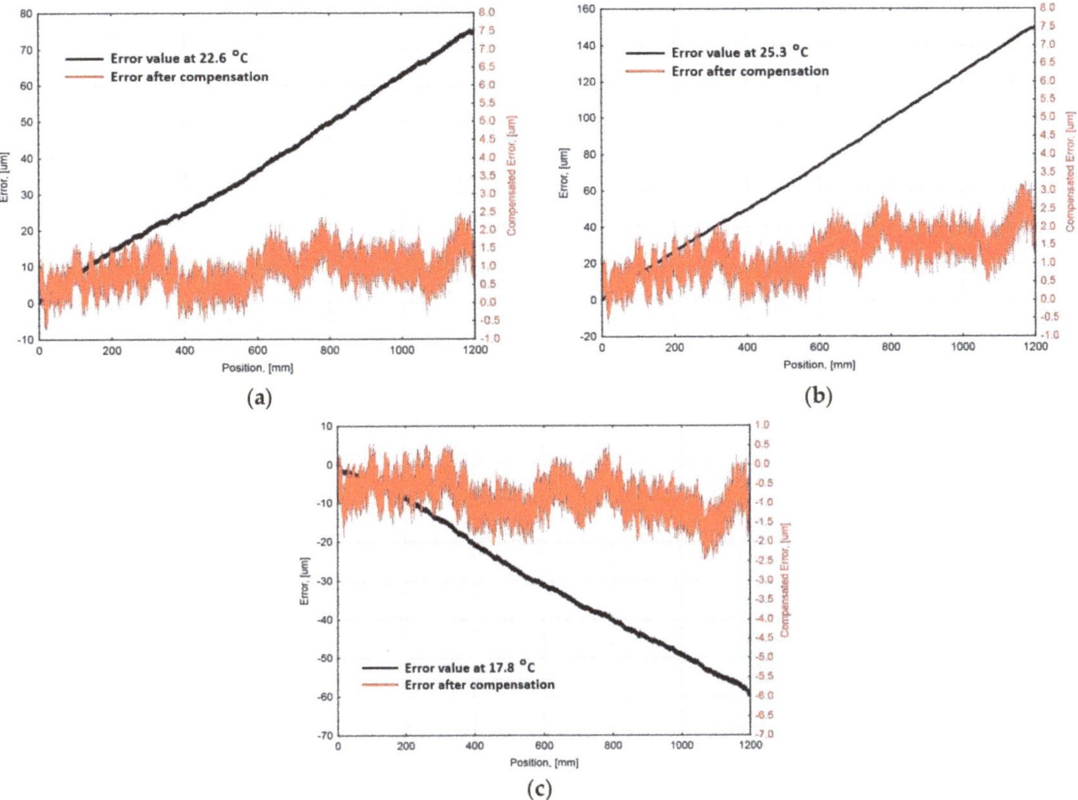

Figure 8. Compensated linear encoder average accuracy graphs (red line/right Y-axis units) with corresponding uncompensated error values (black line/left Y-axis units), at different ambient temperatures: (**a**) at 22.6 °C; (**b**) at 25.3 °C; and (**c**) at 17.8 °C.

The average accuracy graphs obtained at stable 17.8 °C and 22.6 °C ambient temperatures were taken. For a detailed interpretation, both graphs are represented in Figure 9a. The difference between the values of the graphs was calculated at every particular position and plotted in Figure 9b. The received meanings of the differences were approximated with the linear regression line, whose equation is:

$$y = 0.1111q + 1.011, \tag{9}$$

where y—approximated value of the differences, q—is the position value, number 0.1111 is a constant that represents the slope of the linear regression line, and constant 1.011 is the ordinate at the origin.

The real CTE α_{Real} is estimated as the ratio between the determined approximating line slope and the span of ambient temperatures [38]:

$$\alpha_{Real} = \frac{0.1111}{22.6 - 17.8} = 23.15 \left[\frac{\mu m}{m \cdot °C}\right], \tag{10}$$

The calculated value of the CTE is slightly different compared to the simulated $\alpha_{corrected}$. This determined value is greater than the theoretical thermal coefficient, which was used in FEM simulation as aluminum extrusion material property (23×10^{-6} m/(m°C). This suggests that the real CTE of aluminum extrusion is greater. In the literature, the CTE of the aluminum varies from 23×10^{-6}m/(m°C) to 24×10^{-6}m/(m°C). Such an experimental result allows to improve the computer model and the precision of the presented compensation algorithm.

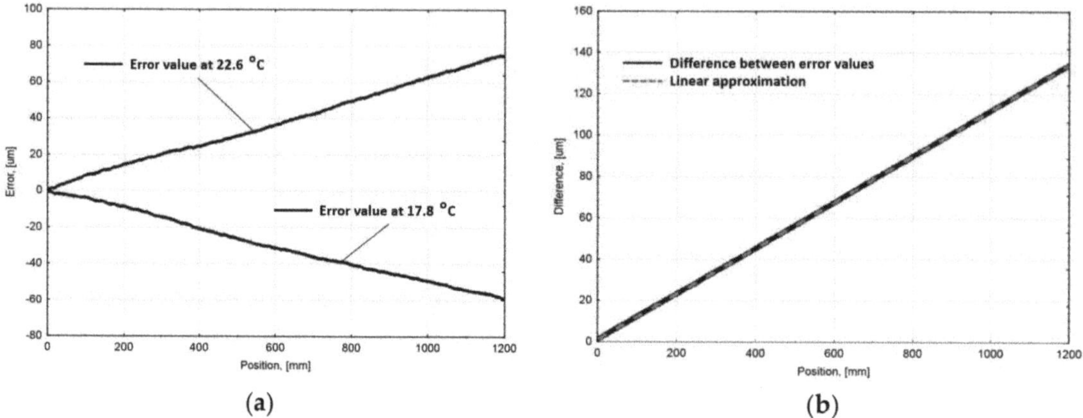

Figure 9. (**a**) Accuracy graphs of uncompensated linear encoder at 17.8 °C and 22.6 °C; (**b**) Differences between accuracy graphs and linear fitting line to the difference.

4.2. Recalculation of the Compensated Results According to the Real CTE

The presented compensation algorithm was adjusted by including the estimated real thermal coefficient into the Equation (6). The compensated error values at different ambient temperatures were calculated and subtracted from the average uncompensated encoder values, recorded during the experiments. This allowed to determine the influence of the introduced changes and compare it to the performed test data. The recalculated compensated error graphs are shown in Figure 10.

The accuracy of approximating compensation functions based on a standard deviation and the parameters of compensated encoder errors are listed in Table 4.

Table 4. Parameters of approximating function and theoretically calculated encoder accuracy graphs (including experimentally estimated real thermal coefficient).

Parameter	Ambient Temperature		
	17.8 °C	**22.6 °C**	**25.3 °C**
Accuracy of approximating function (by mean of standard deviation)	±0.69 µm	±0.67 µm	±0.65 µm
Std. dev. of ~96% measurements (Std. dev. multiplied by 2.1)	±1.45 µm	±1.41 µm	±1.35 µm
Maximal error value (Compensated encoder)	0.95 µm	1.88 µm	1.92 µm
Minimal error value (Compensated encoder)	−1.89 µm	−1.04 µm	−1.04 µm
Compensated encoder accuracy	±1.42 µm	±1.46 µm	±1.48 µm

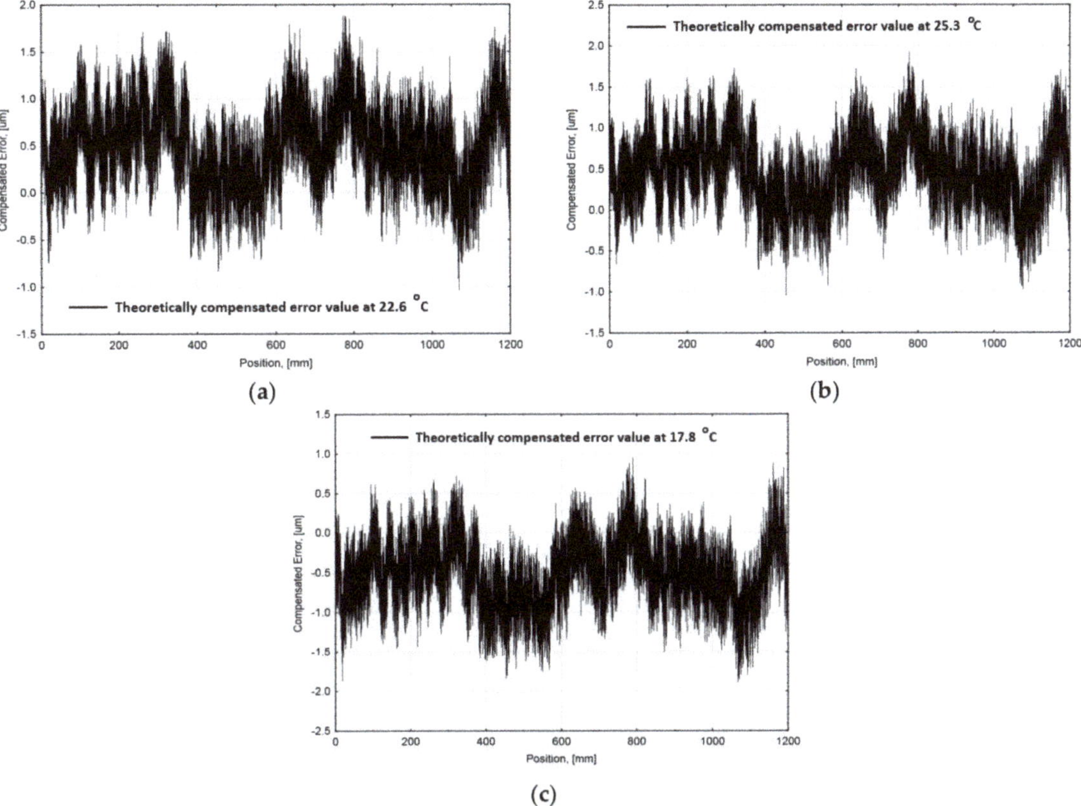

Figure 10. Theoretically recalculated compensated linear encoder accuracy graphs at different ambient temperatures: (**a**) at 22.6 °C; (**b**) at 25.3 °C; and (**c**) at 17.8 °C.

5. Discussion

The proposed mathematical algorithm is based on the approximation of geometric and thermoelastic linear encoder error by a simple parametric function and calculation of linear nature thermal error component in accordance to varying ambient temperature. Based on previous work's [46] accomplished computer modeling results and the practical tests, such a compensation technique was adapted to the selected linear optical encoder.

Performed experiments demonstrated that the designed subsequent electronics are suitable for the realization of the presented compensation approach. An FPGA-based calculation platform properly read tested linear optical encoder position value and compensated it according the installed mathematical function. All processes were performed in real-time; therefore, it could be applied into an industrial, scientific, or other technological application.

The experimental results demonstrate that geometric and introduced thermoelastic linear position measurement error could be drastically reduced:

- At nominal 20 °C ambient temperature, measured encoder average accuracy was ±2.20 μm. After the average position error graph approximation and position compensation, the recorded error was minimized to ±1.08 μm. The approximation accuracy evaluating ~96% measured positions reached ±0.72 μm.
- At different ambient temperatures (17.8 °C; 22.6 °C and 25.3 °C), encoder average accuracy respectively reached ±30.08 μm; ±37.74 μm, and ±75.09 μm without compensation. Applied mathematical algorithm at these temperatures could approximate encoder error correspondingly: ±2.06 μm; ±2.14 μm, and ±2.94 μm. Considering that

the maximal error value reaches up to ~150 µm, the average accuracy of approximation was accepted as reasonable.

- After the position compensation process, the average encoder accuracy at different temperatures was determined as the following: ±1.52 µm (at 17.8 °C); ±1.62 µm (at 22.6 °C), and ±1.95 µm (at 25.3 °C). Considering that the specified accuracy of a standard encoder is ±5 µm per meter, the compensated average encoder accuracy (including the uncertainty of the approximation at different temperatures) was within this range. It can be stated that the performance of the encoder remained under different thermal environmental conditions.
- The presented algorithm could be optimized according to experimentally estimated real CTE value. Embedding this value into the compensation allowed to improve the accuracy of the encoder error approximation which in turn decreased the total error. Theoretical calculations show that the encoder accuracy could reach: ±1.42 µm (at 17.8 °C); ±1.46 µm (at 22.6 °C); and ±1.48 µm (at 25.3 °C).

6. Conclusions

The approach of a thermal and geometric error compensation for a linear encoder is introduced in this article. Having designed the suitable technological equipment, performed experimental research, and analyzed the systematized results, the following conclusions are drawn:

1. The thermoelastic linear encoder deformation caused by external heat sources and changing ambient temperature is significant. Considering the linear thermal expansion coefficient, which greatly depends on an encoder design and used materials, and the working environment conditions, the linear position measurement uncertainty could have a big numerical value. This could lead to undesirable performance of the encoder or even a whole application.
2. The proposed error compensation model is suitable for thermoelastic and geometric error compensation. The performed experiments show that the introduced tested encoder error could be significantly reduced up to 98%. Usage of this kind's compensation might be cheaper and more appropriate solution compared to others, like encoder design including close to zero thermal expansion materials or control of working environment temperature.
3. The compensation algorithm implementation into FPGA-based calculation platform demonstrates the reliable performance. Such hardware selection can ensure an appropriate calculation speed for a real-time application. Due to its flexibility and low cost, it is possible to integrate this device into encoder design or use it like a subsequent electronics module.

However, certain details still exist that require in depth theoretical and experimental research, such as the dynamically changing temperature gradients, different encoder designs, and mounting methods, etc.

Author Contributions: D.G. designed and composed the experimental setup, contributed to the experiments, and wrote the paper; A.K. (Artūras Kilikevičius) performed the experiments and collected and processed the data; A.K. (Albinas Kasparaitis) supervised the research. All authors have read and agreed to the published version of the manuscript.

Funding: This research received no external funding.

Institutional Review Board Statement: Not applicable.

Informed Consent Statement: Not applicable.

Data Availability Statement: Not applicable.

Conflicts of Interest: The authors declare no conflict of interest.

Abbreviations

The following abbreviations are used in this manuscript:

CTE	Coefficient of Thermal Expansion
FPGA	Field-programmable Gate Array

References

1. Zhao, L.; Cheng, K.; Chen, S.; Ding, H.; Zhao, L. An approach to investigate moiré patterns of a reflective linear encoder with application to accuracy improvement of a machine tool. *Proc. Inst. Mech. Eng. Part B J. Eng. Manuf.* **2018**, *233*, 927–936. [CrossRef]
2. Bai, Q.; Liang, Y.; Cheng, K.; Long, F. Design and analysis of a novel large-aperture grating device and its experimental validation. *Proc. Inst. Mech. Eng. Part B J. Eng. Manuf.* **2013**, *227*, 1349–1359. [CrossRef]
3. Liu, C.; Jywe, W.; Hsu, T. The application of the double-redheads planar encoder system for error calibration of computer numerical control machine tools. *Proc. Inst. Mech. Eng. Part B J. Eng. Manuf.* **2004**, *218*, 1077–1089. [CrossRef]
4. Gao, W.; Kim, S.W.; Bosse, H.; Haitjema, H.; Chen, Y.; Lu, X.D.; Knapp, W.; Weckenmann, A.; Estler, T.; Kunzmann, H. Measurement technologies for precision positioning. *Cirp Ann. Manuf. Technol.* **2015**, *64*, 773–796. [CrossRef]
5. Ishii, N.; Taniguchi, K.; Yamazaki, K.; Aoyama, H. Performance improvement of machine tool by high accuracy calibration of built-in rotary encoders. In Proceedings of the 9th International Conference on Leading Edge Manufacturing in 21st Century, Hiroshima City, Japan, 13–17 November 2017.
6. Xie, L.-B.; Qiu, Z.-C.; Zhang, X.-M. Development of a 3-PRR precision tracking system with full closed-loop measurement and control. *Sensors* **2019**, *19*, 1756. [CrossRef]
7. Tang, T.; Chen, S.; Huang, X.; Yang, T.; Qi, B. Combining load and motor encoders to compensate nonlinear disturbances for high precision tracking control of gear-driven Gimbal. *Sensors* **2018**, *18*, 754. [CrossRef]
8. Chong, K.K.; Wong, C.W.; Siaw, F.; Yew, T.; Ng, S.; Liang, S.; Lim, Y.; Liong, L.S. Integration of an on-axis general sun-tracking formula in the algorithm of an open-loop sun-tracking system. *Sensors* **2009**, *9*, 7849–7865. [CrossRef]
9. Algburi, R.N.A.; Gao, H. Health assessment and fault detection system for an industrial robot using the rotary encoder signal. *Energies* **2019**, *12*, 2816. [CrossRef]
10. Han, Z.; Jianjun, Y.; Gao, L. External force estimation method for robotic manipulator based on double encoders of joints. *IEEE Int. Conf. Robot. Biomim.* **2018**. [CrossRef]
11. Peng, L.; Xiangpeng, L. Common sensors in industrial robots: A review. *J. Phys. Conf. Ser.* **2019**, *1267*, 012036.
12. Mikhel, S.; Popov, D.; Mamedov, S.; Klimchik, A. Advancement of robots with double encoders for industrial and collaborative applications. In Proceedings of the 23rd Conference of Open Innovations Association (FRUCT), Bologna, Italy, 13–16 November 2018; pp. 246–252.
13. Rodriguez-Donate, C.; Osornio-Rios, R.A.; Rivera-Guillen, J.R.; Romero-Troncoso, R.J. Fused smart sensor network for multi-axis forward kinematics estimation in industrial robots. *Sensors* **2011**, *11*, 4335–4357. [CrossRef] [PubMed]
14. Kimura, A.; Gao, W.; Hosono, K.; Shimizu, Y.; Shi, L.; Zeng, L. A sub-nanometric three-axis surface encoder with short-period planar gratings for stage motion measurement. *Precis. Eng.* **2012**, *36*, 576–585. [CrossRef]
15. Lee, C.B.; Kim, G.H.; Lee, S.K. Design and construction of a single unit multi-function optical encoder for a six-degree-of-freedom motion error measurement in an ultraprecision linear stage. *Meas. Sci. Technol.* **2011**, *22*, 105901. [CrossRef]
16. Li, Y.T.; Fan, K.C. A novel method of angular positioning error analysis of rotary stages based on the Abbe principle. *Proc. Inst. Mech. Eng. Part B J. Eng. Manuf.* **2018**, *232*, 1885–1892. [CrossRef]
17. Lou, Z.F.; Hao, X.P.; Cai, Y.D.; Lu, T.F.; Wang, X.D.; Fan, K.C. An embedded sensors system for real-time detecting 5-DOF error motions of rotary stages. *Sensors* **2019**, *19*, 2855. [CrossRef]
18. Ramesh, R.; Mannan, M.A.; Poo, A.N. Error compensation in machine tools—A review: Part I: Geometric, cutting-force induced and fixture-dependent errors. *Int. J. Mach. Tools Manuf.* **2000**, *40*, 1235–1256. [CrossRef]
19. Ramesh, R.; Mannan, M.A.; Poo, A.N. Error compensation in machine tools—A review: Part II: Thermal errors. *Int. J. Mach. Tools Manuf.* **2000**, *40*, 1257–1284. [CrossRef]
20. Mareš, M.; Horejš, O.; Havlik, L. Thermal error compensation of a 5-axis machine tool using indigenous temperature sensors and CNC integrated Python code validated with a machined test piece. *Precis. Eng.* **2020**, *66*, 21–30. [CrossRef]
21. Mareš, M.; Horejš, O. Modelling of cutting process impact on machine tool thermal behaviour based on experimental data. *Procedia Cirp* **2017**, *58*, 152–157. [CrossRef]
22. Zaplata, J.; Pajor, M. Piecewise compensation of thermal errors of a ball screw driven CNC axis. *Precis. Eng.* **2019**, *60*, 160–166. [CrossRef]
23. Li, Y.; Zhao, J.; Ji, S.; Liang, F. The selection of temperature-sensitivity points based on K-harmonic means clustering and thermal positioning error modeling of machine tools. *Int. J. Adv. Manuf. Technol.* **2019**, *100*, 2333–2348. [CrossRef]
24. Polyakov, A.N.; Parfenov, I.V. Thermal error compensation in CNC machine tools using measurement technologies. *J. Phys. Conf. Ser.* **2019**, *1333*, 062021. [CrossRef]
25. Yao, X.; Hu, T.; Yin, G.; Cheng, C. Thermal error modeling and prediction analysis based on OM algorithm for machine tools spindle. *Int. J. Adv. Manuf. Technol.* **2020**, *106*, 3345–3356. [CrossRef]

26. Tan, F.; Deng, C.; Xiao, H.; Luo, J.; Zhao, S. A wrapper approach-based key temperature point selection and thermal error modeling method. *Int. J. Adv. Manuf. Technol.* **2020**, *106*, 907–920. [CrossRef]
27. Lei, M.; Yang, J.; Wang, S.; Zhao, L.; Xia, P.; Jiang, G.; Mei, X. Semi-supervised modeling and compensation for the thermal error of precision feed axis. *Int. J. Adv. Manuf. Technol.* **2019**, *104*, 4629–4640. [CrossRef]
28. Shi, X.; Wang, W.; Mu, Y.; Yang, X. Thermal characteristics testing and thermal error modeling on a worm gear grinding machine considering cutting fluid thermal effect. *Int. J. Adv. Manuf. Technol.* **2019**, *103*, 4317–4329. [CrossRef]
29. Alejandre, I.; Artes, M. Machine tool errors caused by optical linear encoders. *Proc. Inst. Mech. Eng. Part B J. Eng. Manuf.* **2004**, *218*, 113–122. [CrossRef]
30. Lopez, J.; Artes, M.; Alejandre, I. Analysis of optical linear encoder's errors under vibration at different mounting conditions. *Measurement* **2011**, *44*, 1367–1380. [CrossRef]
31. Lopez, J.; Artes, M.; Alejandre, I. Analysis under vibrations of optical linear encoders based on different scanning methods using an improved experimental approach. *Exp. Tech.* **2012**, *36*, 35–47. [CrossRef]
32. Alejandre, I.; Artes, M. Method for the evaluation of optical encoders performance under vibration. *Precis. Eng.* **2007**, *31*, 114–121. [CrossRef]
33. Albrecht, C.; Klock, J.; Martens, O.; Schumacher, W. Online estimation and correction of systematic encoder line errors. *Machines* **2017**, *5*, 1. [CrossRef]
34. Mendenhall, M.H.; Windover, D.; Henins, A.; Cline, J.P. An algorithm for the compensation of short-period errors in optical encoders. *Metrologia* **2015**, *52*, 685. [CrossRef]
35. Ye, G.; Fan, S.; Liu, H.; Li, X.; Yu, X.; Yu, H.; Shi, Y.; Yin, L.; Lu, B. Design of a precise and robust linearized converter for optical encoders using a ratiometric technique. *Meas. Sci. Technol.* **2014**, *25*, 12. [CrossRef]
36. Yandayan, T.; Geckeler, R.D.; Just, A.; Krause, M.; Akgoz, S.A.; Aksulu, M.; Grubert, B.; Watanabe, T. Investigation of interpolation errors of angle encoders for high precision angle metrology. *Meas. Sci. Technol.* **2018**, *29*. [CrossRef]
37. Wang, Y.; Liu, Y.; Yan, X.; Chen, X.; Lv, H. Compensation of Moire fringe sinusoidal deviation in photoelectric encoder based on tunable filter. In Proceedings of the Engineering, Computer Science 2011 Symposium on Photonics and Optoelectronics (SOPO) NA, Wuhan, China, 16–18 May 2011. [CrossRef]
38. Alejandre, I.; Artes, M. Real thermal coefficient in optical linear encoders. *Exp. Tech.* **2004**, *28*, 18–22. [CrossRef]
39. Alejandre, I.; Artes, M. Thermal non-linear behaviour in optical linear encoders. *Int. J. Mach. Tools Manuf.* **2006**, *46*, 1319–1325. [CrossRef]
40. Yu, L.D.; Bao, W.H.; Zhao, H.N.; Jia, H.K.; Zhang, R. Application and novel angle measurement error compensation method of circular gratings. *Opt. Prec. Eng.* **2019**, *27*, 1719–1726.
41. Jia, H.K.; Yu, L.D.; Jiang, Y.Z.; Zhao, G.N.; Cao, J.M. Compensation of rotary encoders using Fourier expansion-back propagation neural network optimized by genetic algorithm. *Sensors* **2020**, *20*, 2603. [CrossRef]
42. Hu, F.; Chen, X.; Cai, N.; Lin, Y.J.; Zhang, F.; Wang, H. Error analysis and compensation of an optical linear encoder. *IET Sci. Meas. Technol.* **2018**, *12*, 561–566. [CrossRef]
43. Gurauskis, D.; Kilikevičius, A.; Borodinas, S.; Kasparaitis, A. Analysis of geometric and thermal errors of linear encoder for real-time compensation. *Sens. Actuators A Phys.* **2019**, *296*, 145–154. [CrossRef]
44. ISO 5725-1: 1994 (R2018). Accuracy (Trueness and Precision) of Measurement Methods and Results—Part 1: General Principles and Definitions. Available online: https://www.iso.org/standard/11833.html (accessed on 6 January 2021).
45. Fast and Simple Measurement of Position Changes. Available online: https://www.ichaus.de/upload/pdf/WP2en_EncoderInterface_14082012.pdf (accessed on 27 November 2020).
46. BiSS Interface. Available online: https://biss-interface.com/ (accessed on 27 November 2020).

Letter

Temperature Field Boundary Conditions and Lateral Temperature Gradient Effect on a PC Box-Girder Bridge Based on Real-Time Solar Radiation and Spatial Temperature Monitoring

Xiao Lei [1], Xutao Fan [1], Hanwan Jiang [2,*], Kunning Zhu [3] and Hanyu Zhan [4]

1 College of Civil and Transportation Engineering, Hohai University, 1 Xikang Road, Nanjing 210098, China; leix2010@126.com (X.L.); fanxutao11@163.com (X.F.)
2 Civil & Environmental Engineering Department, University of Wisconsin Platteville, Platteville, WI 53818, USA
3 China Design Group Co., Ltd., 9 Ziyun St., Nanjing 210014, China; redkun@163.com
4 Klipsch School of Electrical and Computer Engineering, New Mexico State University, Las Cruces, NM 88003, USA; hzhan@nmsu.edu
* Correspondence: jianghan@uwplatt.edu

Received: 29 July 2020; Accepted: 4 September 2020; Published: 15 September 2020

Abstract: Climate change could impose great influence on infrastructures. Previous studies have shown that solar radiation is one of the most important factors causing the change in temperature distribution in bridges. The current temperature distribution models developed in the past are mainly based on the meteorological data from the nearest weather station, empirical formulas, or the testing data from model tests. In this study, a five-span continuous Prestressed-concrete box-girder bridge was instrumented with pyranometers, anemometers, strain gauges, displacement gauges, and temperature sensors on the top and bottom slabs and webs to measure the solar radiation, wind speeds, strain, displacement, and surface temperatures, respectively. The continuously monitoring data between May 2019 and May 2020 was used to study the temperature distributions caused by solar radiation. A maximum positive lateral temperature gradient prediction model has been developed based on the solar radiation data analysis. Then, the solar radiation boundary condition obtained from the monitoring data and the lateral temperature gradient prediction model were utilized to compute the tensile stresses in the longitudinal and transverse directions. It was demonstrated in this study that the tensile stress caused by the lateral temperature gradient was so significant that it cannot be ignored in structural design.

Keywords: lateral temperature gradient; temperature field; boundary conditions; solar radiation intensity; temperature monitoring

1. Introduction

Most highway bridges are located in open fields and are exposed to the solar radiation, wind, rain, and other environmental changes. Those climate changes can cause substantial temperature variations in bridges, trigger cracking and bearing displacing, and therefore, endanger the serviceability, durability, and safety of the bridges. In extreme cases, the major structural elements could lose their functionality due to the severe damage causing the global structural failure [1,2]. Solar radiation plays an important role in affecting the temperature distribution in the bridge, so that relationship between the solar radiation and the temperature distribution should be properly established. Lee [3] was monitoring the temperature distribution in a five-foot-long prestressed I-shape girder with thermocouples embedded in the flanges and the web for a year. With the continuously measured temperature variations, both vertical and lateral temperature distributions were established. A two-dimensional heat-transfer model was developed to investigate the effects of seasonal variations and girder orientations on temperature distributions. Wang et al. created a finite element model for a concrete box-girder arch bridge in thermal field caused by solar radiation based on meteorological data [4]. Abid et al. monitored air temperature, solar radiation, and wind speeds with sensors and thermocouples for more than a year. They proposed empirical equations to predict the maximum vertical and lateral temperature gradients based on the data collected [5]. The same approach was applied to concrete-encased steel girders [6]. Taysi and Abid investigated the effects of the thermal properties in a full-scale concrete box-girder segment. The individual effect of each of the thermal properties of the concrete box girder were studied with the aid of three-dimensional finite element analysis. Temperature distributions under the extreme thermal loads were compared with AASHTO's (The American Association of State Highway and Transportation Officials) and Bridge Manual's temperature gradient models [7]. Tian et al. conducted a numerical study on temperature effects in the train–bridge interaction system [8]. Lawson et al. used recent meteorological data from two weather stations in Nevada establishing temperature profiles for bridge superstructures and compared the results with AASHTO model [9]. Hagedorn et al. built an I-beam segment and determined vertical and transverse temperature gradients with temperature-monitoring data [10]. Rodriguez et al. implemented a dense array of thermocouples in a box-girder bridge in California and established a Finite Element model to predict bridge internal stresses under the measured temperature variations [11].

In the above-mentioned literature, either meteorological data from the nearest weather station or empirical equations were utilized to develop the temperature distribution resulting from solar radiation. However, oftentimes, the nearest weather station is still far from the bridge, so the meteorological data obtained could still deviate from the local climate conditions. Empirical equations were developed based on the given temperature data, hence they may not be applicable for a different on-site situation. Some studies used pyranometers to measure the intensity of solar radiation and thermocouples to measure the temperature distributions [5–7]. However, only the radiation intensity on the top flanges was investigated, while the effect of the flange shadow casting on the web was not discussed. In reality most box-girder bridges have tilted webs instead of vertical ones. Hence, to study the temperature distributions in box-girder bridges, it is essential to account for the tilt angle of the web and the overhang shadow casting on the web when estimating the solar radiation intensity.

In this study, we instrumented a prestressed concrete box-girder bridge with pyranometers, anemometers, strain gauges, displacement gauges, and temperature sensors on the top and bottom slabs and webs to measure the solar radiation, wind speeds, strain, displacement, and surface temperatures, respectively. The aforementioned data have been collected continuously for over a year and exploited to establish the solar radiation and temperature gradient relationship. In addition, a three-dimensional finite element model was developed to simulate the temperature field in the worst scenario and analyze the corresponding stress distribution based on the real time radiation monitoring data. The novelty of the work lies in the fact that all the boundary conditions have been obtained from real-time monitoring of solar radiation, temperature, and wind speed on an in-service bridge. From the data analysis, the lateral temperature gradient effect has been studied, which has long been ignored in the bridge design practice worldwide. Finally, the finite element analysis has been performed in conjunct with the maximum lateral temperature gradient from the monitoring data. The result indicated that the tensile stresses generated by the lateral temperature gradient alone are so significant that it cannot be ignored in design.

2. Experimental Instrumentation and Data Acquisition

The bridge was built in 2002. It has five continuous spans and each span is 25 m in length. All the sensors were mounted on the fourth span from the south on upstream side. Three optical pyranometers (JMFS-1001) were installed on the top slab, upper corner on the web, and lower position on the web, respectively, for each bridge (Figure 1). The optical pyranometers have spectral range of 0.3~3 μm and can detect the direct radiation from the sun and the reflected radiation from the other objectives as well as the radiation from the incident sunlight with an angle. 58 temperature sensors were embedded inside or on the surface of the box girder in the midspan cross section (Figure 2). Two anemometers for wind speed monitoring were installed on the outer side and at the bottom of the bridge, as shown in Figure 3.

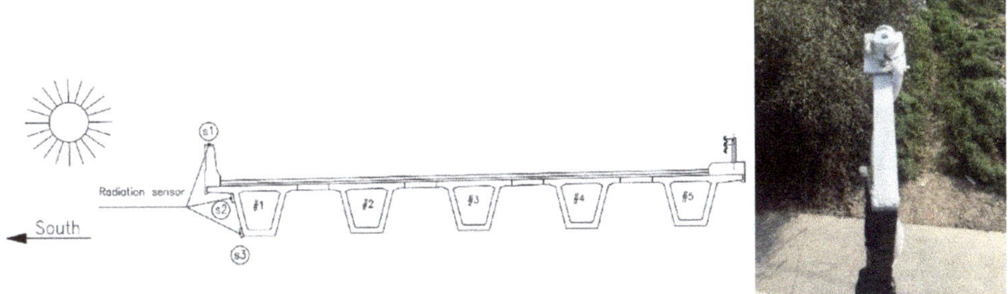

Figure 1. Radiation sensor (optical pyranometer) positions at the midspan.

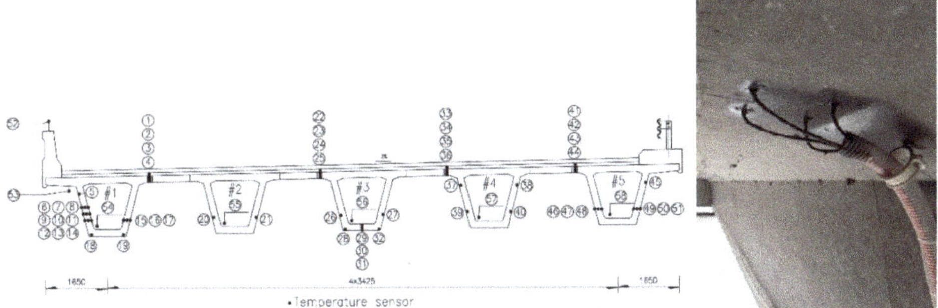

Figure 2. Temperature sensor layout at the midspan.

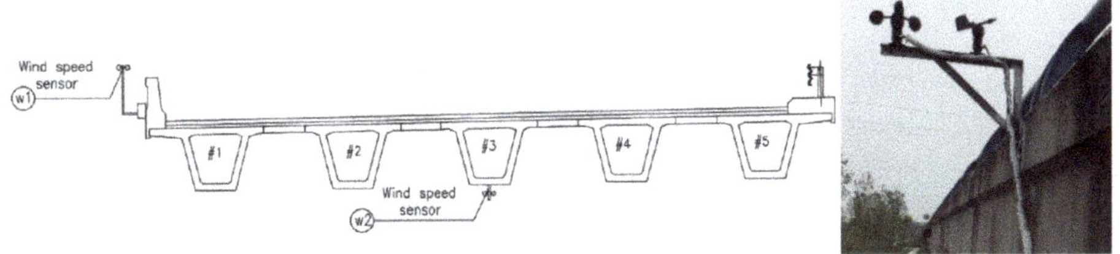

Figure 3. Anemometers for wind speed monitoring positions.

A comprehensive monitoring system was integrated with an all sealed chassis (JMBV-1164), an integrated data acquisition module (JMZX-32A), a temperature acquisition module (JMWT-64RT), and a DTU mobile internet module (JMTX-2017). The system enables 24/7 automatic continuous data acquisition through wire or wireless transmission. The data used in this study were collected from May 2019 to May 2020 at a sampling rate of 1 time per hour.

3. Solar Radiation and Heat Transfer Theory

The Fourier heat transfer differential equation was employed to describe the heat conduction in the concrete bridge [12–15]:

$$\rho C_p \frac{\partial T}{\partial t} = \frac{\partial}{\partial x}(k\frac{\partial T}{\partial x}) + \frac{\partial}{\partial y}(k\frac{\partial T}{\partial y}) + \frac{\partial}{\partial z}(k\frac{\partial T}{\partial z}) \tag{1}$$

where ρ, C_p, and k are density in kg/m^3, specific heat in J/kg °C, and thermal conductivity of concrete W/m °C, respectively. T refers to the temperature at any point in the girder at any time t.

The boundary condition on the surfaces of the box girder can be written as [12–15]:

$$k\frac{\partial T}{\partial x}n_x + k\frac{\partial T}{\partial y}n_y + k\frac{\partial T}{\partial z}n_z + q_c + q_s + q_{re} = 0 \tag{2}$$

where n_x, n_y, and n_z are the direction cosines of the normal vectors; q_c, q_s, and q_{re} are the convection, the total solar radiation and the long-wave radiation in W/m^2. The solar radiation q_s consists of three components including direct solar radiation q_{dr}, solar scattered radiation q_{sr}, and ground reflected radiation q_{gr}, as shown in Equation (3).

$$q_s = q_{dr} + q_{sr} + q_{gr} \tag{3}$$

3.1. Direct Solar Radiation

Direct solar radiation comes straight from the Sun. The direct solar radiation energy I_m can be expressed in the form of Equation (4) [13]:

$$I_m = I_0 \frac{\sin h}{\sin h + \frac{1-p}{p}} \tag{4}$$

where h is solar altitude angle; p is atmospheric transparency coefficient, and I_0 is solar constant.

Solar altitude angle h is generally written as Equation (5) [16].

$$\sin(h) = \cos\varphi\cos\delta\cos\tau + \sin\varphi\sin\delta \tag{5}$$

where ϕ is the latitude of the location; δ is solar declination, and δ = 23.45 sin [360 (284 + N)/365] in which N is the day of the year; τ is the solar hour angle, which is zero at the noon, negative before noon, and positive afternoon converting each hour to 15°.

The direct solar radiation projected on the box-girder web q_{dr} in Equation (3) is calculated as:

$$q_{dr} = I_{\mathrm{m}} \cos\theta \tag{6}$$

where θ is the angle between the Sun rays and the normal line to the web surface and $\cos\theta = \cos\beta \sin h + \sin\beta \cos h \cos(\gamma_z - \gamma)$ in which β is the angle between web and the horizontal direction, γ is the surface azimuth angle, and γ_z is the azimuth angle of the sun.

In a box-girder bridge, the overhang casts shadow on the web and the solar radiation in the shadow should be deducted from the total. The equation of shadow length is defined by Equation (7) [16]:

$$l_s = l_c \frac{\tan h}{\sin(90 + \gamma - \gamma_z)\sin\beta - \cos\beta\tan h} \tag{7}$$

where l_c is the length of the overhang.

3.2. Scattered Solar Radiation

Solar heat scatters on the structures in all directions. Jain assumed that the scattering intensity of the horizontal plane received is linearly related with the transmission coefficient and the averaged radiation. Scattering radiation I_d in the atmosphere can be obtained with Equation (8) [17]:

$$I_d = (1 - 1.13\, k_T)\, I_m \tag{8}$$

where k_T is the transmission coefficient accounting for the attenuation of solar radiation in the atmosphere, which takes a value in between 0.3~0.8 for latitudes of 30~40°.

The scattered solar radiation q_{sr} received by an arbitrary wall is:

$$q_{sr} = I_d\,(1 + \cos\beta)/2 \tag{9}$$

3.3. Ground Reflected Radiation

The bottom slab of the box girder receives the short-length wave radiation reflected from the ground. The reflected solar radiation can be calculated by Equation (10) [18]:

$$q_{gr} = \rho^*(I_m + I_d)(1 - \cos\beta)/2 \tag{10}$$

where ρ^* is the ground reflection coefficient and $\rho^* = 0.1$ for ground reflection and $\rho^* = 0.2$ for water reflection.

4. Theoretical and Measured Solar Radiation Intensity

It is evident from the monitoring data shown in Figure 4 that the solar radiation intensity on the top slab reached maximum in summer due to the direct solar exposure, while the intensity on the lower web reached maximum in winter. Solar radiation intensity on the lower web had maximum value in winter is because the top slab casts shadow on the web and the sun altitude angle gets smaller leading to higher solar radiation in winter.

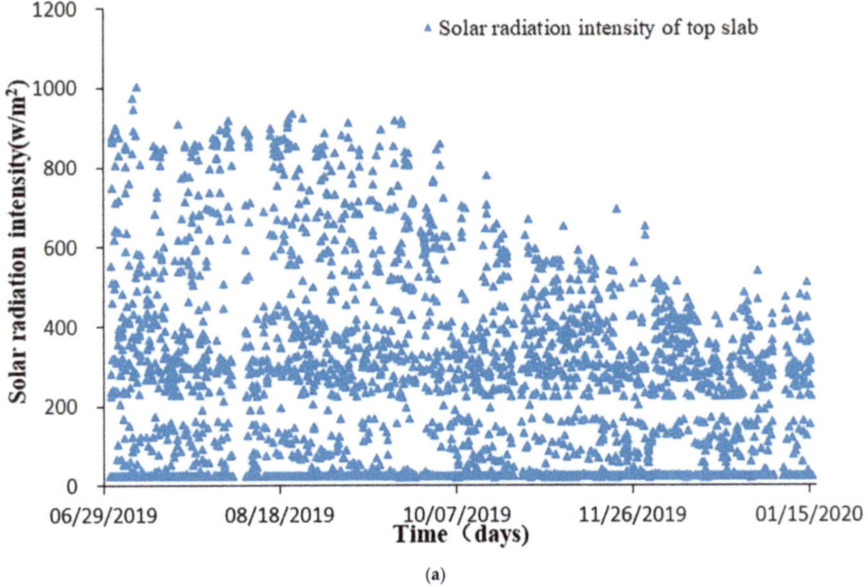

(a)

Figure 4. *Cont.*

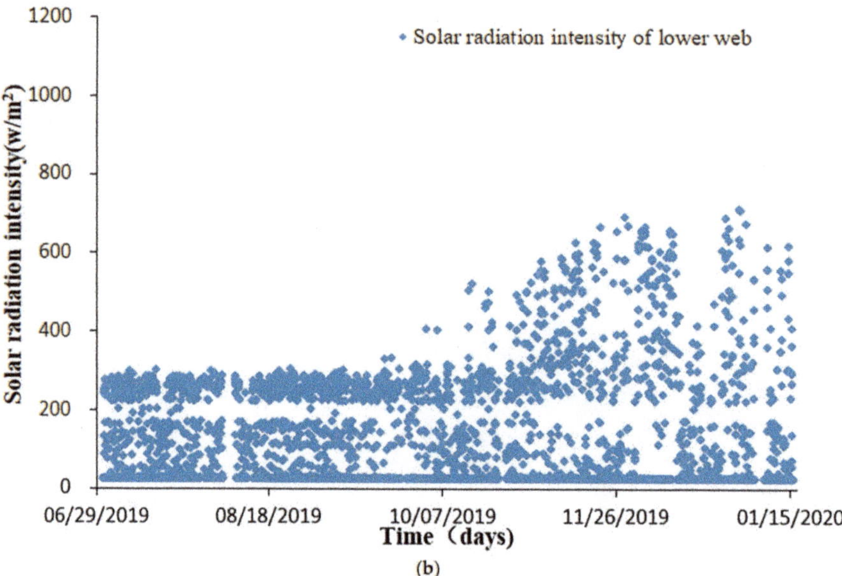

(**b**)

Figure 4. Solar radiation intensity on different parts of the bridge (sampling frequency: 1 time/h). (**a**) Solar radiation variations on the top slab from 29 June 2019 to 15 January 2020; (**b**) solar radiation variations on the lower web from 29 June 2019 to 15 January 2020.

From the aforementioned solar radiation theory, the theoretical solar radiation intensity was calculated and compared with the measured intensity, as shown in Figure 5. During the day of 31 December 2019, the measured solar radiation intensities on the top slab and upper portion of the web were smaller than the theoretical ones, while the measured solar radiation intensities on the lower portion of the web were close to the theoretical values. In addition, note that in Figure 5a,c there was a sudden drawdown around 2 pm, which may be attributed to the temporary cloudiness. The other reason for the underestimation of the theoretical equations may ascribe to the ignorance of the atmosphere counter radiation and earth surface radiation at night.

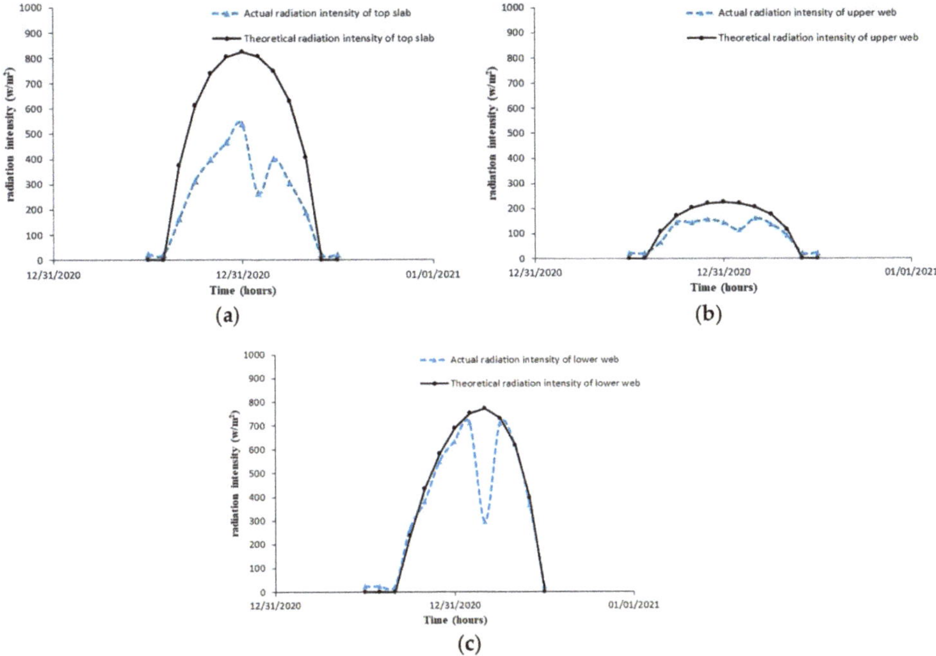

Figure 5. Comparison between measured and theoretical solar radiation intensities. (**a**) Solar radiation on the top slab; (**b**) solar radiation on the upper web; (**c**) solar radiation on the lower web.

5. Solar Radiation and Temperature Gradient Relationship

The box-girder surface temperature increases rapidly under the solar radiation forming a large temperature gradient. The maximum daily temperature gradients were obtained from maximum temperature during the day at #12, #13, and #14 sensor subtracted by the lowest temperature at #15, #16, #17, #20, #21, #26, #27, #39, #40, #46, #47, #48, #49, #50, and #51. It was found that the daily maximum lateral temperature gradient was correlated with daily maximum solar radiation. The Pearson correlation coefficient for each location on the box girder was calculated, as shown in Table 1.

Table 1. Pearson correlation coefficients between daily maximum temperature gradient and daily maximum solar radiation.

Item	Daily Maximum Solar Radiation		
Position	Top slab (s1)	Upper web (s2)	Lower web (s3)
Pearson correlation coefficients	0.124	0.276	0.879

It is indicated in Table 1 that the correlation between maximum daily positive lateral temperature gradient and maximum daily solar radiation on the lower web is the highest. In Figure 6, a linear regression was utilized to best fit the data described in Equation (11).

$$T^{+}_{\text{max-la}} = 0.01377 I_{\text{max-lw}} + 0.4838 \quad (11)$$

where $T^{+}_{\text{max-la}}$ is the maximum daily positive lateral temperature gradient in °C and $I_{\text{max-lw}}$ is the maximum daily solar radiation intensity on the lower web. The standard deviation of $T^{+}_{\text{max-la}}$ d is equal to 2.196 °C. The dashed line in Figure 6 represents the $T^{+}_{\text{max-la}} + 2d$, and it is just above all the temperature gradient data. The dashed line is the envelope for the temperature gradient data and can be used for maximum daily positive lateral temperature gradient prediction through the given maximum daily solar radiation intensity on the lower web (Equation (12)).

$$T^{+}_{\text{max-la}} + 2d = 0.01377 I_{\text{max-lw}} + 4.8752 \quad (12)$$

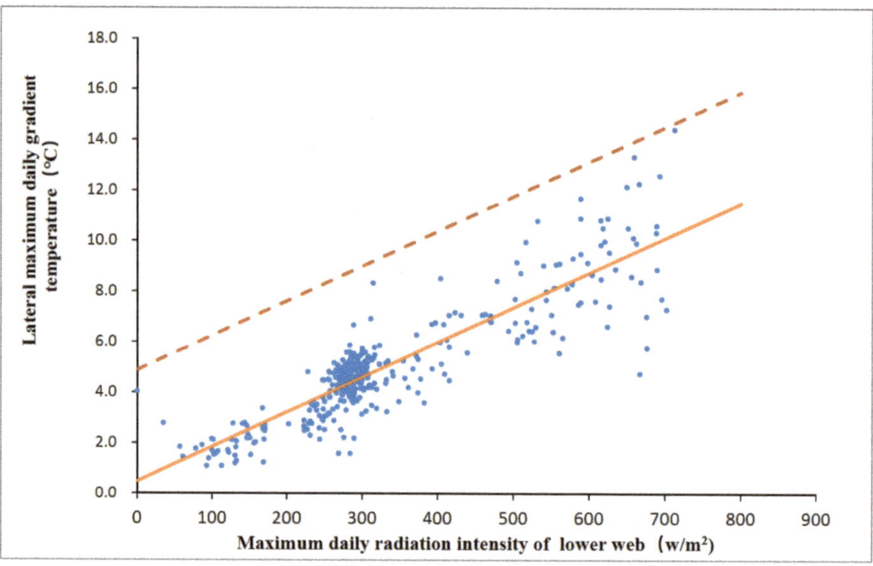

Figure 6. Relation between maximum daily positive lateral temperature gradient and maximum daily solar radiation intensity on the lower web.

6. Simulation of Temperature Field with Finite Element Method

6.1. Finite Element Model

A five-span continuous bridge model was created with ANSYS software. Since the bridge is symmetric in centerline, half of the five spans (2.5 spans) were modeled in ANSYS. The three-dimensional thermal element Solid 90 was selected for the structural analysis. The 3D model has 49,848 elements and 205,400 nodes, as shown in Figure 7.

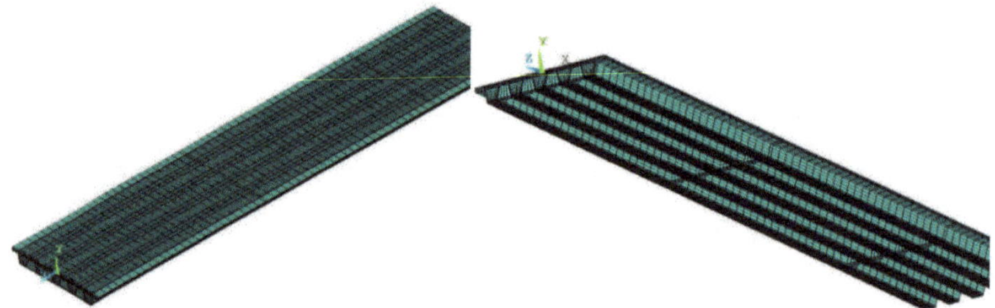

Figure 7. The 3D Finite Element model of the PC box girder.

6.2. Parameters Used in Finite Element Modeling

6.2.1. The Thermo-Physical Properties of the Box Girder

Note that the asphalt pavement was built into the FE model together with the PC box girder. The thermo-physical properties for both of the asphalt and concrete are listed in Table 2.

Table 2. The thermo-physical properties of the box girder and its asphalt pavement.

Material	Density (kg/m^3)	Specific Capacity (J/kg °C)	Heat Thermal Conductivity (W/m °C)	Absorption Rate
Concrete	2500	880	2.5	0.4
Asphalt concrete	1700	1000	1.403	0.8

6.2.2. Boundary Conditions

It was observed during the one-year long monitoring, maximum daily positive lateral temperature gradient varied with solar altitude angle h. In summer, solar altitude angle h is larger than in winter, so that the solar radiation duration in the day is shorter leading to a smaller maximum daily positive lateral temperature gradient. It is evident in Figure 8 that December 31st witnessed the largest maximum daily positive lateral temperature gradient in the year.

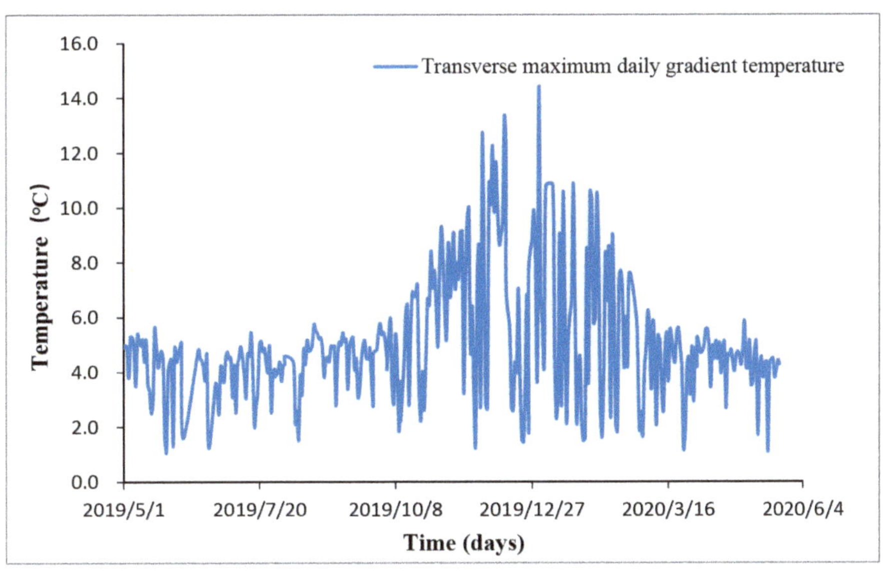

Figure 8. Maximum daily positive lateral temperature gradient variations.

Taking the monitoring data of the wind speed, solar radiation, and temperature within and out of the box girder on 30–31 December 2019 (Figure 9), it was observed that the ambient temperature on the top of the box girder and radiation intensity almost doubled on December 31 compared with that of December 30. As the ambient temperature increased on December 31, the outer surface of the web had been exposed to solar radiation longer in winter than the other time during the year. Meanwhile, the temperature within the box girder had less fluctuation and stayed low. Hence, the combining effects of the ambient temperature and solar radiation increase, and the not much changed inbox temperature, the maximum temperature gradient in the year, was generated in those days.

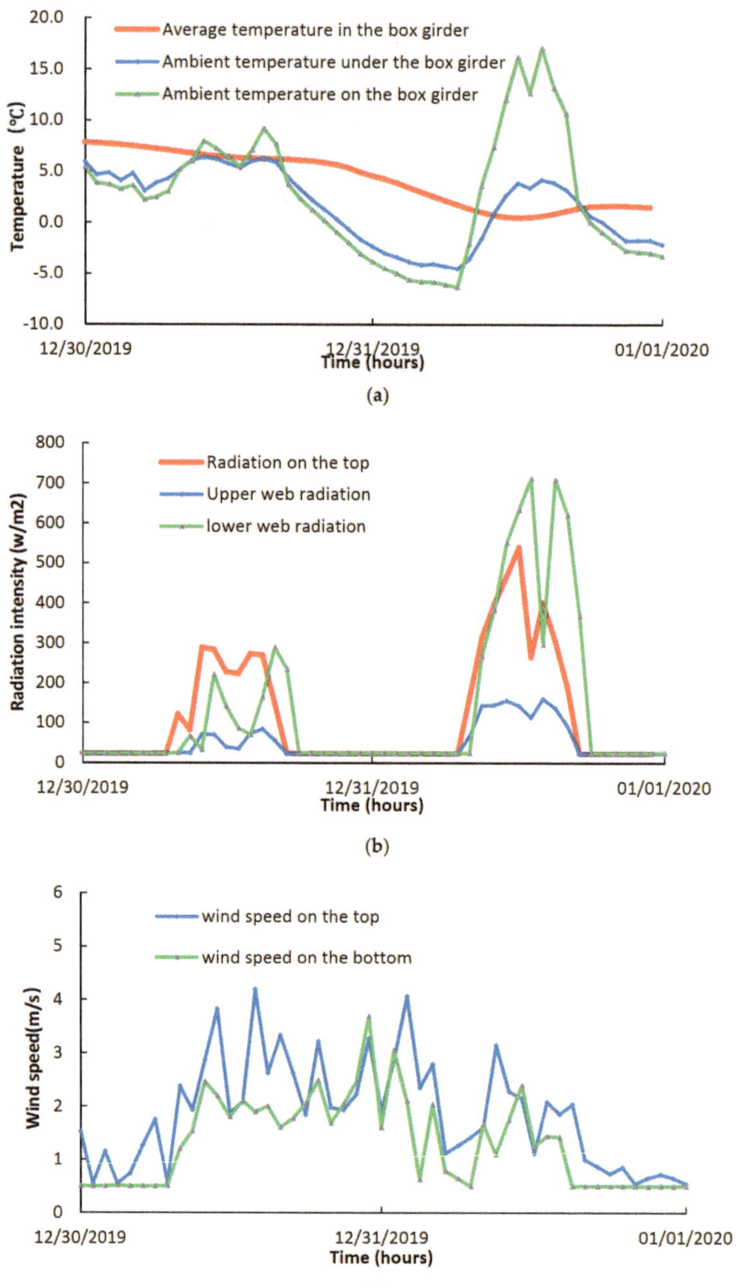

Figure 9. Variations on the boundaries of the box girder. (**a**) Temperature variations within and outside of the box girder; (**b**) solar radiation intensity variations on the top and web surface of the box girder on 30–31 December 2019; (**c**) wind speed variations on the top and web surface of the box girder on 30–31 December 2019.

6.2.3. Comparison between the FE Analysis and Experimental Results

Figure 10 gives the comparison between the measured temperatures on 31 December 2019 and the simulated temperatures in FE analysis for nine temperature test points. The nine test points were grouped into three with each consisting one point on the top slab, one on the web, and one on the bottom slab. It can be seen from the comparison in Figure 10 that the differences between the measured and simulated temperatures are less than 2 °C except those on the top slab. The maximum measured and simulated temperature difference on the top slab reached 4.0 °C. Overall, the 3D FE model can be considered suitable for simulation of thermal conductivity, heat convection, ambient temperatures, and solar radiation in the box girder [6,8,11,14]. The difference between the measured and FE simulated temperature values may be attributed to the assumptions made in FE analysis that the heat thermal conductivity and the absorption rate are uniform in the structure. However, in really, those parameters could not be evenly distributed in heterogeneous materials like concrete and asphalt.

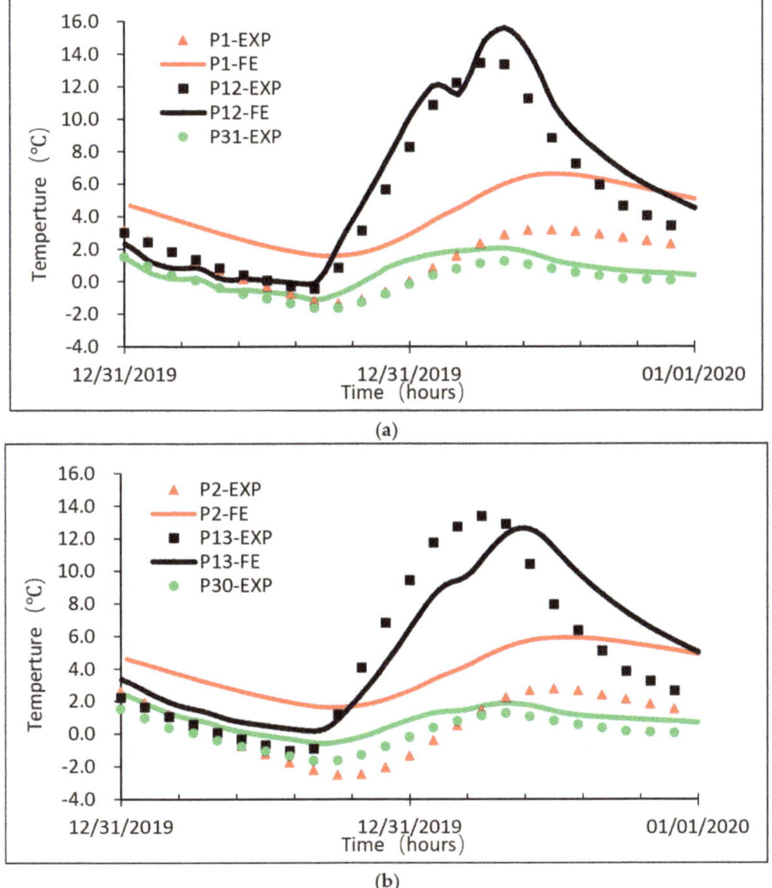

Figure 10. *Cont.*

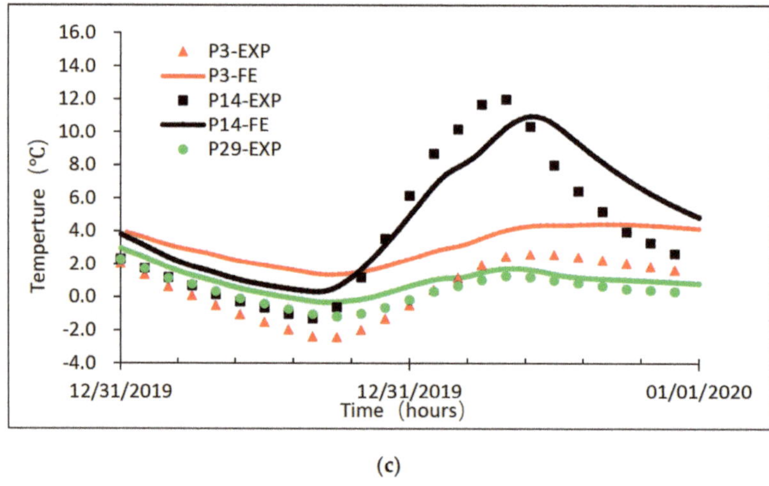

(c)

Figure 10. Measured and simulated temperature values comparisons on 31 December 2019. (**a**) Measured and simulated temperature values at the test point 1, 12, and 31 on 31 December 2019; (**b**) Measured and simulated temperature values at the test point 2, 13, and 30 on 31 December 2019; (**c**) Measured and simulated temperature values at the test point 3, 14, and 29 on 31 December 2019.

6.2.4. Stresses Caused by the Spatial Temperature Gradients in FE Analysis

Solar radiation causes differential temperature distributions that result in a vertical (y axis in Figure 7) and lateral (x axis in Figure 7) temperature gradient in the box girder. In bridge design practice, only the vertical temperature gradient is considered to be attributed to significant stress changes, and the lateral temperature gradient effect is generally ignored. In this paper, the lateral temperature gradient in the x axis direction is computed. Then, the maximum positive lateral temperature gradient was input in the FE model, and the corresponding normal stress in z axis direction and the transverse stress in y axis direct have been calculated, as shown in Figures 11–13.

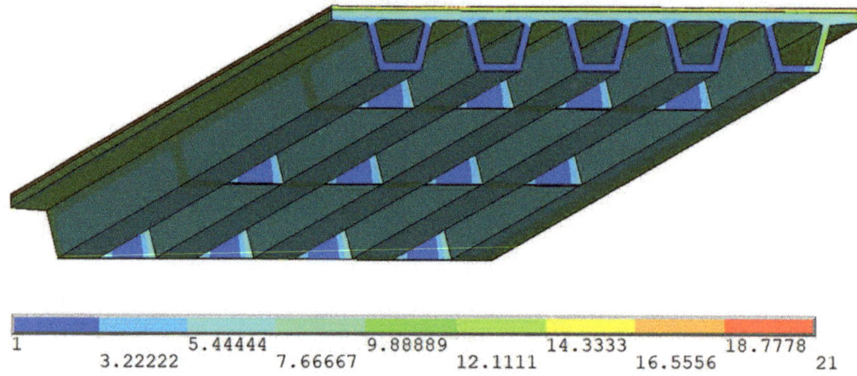

Figure 11. Simulated temperature distribution in the box girder at 15:00 on 31 December 2019 (unit: °C).

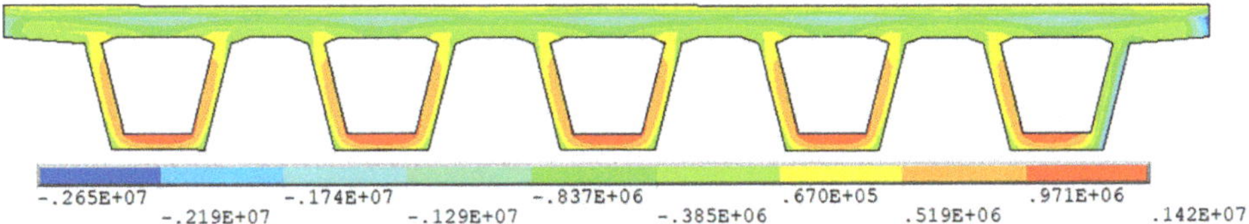

Figure 12. Longitudinal normal stress distribution under maximum positive lateral temperature gradient (unit: Pa).

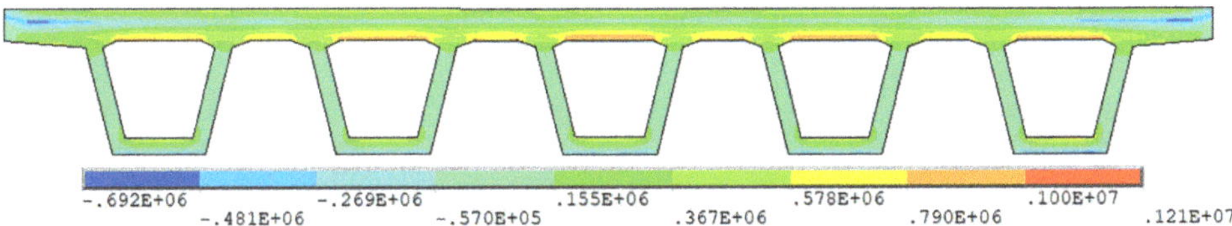

Figure 13. Transverse normal stress distribution under maximum positive lateral temperature gradient (unit: Pa).

Figure 11 shows the temperature distribution at the time of 15:00 on 31 December 2019 when the lateral temperature gradient reached maximum. It is clear that the temperature on the top and right web surface reached maximum 21 °C, while the temperatures on the bottom surface and inside the box remain relatively low.

The positive lateral temperature gradient shown in Figure 11 was exerted onto the FE model for structural analysis and the longitudinal and transverse stress distributions due to the lateral temperature gradient were obtained. Figure 12 shows that the longitudinal normal stress reached maximum 2.65 MPa on the left web in compression and 1.42 MPa in tension on the other webs and bottom slabs. The longitudinal normal tensile stress caused by lateral temperature gradient alone is significant and large enough to create cracks in concrete. The transverse normal stress was also calculated, as shown in Figure 13. The bottom edge of the top slab experienced maximum transverse tensile stress 1.21 MPa, while the all the webs and bottom slabs were in compression, and the maximum compressive stress was 0.69 MPa. It is indicated that the maximum transverse tensile stress is comparable to the longitudinal one. Hence, the lateral temperature gradient effects cannot be ignored in structural design.

7. Conclusions

The authors performed an experimental study on temperature effects caused by solar radiation on a 5-span continuous PC box-girder bridge. Sensors measuring solar radiation, temperatures, strain, wind speed and displacements were installed at various cross sections. The continuously acquired data from May 2019 to May 2020 was utilized to determine the daily lateral maximum positive temperature gradient for the PC box-girder bridge. It was found from the data analysis that the Pearson's correlation coefficient between the daily maximum positive lateral temperature gradient on the lower web and the daily maximum lateral solar radiation intensity reached maximum value of 0.879. Hence, the solar radiation on the web can be considered as the key factor that cause the daily maximum positive lateral temperature gradient. Then, a prediction equation for the daily maximum positive lateral temperature gradient was developed using solar radiation intensity measured on the lower web. Meanwhile, the comparison made between the simulated solar radiation intensities and the measured values indicated that there were differences between the simulated and measured values. It may be attributed to the fact that the uniform heat thermal transfer conductivity and absorption rate are used in the model. On the other hand, a linear elastic material model was utilized in the FE analysis for simplification, which may be attributed to causing the difference. In the future, an improved FE model accounting for the non-linear material properties, cracking, and stress re-distribution will be developed to generate more realistic stress distribution result. In addition, it is suggested that actual monitoring data be used to establish the temperature field boundary conditions when possible.

It was discovered that the daily maximum positive lateral temperature gradient took place at 15:00 on 31 December 2019 over the one-year monitoring. Entering the obtained daily maximum positive lateral temperature gradient into the FE model, the maximum longitudinal tensile stresses obtained were 1.42 MPa on the inner and farther webs and bottom slabs, and the maximum transverse ones were 1.21 MPa at the bottom of the top slab. The study demonstrated that the positive lateral temperature gradient effects were so significant that they should be taken into account in structural design.

Our future work will focus on developing an advanced three-dimensional solar radiation model accounting for modifications based on field testing data and comparing it with the vertical temperature gradient model in AASHTO bridge design specification.

Author Contributions: Conceptualization, X.L.; methodology, X.L.; formal analysis, X.F.; investigation, K.Z.; data curation, X.L. and X.F.; writing—original draft preparation, H.Z., H.J.; writing—review and editing, X.L.; funding acquisition, X.L. All authors have read and agreed to the published version of the manuscript.

Funding: This research was funded by National Natural Science Foundation of China (No. 51108152).

Acknowledgments: We would like to acknowledge the reviewers who provided us constructive suggestions that helped us improve our work.

Conflicts of Interest: The authors declare no conflict of interest.

References

1. Hossain, T.; Segura, S.; Okeil, A.M. Structural effects of temperature gradient on a continuous prestressed concrete girder bridge: Analysis and field measurements. *Struct. Infrastruct. Eng.* **2020**, 1–12. [CrossRef]
2. Maguire, M.; Roberts-Wollmann, C.L.; Cousins, T. Live-load testing and long-term monitoring of the Varina-Enon Bridge: Investigating thermal distress. *J. Bridge Eng.* **2018**, *23*, 04018003. [CrossRef]
3. Lee, J. Investigation of extreme environmental conditions and design thermal gradients during construction for prestressed concrete bridge girders. *J. Bridge Eng.* **2012**, *17*, 547–556. [CrossRef]
4. Wang, Y.; Zhan, Y.; Zhao, R. Analysis of thermal behavior on concrete box-girder arch bridges under convection and solar radiation. *Adv. Struct. Eng.* **2016**, *19*, 1043–1059. [CrossRef]
5. Abid, S.R.; Taysi, N.; Ozakca, M. Experimental analysis of temperature gradients in concrete box-girders. *Constr. Build. Mater.* **2016**, *106*, 523–532. [CrossRef]
6. Abid, S.R.; Mussa, F.; Taysi, N.; Özakça, M. Experimental and finite element investigation of temperature distributions in concrete-encased steel girders. *Struct. Control Health Monit.* **2018**, *25*, e2042. [CrossRef]
7. Taysi, N.; Abid, S.R. Temperature distributions and variations in concrete box-girder bridges: Experimental and finite element parametric studies. *Adv. Struct. Eng.* **2015**, *18*, 469–486. [CrossRef]
8. Tian, Y.; Zhang, N.; Xia, H. Temperature effect on service performance of high-speed railway concrete bridges. *Adv. Struct. Eng.* **2017**, *20*, 865–883. [CrossRef]
9. Lawson, L.; Ryan, K.L.; Buckle, I.G. Bridge Temperature Profiles Revisited: Thermal Analyses Based on Recent Meteorological Data from Nevada. *J. Bridge Eng.* **2020**, *25*, 04019124. [CrossRef]
10. Hagedorn, R.; Martivargas, J.R.; Dang, C.N.; Hale, W.M.; Floyd, R.W. Temperature Gradients in Bridge Concrete I-Girders under Heat Wave. *J. Bridge Eng.* **2019**, *24*, 04019077. [CrossRef]
11. Rodriguez, L.E.; Barr, P.J.; Halling, M.W. Temperature Effects on a Box-Girder Integral-Abutment Bridge. *J. Perform. Constr. Facil.* **2014**, *28*, 583–591. [CrossRef]
12. Kehlbeck, F. *Effect of Solar Radiation on Bridge Structure*; Liu, X., Translator; Railway Publishing House: Beijing, China, 1981.
13. Ghali, A.; Favre, R.; Elbadry, M. *Concrete Structures: Stresses and Deformation*, 3rd ed.; E & FN Spon: London, UK, 2002.
14. Lei, X.; Jiang, H.; Wang, J. Temperature effects on horizontally curved concrete box-girder bridges with single-column piers. *J. Aerosp. Eng.* **2019**, *32*, 04019008. [CrossRef]
15. Liu, X.F. *Temperature Induced Stress Analysis for Concrete Structures*; Communications Press: Beijing, China, 1991.

16. Elbadry, M.; Ghali, A. Temperature Variations in Concrete Bridges. *J. Struct. Eng.* **1983**, *109*, 2355–2374. [CrossRef]
17. Jain, P.C. A method for diffuse and global irradiation of horizontal surfaces. *Solar Energy* **1990**, *44*, 301–308. [CrossRef]
18. Page, J.K. The estimation of monthly means values of daily total short-wave radiation oil vertical and inclined surfaces from sunshine records for latitudes 40° N–40° S. In Proceedings of the UN Conference on News Sources of Energy, Rome, Italy, 16 May 1961; pp. 1–16.

Letter

Internal Cylinder Identification Based on Different Transmission of Longitudinal and Shear Ultrasonic Waves

Wen-Bei Liu [1,†], Wen-Bo Yan [2,†], Huan Liu [2], Cheng-Guo Tong [1,3], Ya-Xian Fan [1,3] and Zhi-Yong Tao [1,3,*]

1 Guangxi Key Laboratory of Wireless Wideband Communication and Signal Processing, Guilin University of Electronic Technology, Guilin 541004, China; lwb2019@guet.edu.cn (W.-B.L.); tcg@guet.edu.cn (C.-G.T.); yxfan@guet.edu.cn (Y.-X.F.)
2 Key Lab of In-Fiber Integrated Optics, Ministry Education of China, Harbin Engineering University, Harbin 150001, China; yanwenbo83@163.com (W.-B.Y.); liuhuana@hrbeu.edu.cn (H.L.)
3 Academy of Marine Information Technology, Guilin University of Electronic Technology, Beihai 536000, China
* Correspondence: zytao@guet.edu.cn
† These authors contributed equally to this work.

Abstract: We have built a Fizeau fiber interferometer to investigate the internal cylindrical defects in an aluminum plate based on laser ultrasonic techniques. The ultrasound is excited in the plate by a Q-switched Nd:YAG laser. When the ultrasonic waves interact with the internal defects, the transmitted amplitudes of longitudinal and shear waves are different. The experimental results show that the difference in transmission amplitudes can be attributed to the high frequency damping of internal cylinders. When the scanning point is close to the internal defect, the longitudinal waves attenuate significantly in the whole defect area, and their amplitude is always smaller than that of shear waves. By comparing the transmitted amplitudes of longitudinal and shear waves at different scanning points, we can achieve a C scan image of the sample to realize the visual inspection of internal defects. Our system exhibits outstanding performance in detecting internal cylinders, which could be used not only in evaluating structure cracks but also in exploring ultrasonic transmission characteristics.

Keywords: elastic waves; laser ultrasound; Fizeau fiber interferometer; defect identification

Citation: Liu, W.-B.; Yan, W.-B.; Liu, H.; Tong, C.-G.; Fan, Y.-X.; Tao, Z.-Y. Internal Cylinder Identification Based on Different Transmission of Longitudinal and Shear Ultrasonic Waves. *Sensors* **2021**, *21*, 723. https://doi.org/10.3390/s21030723

Academic Editors: Daria Wotzka and Luca De Marchi

Received: 6 November 2020
Accepted: 20 January 2021
Published: 21 January 2021

1. Introduction

In the aeronautics and automotive industry, there is a high demand for the quality of mechanical structures and materials. The cavities, grooves, and other types of defects generated during processing can affect the performance of the equipment and cause potential safety hazards. Therefore, harmful defects should be detected and be properly treated in time. In order to achieve safe and efficient detection for different kinds of manufacturing damages, the non-destructive evaluation (NDE) technology has been rapidly developed in various fields [1–7]. In NDEs, the detection of surface cracks has been paid much attention. For instance, Hong et al. used piezoelectric ultrasonic transducers to detect defects in plastic pipes [8], whereas Edwards et al. studied the relationship between depth of defects and ultrasonic signals by using the electro-magnetic acoustic transducers [9]. Campman et al. imaged surface defects in aluminum plates with laser ultrasonic methods [10] and developed the related methods and theories of wave propagation in marine seismic waves for applications [11,12]. In these works, researchers focused on surface defects and usually analyzed the reflected or transmitted ultrasound to obtain information about the cracks. Compared to surface defects, internal defects cannot be observed visually. In addition, influenced by the internal defects, the ultrasonic waves would experience amplitude attenuation, reflection, and scattering, which makes the detected signals too complex to be analyzed [13,14]. Therefore, the way to effectively detect and evaluate the internal defects is of great significance.

Due to the limitation of sensor size, the traditional piezoelectric transducer has limited spatial resolution, which makes it difficult to work normally in harsh environments such as high temperature, high radioactivity, and complex surfaces. With the development of NDEs, laser ultrasonic technology, as an effective method to solve these problems, has been widely used for years [15–20]. According to different detection requirements, laser ultrasonic technology can excite ultrasound through the thermo-elastic mechanism or ablation mechanism, which has the advantages of non-contact, high spatial resolution, and remote diagnosis. Based on a finite element method, Dai et al. obtained surface displacements of a 20 mm thick aluminum plate with various surface notch orientations [21] while Guan et al. modeled a surface notch with a depth of 200 μm, which is smaller than the center wavelength of Rayleigh waves [22]. Sohn et al. used the scanning laser-line source technique to investigate the interaction of surface-breaking flaws with ultrasonic waves [23]. Li et al. analyzed the ultrasonic frequency characteristics of a partially closed surface crack [24]. Zhou et al. numerically and experimentally investigated the reflection and transmission properties of a surface notch on an aluminum plate [25]. Wang et al. measured surface slot width using laser-generated Rayleigh waves [26]. Harb et al. proposed a fast imaging method for metal plate damage using zero-lag cross-correlation imaging conditions in the frequency domain [27]. In the above studies, ultrasonic signals were detected by spatial light paths system, laser interferometer, laser vibration meter, or other devices. Compared with these detection devices, the optical fiber interferometer not only maintains the advantages of high spatial resolution and high sensitivity but also overcomes the disadvantages of cumbersome and difficult adjustment of the spatial optical path, which can adapt to various detection conditions with high stability [28–30]. Moreover, the fiber interferometer can be built simply by using several kinds of optical fiber devices. The convenience of the system allows users to quickly master the operation method, which has good application potential in the field of NDE, especially for internal defects.

In this paper, we build a Fizeau fiber interferometer system to detect the size and shape of internal cylindrical defects. By using a precisely motorized positioning system, we obtain the ultrasonic signals at different positions within a two-dimensional area of the sample. We have found that the amplitudes of ultrasonic waves can be evidently affected by the internal defects and that the longitudinal wave attenuation is much larger than that of shear waves. Based on the relative intensity of the transmitted shear and longitudinal elastic waves, the shapes of internal cylinders have been imaged by carefully scanning the two-dimensional areas. In the following section, we build the proposed Fizeau fiber interferometer and introduce the experimental equipment. The waveform signals of transmitted ultrasonic waves are analyzed to illustrate the different attenuation effects of the internal defects on the longitudinal and shear waves in Section 3. In Section 4, by observing the trend of the shear wave amplitudes, we propose a data processing method to achieve the C scan images of the internal defects. Then, we spatially estimate the location, size, and shape of the prefabricated internal cylinders. Finally, the major results on internal defect identification based on the proposed Fizeau fiber interferometer and data processing method are summarized in Section 5.

2. Laser Ultrasonic Detection

2.1. Fizeau Fiber Interferometer

The Fizeau fiber interferometer [31] was designed to study the ultrasonic field on the relevant surface of the sample with internal defects. The schematic diagram of the interferometer is shown in Figure 1. The structure of Fizeau fiber interferometer is composed of single-mode fibers used to transmit signal light and reference light. The narrow linewidth tunable laser (TLT-1500 LaseGen, CA, USA) with output wavelength of 1550 nm and power of 15 mW is used as the detection light source of the interferometer. The continuous detection light is transmitted to the circulator through the isolator. Here, the isolator is used to prevent the detection light returning to the laser to ensure the stability of the interferometer.

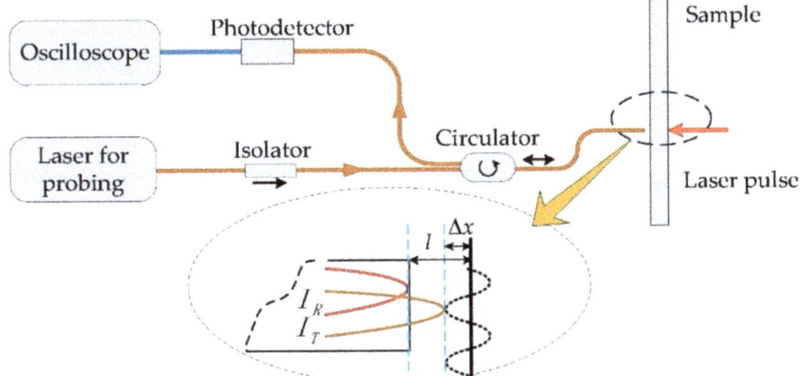

Figure 1. Schematic diagram of Fizeau fiber interferometer and its working principle.

When the detection laser outputs from the circulator, 4% of the energy at the end of the fiber would be reflected as the reference light. Some of the transmitted light would radiate on the surface of the sample and then be reflected by the surface of the sample. Most of the light would enter the fiber again. The two beams satisfy the coherent condition, and the vibration on the surface of the sample can change the optical path difference of the two beams. The coherent light intensity satisfies the following formula:

$$I = I_R + I_T + 2\sqrt{I_R I_T}\cos\Delta\varphi \tag{1}$$

$$\Delta\varphi = \frac{4\pi n_0(l + \Delta x)}{\lambda} \tag{2}$$

where I is the interference light intensity, $\Delta\varphi$ is the phase difference, Δx is the vibration displacement, and n_0 is the air refractive index. By adjusting the length of the interference cavity l, we can make $4n_0 l = (2k + 0.5)\lambda$, where k is an integer. Then, the coherent light intensity has a linear relationship with the very small vibration displacement. The intensity of interference light is converted into electrical signal by the photoelectric detector and displayed on the oscilloscope. We can thus get the ultrasonic signals detected by the Fizeau fiber interferometer.

2.2. Experimental Processes

The experimental setup is schematically shown in Figure 2. In the experiments, a pulsed laser with a wavelength of 1064 nm and a duration of 6 ns was generated by an Nd:YAG laser (DAWA-200 from Beamtech Optronics Co., Ltd., Beijing, China), and the repetition rate of the laser source was selected as 3 Hz. The average laser pulse energy was controlled around 20 mJ to guarantee that the ultrasonic wave was generated. The splitter lens was used to divide the pulse laser into two beams. One beam of the laser was focused into a spotlight through a focus lens with a 100 mm focal length to excite the ultrasonic waves, and the other beam was received by a space photodetector (KG-PR-1G-A-PC from Beijing Conquer Optics Science & Technology Co., Ltd., Beijing, China) as the trigger signal to the digital oscilloscope (MOS-S-204A from Keysight Technologies, Inc., Santa Rosa, CA, USA). The position of the probe fiber and the pulsed laser relative to the sample is also depicted in the inset of Figure 2.

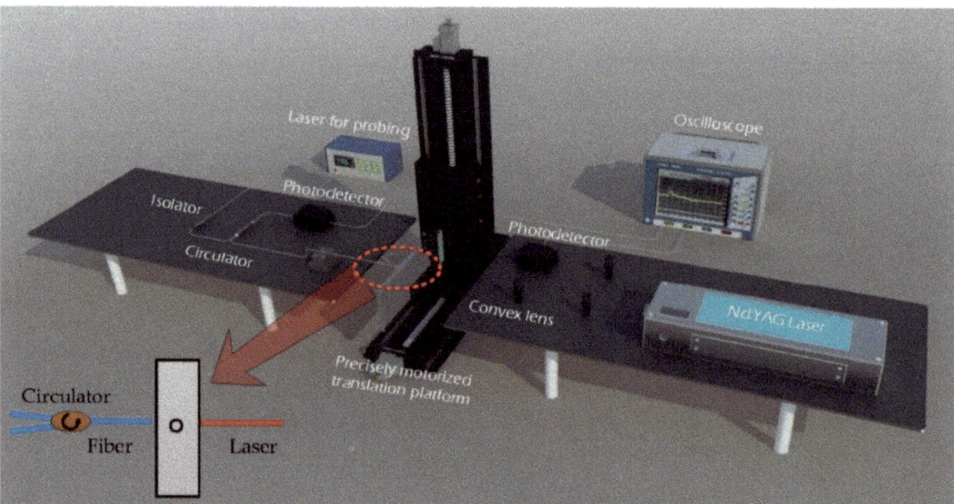

Figure 2. Schematic diagram of experimental setup. The inset also shows the relative position of the fiber and the pulse laser on the sample.

The two aluminum plates with length of 300 mm, width of 100 mm, and thickness of 10 mm were selected and held alternately by a two-dimensional precisely motorized translation stage (Beijing Optical Century Instrument Co., Ltd., Beijing, China) as shown in Figure 2. The two aluminum plate samples with the thickness h = 10 mm were prepared for the test as shown in Figure 3a. We drilled a 2.0 mm diameter hole vertically on the upper side of one aluminum plate, while on the other aluminum plate, we used a 1.5 mm diameter drill bit to create a vertical hole, and then reprocessed it with a 1 mm diameter drill at the same location to create the hole with the diameter varying from 1.5 mm to 1 mm. The precise electric translation table could move in a two-dimensional direction, and the repetition position accuracy was less than 5 μm. Before the test, we adjusted the relative position of the sample and the optical fiber probe so that the bottom of the internal defect was located in the center of the scanning area. As shown in Figure 3b, the scanning area with the size of 5 mm × 5 mm was 5 mm lower than the upper edge of the aluminum plate, the scanning step size was 0.2 mm, and the red arrows denote the scanning direction. Then, the probe laser was focused on the sample surface with a fiber probe. During the scanning process, we kept the position of the optical fiber probe and laser focus unchanged and only used the translation stage to move the samples. This scanning method can avoid the additional phase change and laser energy fluctuation caused by the movement of the fiber probe, which effectively improves the stability and efficiency of the system.

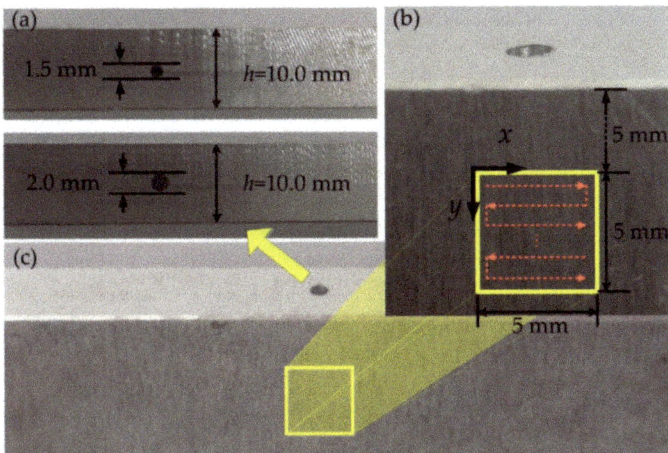

Figure 3. Aluminum plate samples with the different holes on one side. (**a**) Two holes with 1.5 mm and 2.0 mm diameters. (**b**) Scanning area and path. (**c**) Aluminum plate with the hole and scanning area.

3. Transmission of Longitudinal and Shear Waves

To illustrate the advantages of the proposed fiber interferometer, we compared our interferometer with the traditional ultrasonic devices. The fiber end face of the Fizeau fiber interferometer is presented in Figure 4a, which is the detailed enlarged view of the fiber probe in the actual picture of the detection as shown in Figure 4b. Figure 4c shows the detection by using an ultrasonic transducer with a center frequency of 1 MHz. The ultrasonic transceiver model of the ultrasonic probe is Olympus 5077PR.

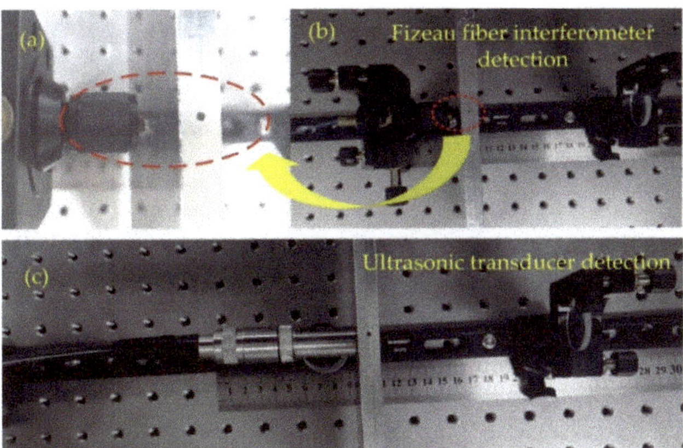

Figure 4. Fiber end face of the Fizeau fiber interferometer (**a**), and the interferometer (**b**) and ultrasonic transducer (**c**) detection.

The typical ultrasonic time and frequency domain signals obtained by the proposed interferometer are shown in Figure 5a–c. The detected ultrasonic signal without defects is shown in Figure 5a. By analyzing the signal, we can find that the Fizeau fiber interferometer can detect the two ultrasonic modes: longitudinal and shear waves. A longitudinal wave is a kind of compression wave whose vibration direction is parallel to the propagation direction. The interference optical path difference caused by its vibration will be the largest, so the amplitude change of longitudinal wave signal detected by Fizeau fiber interferometer is the strongest. The thickness of the aluminum plate is 10 mm. The propagation velocity of longitudinal waves in an aluminum plate is about 6400 m/s [32]. According to the velocity formula, the time of receiving longitudinal waves by Fizeau fiber interferometer can be calculated by

$$t_P = \frac{v_P}{h}, \tag{3}$$

where t_p and v_p are the receiving time and the propagation velocity of longitudinal waves, respectively. The results show that the peak at t_p = 1.56 μs was the longitudinal wave signal (denoted by P). In addition, due to the influence of the sample boundaries, the reflected signals of longitudinal waves (denoted by RP) were also detected. The propagation velocity of shear waves in the aluminum plate was about 3150 m/s [32]. According to the following velocity formula, the shear wave receiving time can be calculated.

$$t_S = \frac{v_S}{h}, \tag{4}$$

where t_s and vs. are the receiving time and the propagation velocity of shear waves, respectively. It was calculated that the lower peak at t_s = 3.21 μs is the shear wave signal (denoted by S).

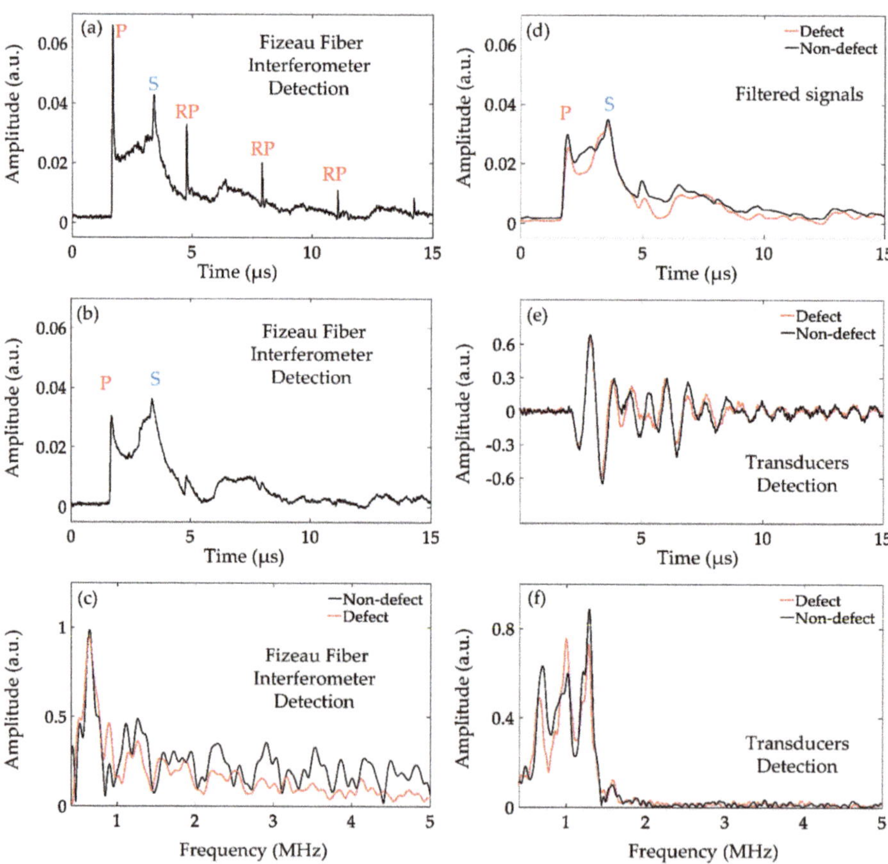

Figure 5. Detected time and frequency domain ultrasonic signals: Fizeau fiber interferometer signals without (**a**) and with (**b**) defects, as well as their frequency spectra (**c**) and the low-pass filtered signals (**d**). Signals (**e**) and their frequency spectra (**f**) detected by the ultrasonic transducers.

When defects are detected by the optical fiber interferometer, the obtained ultrasonic signals were different as shown in Figure 5b. Compared with the defect-free results, we find that the amplitudes of P and S waves are both attenuated. However, it can be also observed that the attenuation degrees of these two kinds of ultrasonic signals are different. The amplitude attenuation of the P wave is more severe. The frequency domain expansions of ultrasonic signals measured with and without defects are depicted in Figure 5c by the red dashed and black solid lines, respectively. We find that the high-frequency components of the signal without defects are much larger. As shown in Figure 5a,b, the P wave attenuation is severe. We suppose that the P wave contains more frequency components higher than 2 MHz, because when defects are detected, they are severely attenuated in the time and frequency domain, respectively. In order to verify this point of view, we chose a Butterworth low-pass filter with a cutoff frequency of 2 MHz to the ultrasonic signals measured with and without defects as shown in Figure 5d. The black solid line represents the signal processed by filtering when there is no defect. It can be found that both P wave and RP waves are filtered out. This proves that P waves contain more high-frequency components than S waves. The red dashed line indicates the signal processed by filtering when there are defects. It can be found that the signal becomes smooth but does not change dramatically from the nonfiltered signal. This not only proves that P waves contain more high-frequency components but also confirms that the attenuation of P waves is more severe than that of S waves when the ultrasound passes through defects. The internal defect does affect the ultrasonic wave transmission, and the effects on the P and S waves are quite different. This is because the angle of the longitudinal and transverse wave propagation is different when the ultrasound is excited by lasers [32]. The propagation direction of P waves is always perpendicular to the aluminum plate surface, whereas the S wave propagates in a certain angle; thus it can bypass the defect. Therefore, it seems very promising that we can identify the internal defect by carefully testing the different transmissions of P and S waves.

At the same time, we also used the ultrasonic transducers with the center frequency of 1 MHz to detect the samples with and without defects as shown in Figure 5e by the red dashed and black solid lines, respectively, as well as their frequency spectra in Figure 5f. It is difficult to directly observe the difference between the two ultrasound signals received by the transducer due to its much larger size. In comparison, the proposed optical fiber interferometer has more advantages in the detection of small internal defects.

In order to further investigate the effects of internal defects on the transmission of ultrasonic waves, we selected the same horizontal scanning in the non-defective and defective regions and obtained the B scan images along the x-axis by data processing, shown respectively as Figure 6a,b. We have normalized the ultrasonic waveform by the P wave intensity without defects as the standard to facilitate the observation of the amplitude changes of ultrasonic waves. The intensity of the amplitude is represented by the color from blue to green: the greener the color, the greater the intensity of the signal. Referring to the ultrasonic mode analysis above, we can clearly observe the P wave and the S wave, and a series of boundary reflected waves in Figure 6a, when the scanning does not encounter any defects. Moreover, it has been proved that the amplitude intensity of recorded ultrasonic signals is very stable in the scanning process. In contrast, the ultrasonic waveform would change significantly when the interferometer was scanning in the defective region shown in Figure 6b. It can be seen that the amplitude of P waves will decrease as the scanning position gradually approaches the internal defect, while the amplitude of S waves increases due to the high-frequency damping phenomenon. Therefore, we have observed a bright region appearing around 3.3 μs as the scanning position moves to the location of the internal defect. As the scanning point moves out of the defective region, the ultrasonic waveform is restored to a non-defective state and remains stable again. In addition, we also found the arc-shaped ultrasonic from the diagram due to the cylindrical shape of the predrilled internal defect, indicating the possibility of defect shape identification based on the recorded ultrasonic signals by the interferometer.

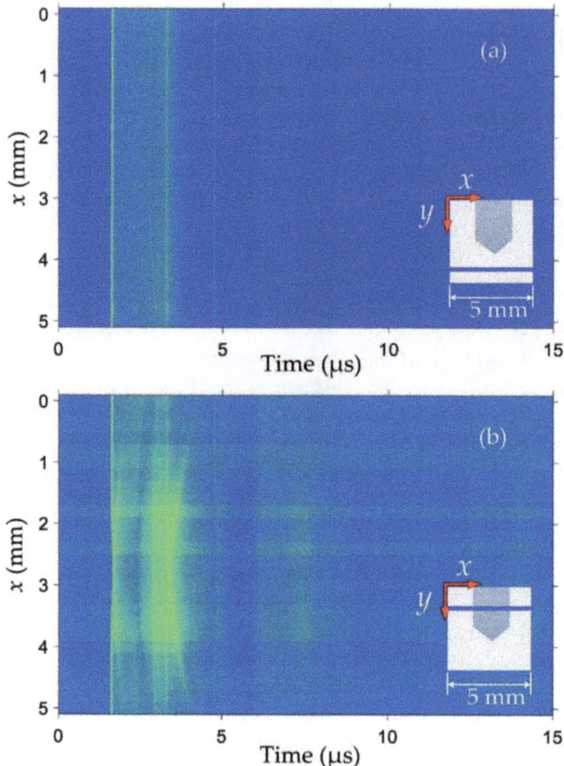

Figure 6. B scan image along the horizontal direction in the non-defect (**a**) and defective (**b**) regions. The scanning route is also depicted by the bold blue line in each insect. To compare the attenuation of P waves and S waves, the signal amplitudes are both normalized by their own P wave intensities.

4. Visualization of Internal Cylinders

Based on the B scan results mentioned above, the existence of internal defects can be determined by the interferometer received ultrasonic waves. However, since the amplitude change of ultrasonic waves is a gradual process, the boundary of the defect cannot be accurately determined yet. Therefore, it is also necessary to investigate the C scan images to inspect the internal defects more intuitively. We found that the amplitudes of P and S waves have different degrees of attenuation due to the influence of internal defects, and the attenuation of P waves is more significant. Near the defect boundary, the amplitude relationship between the two types of ultrasonic waves changes, and the S wave amplitude will gradually approach and exceed that of the P waves. In order to determine the defect location, the maximum value of P wave amplitude and S wave amplitude are selected as $P_{\max}$ and $S_{\max}$, respectively. We set the parameter K, which is given by the following formula.

$$K = \frac{S_{\max}}{P_{\max}}. \tag{5}$$

The defect can change the value of K by attenuating P waves. Therefore, when scanning the area shown in Figure 3b, the value of K can be used to determine whether there are defects in the scanning position. A scanning track in the x-axis direction is selected, and the relationship between K and x-axis displacement is shown in Figure 7a. It can be seen that when the detection position moves into the location of the defect, K begins to increase significantly. In order to determine the defect boundary more accurately, we introduce the following formula.

$$L = \tanh[A(K - K')] \tag{6}$$

where A and K' are both the measurement parameters, which can be calibrated by a hole with standard sizes. The relationship between L and x-axis displacement is shown in Figure 7b, where A = 100 and K' = 0.95 are selected.

With the measured data from a standard hole, we can optimize the measurement parameters A and K'. The value of A can affect the change degree of L at the boundary. Choosing A = 100 satisfies the change judgment of L at the boundary and K' is the criterion of defect boundary. When we select A = 100 and K' = 0.95, L = 1 inside the defect, while $L = -1$ outside. It can be seen from the results that the boundary of defects can be accurately determined. The final imaging results of the whole scanning area are shown in Figure 7c,d for the two samples with 2 mm and 1.5 mm diameters, respectively. The yellow is for L = 1 and the blue is for $L = -1$. Thus, we can figure out the internal cylinders according to the different colors. The experimental results show that the scanning image obtained by this method is basically consistent with the characteristics of cylindrical defects in the samples, even for the cylinder with varying diameters. In addition, there were still some spots at the edges of the defects in the C scan images due to the measuring errors. Because our scanning step is 0.2 mm, the edge of the defect image has an error within 0.2 mm. This error can be further reduced by decreasing the scanning step. This method has nothing to do with the propagating media of ultrasounds, so it can also be easily applied to other metals.

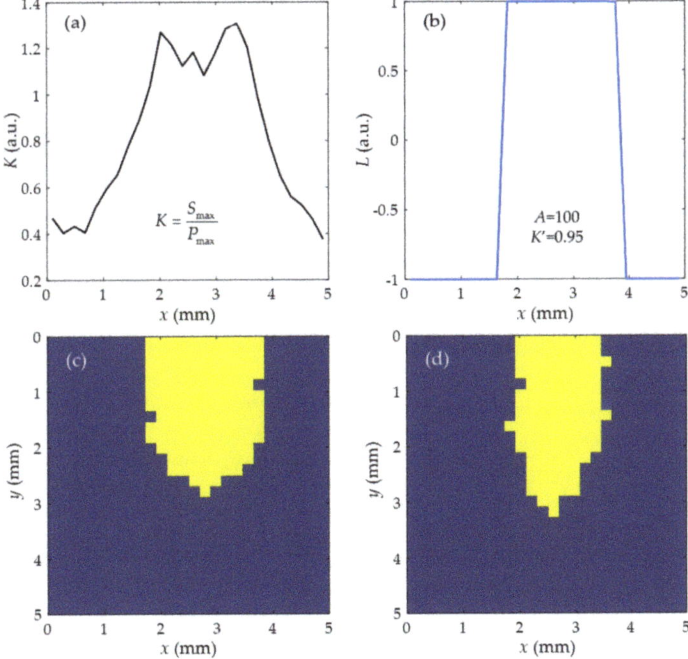

Figure 7. K (**a**) and L (**b**) parameters for the scanning signals along the x-axis and the C scan images of two samples with 2 mm (**c**) and 1.5 mm (**d**) diameter cylinders, where the yellow is for L = 1 and the blue is for $L = -1$.

5. Conclusions

We built a Fizeau fiber interferometer to detect the internal defects in aluminum plates. This system has the advantages of simple structure, high sensitivity, and high spatial resolution. Through experimental and theoretical analysis, we studied the interaction between internal defects and ultrasonic waves. We observed that the variation of the S waves and P waves is affected by internal defects based on the B scan images with different y-axis positions, and the phenomenon of different attenuation degree of waveforms in different modes are explained by the damping of high frequencies. Finally, by comparing the S wave amplitude of different scanning points with the set measurement parameters A and K', the C scan images of the predrilled cylinders were obtained to implement the visual inspection of internal defects. It is very impressive that the C scan images are consistent with the actual features of the internal defects based on the Fizeau fiber interferometer system, which provides a more simple way to measure the size, position, and shape of internal defects. The measurement error is dependent on the step of scanning, which can be decreased according to the resolution of translation stage.

In summary, the Fizeau fiber interferometer system can be effectively used to investigate the internal defects, and the measured results reveal the propagation characteristics of ultrasonic waves and their interaction mechanism with internal defects. The proposed system and methods not only enrich our knowledge of the interaction between defects and ultrasound, but also find a simple way to detect the structures in high resolution, which have good application prospects in the field, such as NDE, ultrasonic measurements, and biomedical imaging.

Author Contributions: Conceptualization, Y.-X.F. and Z.-Y.T.; methodology, H.L. and W.-B.Y.; validation, W.-B.L. and W.-B.Y.; formal analysis, C.-G.T., Y.-X.F., and Z.-Y.T.; writing—original draft preparation, W.-B.L. and W.-B.Y.; writing—review and editing, Y.-X.F. and Z.-Y.T.; visualization, W.-B.L.; supervision, Y.-X.F. and Z.-Y.T. All authors have read and agreed to the published version of the manuscript.

Funding: This research was funded by the National Natural Science Foundation of China (12064005), the Dean Project of Guangxi Key Laboratory of Wireless Wideband Communication and Signal Processing, and the Natural Science Foundation of Heilongjiang Province, China (A2018004).

Institutional Review Board Statement: Not applicable.

Informed Consent Statement: Not applicable.

Data Availability Statement: The data presented in this study are available on request from the corresponding author.

Conflicts of Interest: The authors declare no conflict of interest.

References

1. Monchalin, J.-P. Optical detection of ultrasound. *IEEE Trans. Ultrason. Ferroelectr. Freq. Control* **1986**, *33*, 485–499. [CrossRef] [PubMed]
2. Blitz, J.; Simpson, G. *Ultrasonic Methods of Non-Destructive Testing*; Chapman & Hall: London, UK, 1996.
3. Arias, I.; Achenbach, J.D. A model for the ultrasonic detection of surface-breaking cracks by the scanning laser source technique. *Wave Motion* **2004**, *39*, 61–75. [CrossRef]
4. Drinkwater, B.W.; Wilcox, P.D. Ultrasonic arrays for non-destructive evaluation: A review. *NDT Int.* **2006**, *39*, 525–541. [CrossRef]
5. Pelivanov, I.; Shtokolov, A.; Wei, C.-W.; O'Donnell, M. A 1 kHz A-Scan Rate Pump-Probe Laser- Ultrasound System for Robust Inspection of Composites. *IEEE Trans. Ultrason. Ferroelectr. Freq. Control* **2015**, *62*, 1696–1703. [CrossRef] [PubMed]
6. Lee, F.W.; Chai, H.K.; Lim, K.S. Assessment of Reinforced Concrete Surface Breaking Crack Using Rayleigh Wave Measurement. *Sensors* **2016**, *16*, 337. [CrossRef]
7. Hong, X.; Lin, X.; Yang, B.; Li, M. Crack detection in plastic pipe using piezoelectric transducers based on nonlinear ultrasonic modulation. *Smart Mater. Struct.* **2017**, *26*, 104012. [CrossRef]
8. Sun, X.; Zeng, K.; He, X.; Zhang, L. Ultrasonic C-scan imaging and analysis of the mechanical properties of resistance spot-welded joints of stainless steel. *Nondestruct. Test. Eval.* **2017**, *32*, 242–254. [CrossRef]
9. Edwards, R.S.; Dixon, S.; Jian, X. Depth gauging of defects using low frequency wideband Rayleigh waves. *Ultrasonics* **2006**, *44*, 93–98. [CrossRef]

10. Campman, X.; van Wijk, K.; Riyanti, C.D.; Scales, J.; Herman, G. Imaging scattered seismic surface waves. *Near Surf. Geophys.* **2004**, *2*, 223–230. [CrossRef]
11. Campman, X. Imaging and Suppressing Near-Receiver Scattered Seismic Waves. Ph.D. Thesis, Delft University of Technology, Delft, The Netherlands, 31 May 2005.
12. Campman, X. Marine seismic acquisition: What's next? In Proceedings of the New Advances in Marine Geophysical Acquisition Technologies, Abu Dhabi, UAE, 7–9 May 2018.
13. Tanaka, T.; Izawa, Y. Nondestructive Detection of Small Internal Defects in Carbon Steel by Laser Ultrasonics. *Jpn. J. Appl. Phys.* **2001**, *40*, 1477–1481. [CrossRef]
14. Sun, K.; Shen, Z.; Shi, Y.; Xu, Z.; Yuan, L.; Ni, X. Non-destructive Detection of Small Blowholes in Aluminum by Using Laser Ultrasonics Technique. *Int. J. Thermophys.* **2015**, *36*, 1181–1188. [CrossRef]
15. Cooper, J.A.; Crosbie, R.A.; Dewhurst, R.J.; Mckie, A.D.W.; Palmer, S.B. Surface acoustic wave interactions with cracks and slots: A noncontacting study using lasers. *IEEE Trans. Ultrason. Ferroelectr. Freq. Control* **1986**, *33*, 462–470. [CrossRef] [PubMed]
16. Hutchins, D.A.; Hauser, F.; Goetz, T. Surface waves using laser generation and electromagnetic acoustic transducer detection. *IEEE Trans. Ultrason. Ferroelectr. Freq. Control* **1986**, *33*, 478–484. [CrossRef] [PubMed]
17. Jian, X.; Fan, Y.; Edwards, R.S.; Dixon, S. Surface-breaking crack gauging with the use of laser-generated Rayleigh waves. *J. Appl. Phys.* **2006**, *100*, 064907. [CrossRef]
18. Lomonosov, A.M.; Grigoriev, P.V.; Hess, P. Sizing of partially closed surface-breaking microcracks with broadband Rayleigh waves. *J. Appl. Phys.* **2009**, *105*, 084906. [CrossRef]
19. Rinkevich, A.B.; Perov, D.V.; Korkh, Y.V. Laser detection of elastic waves diffraction by the crack's edge. *Appl. Acoust.* **2012**, *73*, 803–808. [CrossRef]
20. Faëse, F.; Jenot, F.; Ouaftouh, M.; Duquennoy, M.; Ourak, M. Fast slot characterization using laser ultrasonics and mode conversion. *Meas. Sci. Technol.* **2013**, *24*, 095602. [CrossRef]
21. Dai, Y.; Xu, B.Q.; Luo, Y.; Li, H.; Xu, G.D. Finite element modeling of the interaction of laser-generated ultrasound with a surface-breaking notch in an elastic plate. *Opt. Laser Technol.* **2010**, *42*, 693–697. [CrossRef]
22. Guan, J.; Shen, Z.; Lu, J.; Ni, X.; Wang, J.; Xu, B. Finite Element Analysis of the Scanning Laser Line Source Technique. *Jpn. J. Appl. Phys.* **2006**, *45*, 5046–5050. [CrossRef]
23. Sohn, Y.; Krishnaswamy, S. Interaction of a scanning laser-generated ultrasonic line source with a surface-breaking flaw. *J. Acoust. Soc. Am.* **2004**, *115*, 172–181. [CrossRef]
24. Li, J.; Zhang, H.C.; Ni, C.Y.; Shen, Z.H. Analysis of laser generated ultrasonic wave frequency characteristics induced by a partially closed surface-breaking crack. *Appl. Opt.* **2013**, *52*, 4179–4185. [CrossRef] [PubMed]
25. Zhou, Z.G.; Zhang, K.S.; Zhou, J.H.; Sun, G.K.; Wang, J. Application of laser ultrasonic technique for non-contact detection of structural surface-breaking cracks. *Opt. Laser Technol.* **2015**, *73*, 173–178. [CrossRef]
26. Wang, C.Y.; Sun, A.; Xue, M.; Ju, B.F.; Xiong, J.; Xu, X. Width gauging of surface slot using laser-generated Rayleigh waves. *Opt. Laser Technol.* **2017**, *92*, 15–18. [CrossRef]
27. Harb, M.S.; Yuan, F.-G. Damage imaging using non-contact air-coupled transducer/laser Doppler vibrometer system. *Struct. Health Monit.* **2016**, *15*, 193–203. [CrossRef]
28. Imai, M.; Ohashi, T.; Ohtsuka, Y. Fiber-optic Michelson inter-ferometer using an optical power divider. *Opt. Lett.* **1980**, *5*, 418–420. [CrossRef]
29. Alcoz, J.J.; Lee, C.E.; Taylor, H.F. Embedded fiber-optic Fabry-Perot ultrasound sensor. *IEEE Trans. Ultrason. Ferroelectr. Freq. Control* **1990**, *37*, 302–306. [CrossRef]
30. Yuan, L.B.; Zhou, L.M.; Jin, W. Long-gauge length embedded fiber optic ultrasonic sensor for large-scale concrete structures. *Opt. Laser Technol.* **2004**, *36*, 11–17. [CrossRef]
31. Drake, A.D.; Leiner, D.C. Fiber-optic interferometer for remote subangstrom vibration measurement. *Rev. Sci. Instrum.* **1984**, *55*, 162–165. [CrossRef]
32. Scruby, C.B.; Drain, L.E. *Laser Ultrasonics: Techniques and Applications*; Adam Hilger: Bristol, UK, 1990; pp. 262–274.

Article

Ultrasonic Propagation in Highly Attenuating Insulation Materials

David A. Hutchins [1], Richard L. Watson [1], Lee A.J. Davis [1], Lolu Akanji [1], Duncan R. Billson [1], Pietro Burrascano [2], Stefano Laureti [3,*] and Marco Ricci [3]

[1] School of Engineering, University of Warwick, Coventry CV4 7AL, UK; D.A.Hutchins@warwick.ac.uk (D.A.H.); r.watson.2@warwick.ac.uk (R.L.W.); lee.davis@warwick.ac.uk (L.A.J.D.); omololu@akanji.com (L.A.); D.R.Billson@warwick.ac.uk (D.R.B.)
[2] Department of Engineering, University of Perugia, Polo Scientifico Didattico di Terni, Via di Pentima 4, 05100 Terni, Italy; pietro.burrascano@unipg.it
[3] Department of Informatics, Modelling, Electronics and Systems Engineering, University of Calabria, Via Pietro Bucci, 87036 Arcavacata, Rende (CS), Italy; marco.ricci@unical.it
* Correspondence: stefano.laureti@unical.it

Received: 6 March 2020; Accepted: 15 April 2020; Published: 17 April 2020

Abstract: Experiments have been performed to demonstrate that ultrasound in the 100–400 kHz frequency range can be used to propagate signals through various types of industrial insulation. This is despite the fact that they are highly attenuating to ultrasonic signals due to scattering and viscoelastic effects. The experiments used a combination of piezocomposite transducers and pulse compression processing. This combination allowed signal-to-noise levels to be enhanced so that signals reflected from the surface of an insulated and cladded steel pipe could be obtained.

Keywords: pulse compression; scattering; attenuation; insulation; ultrasonic testing

1. Introduction

Corrosion under insulation (CUI) is a significant industrial problem that has been identified as one of major concern [1]. The basic cause is the leakage of water through the insulation and subsequent contact with the surface of the underlying structure (e.g., a steel pipe), which then corrodes. Water can reach the surface of the underlying metal through two different mechanisms. The first is damage caused to the insulation layers, which creates a path to the metallic material underneath. Alternatively, water could diffuse into the insulation and hence reach the metal surface, initiating the corrosion process. CUI tends to occur at low-lying sections where water can concentrate and/or where there are junctions in structures or pipework. This accelerates corrosion of the outer metal surface, leading to pitting, cracking, and the possibility of subsequent failure.

There is an increasing problem with CUI within ageing infrastructure across numerous existing industrial sites. Sudden failure of pipes or large vessels due to CUI is a serious health and safety issue which, depending on the application, could lead to environmental concerns. There is often a need to strip off the insulation for a manual inspection if CUI is suspected, and additional manpower and costs are associated with this. Corrosion and material failure together are thought to contribute towards 50% of all significant incidents involving hazardous liquid in industrial pipelines [2]. Corrosion processes that occur in CUI include galvanic, alkaline, acidic, and chloride corrosion due to chemicals leaching out from wet insulation [3]. Various techniques exist for the monitoring and inspection of corrosion, and these have been reviewed [4]. They include ultrasound, radiography, eddy currents, and thermography. In many such cases, it is generally assumed that direct access to the metal pipe surface is available, so that the techniques are not designed to be used through insulation layers.

One example, the use of ultrasonic guided waves for the detection of pitting due to corrosion, shows promise in the remote detection of corrosion defects [5–7], but requires access to a bare pipe and the clamping on of transducer arrays for ultrasonic generation and detection. For insulated pipes, this could be inconvenient and expensive. Radio frequency (RF) signals can also be transmitted along the insulation-filled waveguide formed by a pipe wall and external metallic cladding, and has been used to detect water ingress [8], but suitable access could be a problem in this case also. Low frequency pulsed eddy-current inspection is a through-cladding Nondestructive Evaluation (NDE) technique that has recently been adopted for the NDE of insulted metallic pipes [9–11], and has been developed for practical use. The metal magnetic memory (MMM) technique also shows promise for detecting buried pipelines using the earth's magnetic field, but this is a small effect which would be difficult to use on complicated metallic pipework within an industrial plant [12].

Ultrasound does not seem to have received much interest as an NDE tool for CUI detection through thick sections of insulation. This is probably due to high transmission losses for many common insulation materials, and the wide range of acoustic properties which can vary substantially from one insulation material to the next. This paper thus investigates the possibility that frequencies in the 100–400 kHz range could be used for this task. This would use a combination of transducers with the appropriate bandwidth and matched-filter techniques to improve signal to noise ratios (SNRs). It might then be possible to penetrate significant distances into very attenuating insulation materials. If so, then ultrasound could potentially measure surface topographies caused by pitting of the outer pipe wall, and could also identify the presence of water via a change in propagation velocity within the insulation itself. The present paper investigates whether ultrasound could penetrate through common insulation materials, and recommends how future development of a practical industrial NDE technique for CUI could be achieved.

2. Pulse Compression

Pulse compression (PuC) is the chosen method for dealing with the propagation through insulation-covered pipes. The main aim is to output an estimate of the impulse response close to that retrievable by pulsed excitation, but with a much-improved SNR. PuC methods have been used previously in the context of highly attenuating materials [13,14], but experiments in common industrial insulation materials have not previously been reported. This is because of their highly attenuating and scattering nature, and their low acoustic impedances (arising from their naturally low acoustic velocities and densities, making matching to conventional transducers difficult). This necessitated the use of low ultrasonic frequencies and piezocomposite transducers in the present work.

PuC relies on the use of matched filtering techniques, where a coded signal is transmitted into the insulation across a pre-determined bandwidth, and the pulse compression algorithm is then applied to the recorded output. It is essential that the characteristics of this signal are chosen carefully. There are many options available in terms of the signal type, windowing function and processing algorithm. These have been reviewed for use with ultrasound, for example when used for air-coupling [15]. The choice of waveform is associated with several features: the bandwidth required, whether or not the device has a known centre frequency of operation, and the type of measurement to be performed (e.g., whether temporal or spatial information more important [16–18]). In the present case, the piezoelectric transducers will operate over a predetermined bandwidth around their frequency of resonance. In addition, the main requirement in the present measurements is to maximize sensitivity, due to the highly-attenuating nature of the insulation materials.

To gain insight into the PuC algorithm, consider a standard ultrasonic test whereby a short time duration yet high voltage amplitude signal feeds the transmitter transducer having a central frequency of (f_0). If the amplitude of the signal is such that the onset of any non-linear behaviour is avoided, the system can be thus considered linear and time-invariant. Further, if the duration is short enough to excite the whole transducer bandwidth, the output signal represents a good estimate of the overall measurement system impulse response. Theoretically, the impulse response carries all the information about the system under test, in practice, the performance of an ultrasonic inspection will be limited due to the bandwidth (B) of the whole measurement system and the achieved value of signal-to-noise ratio (SNR). It then follows that different aspects must be considered to design an optimal input signal to distinguish between different reflectors/scatterers and to optimise the SNR. Note that PuC is also likely to reduce the effects of structural noise, and to allow better separation of multi-interface echoes if they are present.

In a conventional pulsed ultrasonic system, a good time resolution requires the input signal to have a short time duration (T_{pulse}) leading to a broad bandwidth B. This results in a strict constraint for the product between B and T_{pulse} of the signal, such that $T_{pulse} \cdot B \cong 1$. In addition, the conventional way to improve the SNR is to increase the excitation signal amplitude, but this has limitations and may not be suitable in industrial environments where voltage levels are restricted for safety reasons. An alternative solution is the use of a coded signal input where the excited bandwidth B and the duration T are not constrained by $T \cdot B \cong 1$. PuC uses this approach. It is an effective solution for estimating the system impulse response in poor SNR conditions, such as those encountered in high-attenuation insulation. A coded signal $x(t)$ is defined, which distributes its energy over the frequency range of interest and over a relatively long period of time T_{PuC}. This is employed as the input signal to the transmitter. The received signal $y(t)$ is the convolution between the system impulse response $h(t)$ and the coded signal, i.e., $y(t) = h(t) * x(t)$. A matched-filter $\psi(t)$, defined such that $x(t) * \psi(t) \approx \delta(t)$, where $\delta(t)$ is the Dirac delta function, compresses the received signal $y(t)$ allowing an estimation of the system response $\hat{h}(t)$ to be retrieved. If the noise affecting the measurement can be modelled as Arbitrary White Gaussian Noise ($e(t)$), then the PuC process for a single measurement can be described as:

$$\hat{h}(t) = y(t) * \psi(t) = h(t) * \underbrace{s(t) * \psi(t)}_{=\widetilde{\delta}(t)} + \underbrace{e(t) * \psi(t)}_{=\widetilde{e}(t)} = h(t) * \widetilde{\delta}(t) + \widetilde{e}(t) \approx h(t) + \widetilde{e}(t). \tag{1}$$

It follows that if imaging is needed, the PuC algorithm must applied pixelwise for each acquired signal across the scanned area. It can be also noted that the PuC approach can improve SNR with respect to a standard pulsed excitation while assuring the same or even better range resolution. This happens when the energy of the PuC excitation signal is larger than that of a standard pulsed excitation and the same or a larger bandwidth is excited by the coded excitation. As stated above, for a standard pulse of duration T_{pulse}, the excited bandwidth is $B \propto \frac{1}{T_{pulse}}$. If the pulse amplitude is constant and equal to A_{pulse}, the pulse energy is $E_{pulse} = A_{pulse} T_{pulse} = \frac{A_{pulse}}{B}$. Instead, for the PuC case, the duration T_{PuC} is not constrained by B and can be arbitrary. If the amplitude is constant and equal to A_{PuC}, the PuC energy is $E_{PuC} = A_{PuC} T_{PuC}$ from which we can derive the maximum available SNR gain as Equation (2).

$$SNR_{gain} = \frac{SNR_{PuC}}{SNR_{pulse}} = \frac{E_{PuC}}{E_{pulsed-excitation}} = \frac{A_{PuC}}{A_{pulsed}} \cdot T_{PuC} B = k \cdot T_{PuC} B. \tag{2}$$

Thus, the available SNR_{gain} is proportional to the so-called time-bandwidth product of the coded signal [14], and, for a fixed bandwidth B, it can be increased by using a longer duration signal (T_{PuC}), while the voltage drive amplitude, i.e., the k factor in Equation (2), can be chosen to suit the system requirements [19].

A wide number of signal pairs $\{x(t), \psi(t)\}$ are available for use in PuC. These include both binary sequences such as maximum length sequences, Golay complementary sequences, Barker codes, and continuous time signals modulated in frequency such as the various types of chirp signals [15]. The choice of signal pairs for a specific test is linked to many factors: the frequency response of the instrumentation (whether it is based on a device with a resonant frequency for example), the need to bias the signal (as in thermography [19]) and the decision on whether the priority is either SNR or time resolution. Conventional binary codes tend to have bandwidths that are a maximum at DC. This is not effective for piezoelectric devices which have a well-defined centre frequency of operation.

Bandpass-like binary codes could be defined starting from conventional ones to overcome this problem. This is the case of inverse repeated sequences introduced in air-coupled ultrasound in [15]. The main choice then is between such binary codes and a chirp signal, as will be illustrated below. Although a fair comparison of windowing functions to be applied on the matched filter $\psi(t)$ lies beyond the scope of this work, some discussion has been included in the next section to gain more insight on the sidelobe problem arising from PuC.

3. Apparatus

There are many types of insulation that are used industrially, but these are classified by ASTM (American Society for Materials Testing, USA) as being available within three basic categories: fibrous, closed-cell expanded foam, and cellular glass types of insulation [20]. Examples of each are shown respectively in Figure 1. Each is in the form to be used on a cylindrical pipe and it has a different type of structure and widely different acoustic properties. Rockwool™ comprises a mineral or rock wool in the form of fibres which are then formed into rolls or slabs of a known density and thickness. It has good thermal and acoustic insulation properties, but is also permeable to water. Moisture is easily retained within its structure—a potential problem for CUI in certain situations. Polyurethane foam is less dense and has better thermal insulation properties than Rockwool, and is more resistant to water ingress due to the closed-pore foam structure. It is structurally stronger than Rockwool. Finally, Foamglas™ is a lightweight, rigid material with gas-filled closed-pore glass cells. It offers a combination of fire resistance, high compressive strength, and impermeability to water.

Figure 1. Photographs of the three types of insulation investigated in this work.

With reference to the three materials shown in Figure 1, their physical properties vary widely. For example, ultrasonic coupling to Rockwool™ without a liquid intermediate layer introduces losses, and propagation though its fibre/air matrix is complex. Polyurethane is a lossy polymer and the foam contains pores that are gas-filled and scattering. Finally, Foamglas™ has gas pockets surrounded by thin layers of glass which are expected to introduce severe ultrasonic scattering. The use of the PuC technique was designed to deal with these issues.

It is important initially to consider the centre frequency (f_0) at which to perform ultrasonic measurements on these types of insulation samples, noting that their structures vary considerably. Values of f_0 below 100 kHz would not be appropriate for the condition monitoring of insulated pipework (e.g., for identifying pitting or water ingress). This is because the corresponding wavelengths (λ) would be too long for a reasonable estimation of pitting depth at an insulated metal surface. To give an example, λ = 14.8 mm at 100 kHz in water, and while the acoustic velocity in insulation materials can vary (see below), this gives an idea of the wavelengths that could be present in a practical measurement. Conversely, the use of higher frequencies would make ultrasonic transmission more difficult in all three materials because of their composition, and the chosen technique must deal with this fact. The technique would also need to work through the thicknesses of insulation typically used in standard industrial applications. 50 mm has been chosen for study here, but some applications require insulation layers that may be thicker.

Initial experiments used dry contact, to avoid the need for couplant that would be absorbed by these materials. These used the apparatus shown schematically in Figure 2a, with the transducers simply pressed against the insulation. Ultrasonic signals were thus transmitted through the samples of insulation shown in Figure 1 in through-transmission. A pair of piezo-composite transducers (see Figure 2b) was chosen for this work, because of their ability to operate over a reasonable bandwidth at these low frequencies. In addition, each transducer was fitted with a quarter-wavelength matching layer for water, which helped transmission into the insulation layers. The transmitter was excited by a chirp signal generated by the National Instruments PXI-1042, incorporating an Arbitrary Waveform Generator NI-PXI 5412 and NI-PXI 5105 digitizer. The output was subsequently amplified via a custom-built power amplifier to 120 V. The piezocomposite transducers were positioned in a through-transmission configuration, and pressed lightly onto either side of a 50 mm thick sample of insulation. The receiver output was connected to a Cooknell CA6C charge amplifier before being input into the same PXI system for processing. Note that later experiments were also conducted in both a water tank and at the surface of a polymer-cladded pipe—the same transducers and instrumentation as that shown in Figure 2a was used for these, with the matching layer well-suited to such situations.

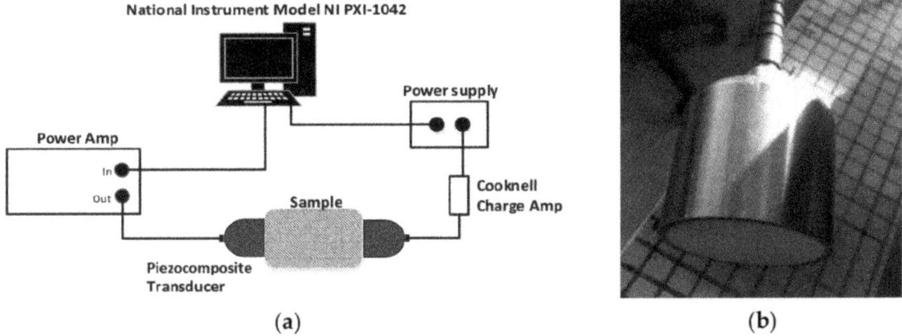

Figure 2. (**a**) Apparatus for low frequency ultrasound at a centre frequency of 150 kHz and a chirp signal swept from 100–200 kHz. (**b**) Piezo-composite transducer.

As stated above, the choice of pulse compression waveform depends in part on the frequency response of the transducers used. To investigate this, the impulse response of the piezo-composite transducers was measured using a transient voltage signal from a Panametrics 5052PRX pulser-receiver, and the results recorded via a Tektronix TDS3032C digital oscilloscope. The impulse response is shown in Figure 3a, whereas Figure 3b depicts the corresponding frequency spectrum (black line). Also shown is the simulated spectrum for two excitation waveforms: an inverse repeated sequence generated from a complementary Golay code and a windowed chirp signal in the 100–700 kHz range. It can be seen that the two types of excitation waveform can be used to produce a response that would excite the piezocomposite transducers efficiently. In the present case, the measurement priorities for a practical NDE tool were for simplicity and improved SNR. However, the use of an inverse repeated complementary Golay code would increase the complexity of a measurement, as two separate signals have to be transmitted at each measurement point. Note also that the complementary Golay code excitation introduces a sideband at frequencies above 750 kHz, which is of lower amplitude in the chirp waveform. For this reason, a set of linear chirp signals with specific values for B and T were used to optimise PuC for the different piezocomposite transducer pairs used in the measurements to be described below.

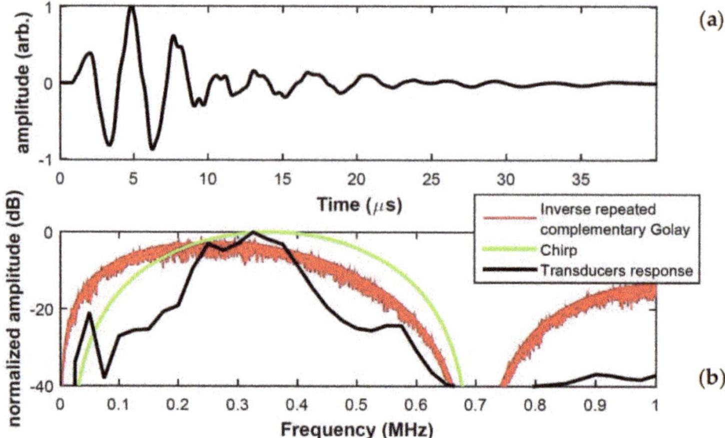

Figure 3. (**a**) The measured impulse response of a 300 kHz piezocomposite transducer. (**b**) Comparison of Inverse repeated complementary Golay and Chirp excitation waveforms for matching to the actual transducer bandwidth.

To gain insight on the choice on the matched filter $\psi(t)$ employed in the PuC algorithm reported in Equation (1), the mathematical definition of a chirp signal is given in Equation (3).

$$s(t) = \cos(\phi(t)), \tag{3}$$

with $\phi(t)$ being the instantaneous signal phase. Both the properties and the design of a chirp strictly depend on the definition of the instantaneous frequency.

$$f_{ist}(t) = \frac{1}{2\pi}\frac{d\phi(t)}{dt}. \tag{4}$$

For a linear chirp signal, the phase is a quadratic function $\phi(t) = 2\pi\left(f_{start}t + \frac{B}{2T}t^2\right)$, leading to an $f_{ist}(t)$ being a linear function of time.

$$f_{ist}(t) = f_{start} + \frac{B}{T}t = f_0 + \frac{B}{T}\left(t - \frac{T}{2}\right). \tag{5}$$

Thus, according to Equation (5) a linear chirp signal is a frequency modulated signal whose instantaneous frequency varies linearly within a chosen time duration and bandwidth. Many studies exist in the literature concerning the use of windowing functions $w(t)$ aimed at mitigating the magnitude of sidelobes inevitably arising from the use of coded signals with PuC [21–26]. In general, the $w(t)$'s are applied to the matched filter, i.e., $\psi(t) = w(t) \cdot x(-t)$, whereby $x(-t)$ is the inverted replica of the input signal. Figure 4 shows a comparison of the effect of some well-known windowing functions after applying Equation (1) with $y(t) = x(t)$, i.e., showing the envelope of the auto-correlation function $x(t) * \psi(t) \approx \delta(t)$.

With the exception of the rectangular window, the application of the other windowing functions widens the main lobe of the resultant $\delta(t)$'s, although all have a positive smoothing effect on sidelobes reduction. However, the main aim here is to obtain the greatest SNR improvement via PuC rather than distinguish among close scatterers, whereby windowing algorithms are extremely useful. For this reason, a rectangular window having unitary amplitude over the whole duration of the chirp signal has been employed, i.e., $w(t) = 1$ with $t \in [0, T]$. Thus, the matched filter is simply the inverted replica of the input signal, i.e., (t) $\psi(t) = x(-t)$, which has been shown to maximise the resulting SNR [21].

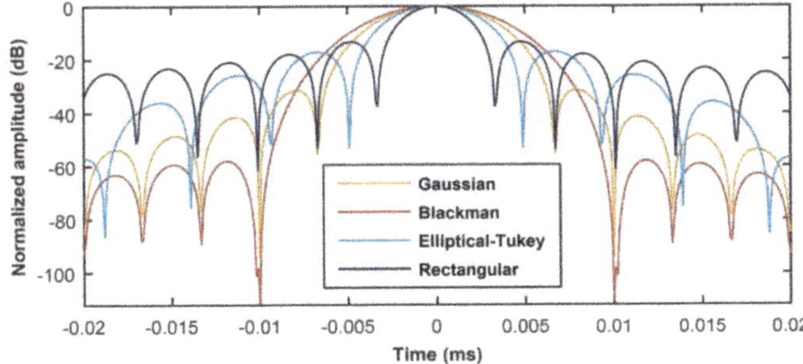

Figure 4. Comparison of the effect of different windowing functions $w(t)$ applied over the matched filter $\psi(t)$ to the PuC output.

4. Results

4.1. Experiments on Ultrasonic Propagation in Insulation

Dry-contact experiments were performed using the apparatus of Figure 2, with a low chirp centre frequency of 175 kHz. The linear chirp excitation signal was swept from 100–250 kHz (B = 150 kHz) with T = 1 ms. The results from these experiments, in terms of a rectified and smoothed pulse compression outputs, are shown for the three insulation samples in Figure 5. Also shown in each case is an estimate of the longitudinal velocity (v) in each material which was derived from these measurements. As might be expected, Rockwool™ had the lowest velocity, polyurethane foam was higher, and Foamglas™ had the highest value at 1000 m/s.

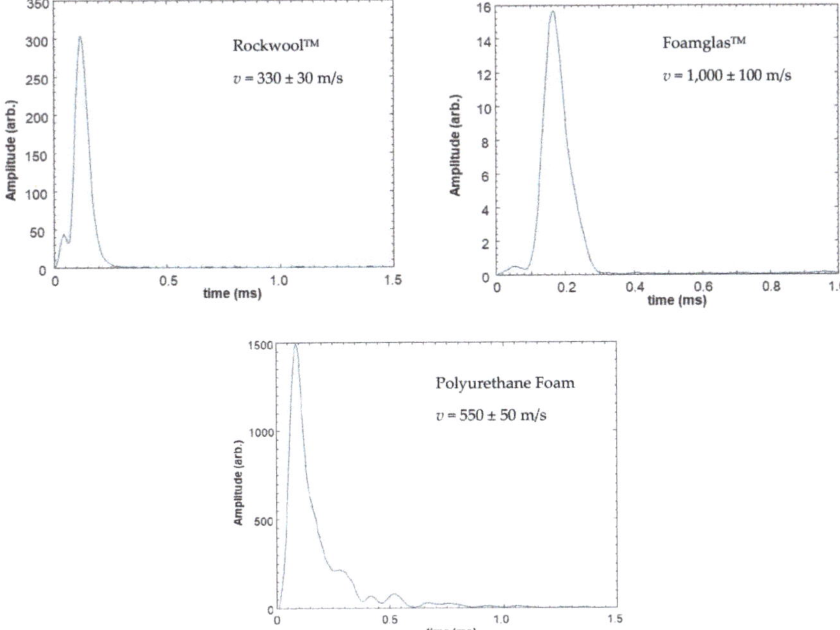

Figure 5. Rectified and smoothed ultrasonic signals transmitted through three samples of 50 mm thick insulation material. The longitudinal velocity (v) is shown in each case. The quoted uncertainties arise from the measurement of the sample thickness in each case.

An experiment was also conducted to determine the effects of water when absorbed by Rockwool™ insulation. To this end, a 50 mm thick sample was left in water for 24 h so as to be totally saturated. It was then left to dry naturally at room temperature (20 °C) and the longitudinal velocity (v) measured using the apparatus of Figure 2a. The results are shown as a function of time in Figure 6. It can be seen that v decreases rapidly initially, but then settles into an approximate linear decrease with time, as indicated by the best fit line shown. While this is only a qualitative result, it does indicate that there is a large change in acoustic velocity with water ingress for Rockwool, and this could thus be used as a good indicator of the presence of water within this insulation material. The structure of the other two materials (polyurethane foam and Foamglas™) is of a closed-pore nature, and hence water retention would not be as marked as that observed for Rockwool™, although this was not measured in this work.

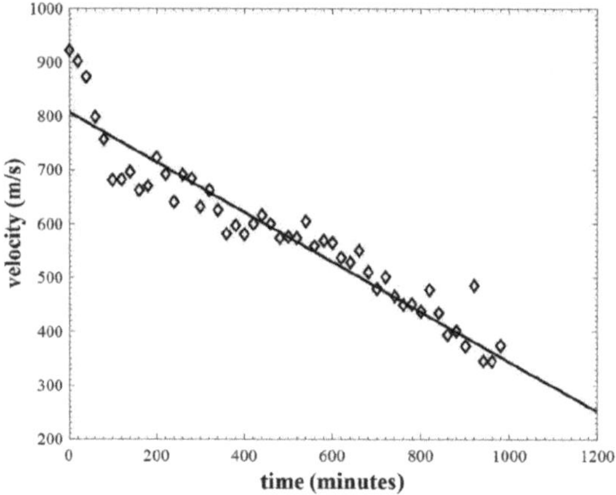

Figure 6. Longitudinal wave velocity decreasing as a Rockwool insulation sample is first saturated with water, and then left to dry out with time at room temperature.

4.2. Experiments in Wet Rockwool Insulation

The aim of these second set of experiments was to demonstrate the use of pulse compression ultrasound in the measurement of pitting of a surface when Rockwool insulation was wet—a common situation that would lead to CUI. While the dry contact results above were obtained at f_0 = 175 kHz, it was thought that a higher frequency would be needed if the detection of pitting was to be successfully accomplished with a reasonable resolution. The measurements below were thus performed using a pair of 25.4 mm diameter piezo-composite transducers with resonant frequency equal to 300 kHz, again fitted with a matching layer for water. As before, a chirp signal was used to drive the transmitter, this time with a frequency sweep from 200–400 kHz (B = 200 kHz and f_0 = 300 kHz) and with T = 1 ms. Steel samples with surface depressions were fabricated, with known surface contours, and a layer of Rockwool placed over the affected area. The transducer pair, positioned side by side at a centre-to-centre distance of 75 mm, could be scanned in unison horizontally over the sample surface, as shown in Figure 7. Similar instrumentation to that shown earlier in Figure 2a was used to record signals reflected from the metal surface in a pitch–catch arrangement, but now the transducers were immersed within a water tank, and their position controlled using an x–y stage under control of an external PC. In this way, either single waveforms could be captured at specific locations, or images of sample surface profiles obtained by measuring changes in delay time from transmitter to receiver. The pulse compression output was typically in the form of a smoothed, rectified response from which this travel time could be estimated. Figure 7 indicates an experiment where a sample of 20 mm thick Rockwool was placed over a 4 mm deep, 20 mm wide flat-bottomed square hole, machined into a flat steel surface.

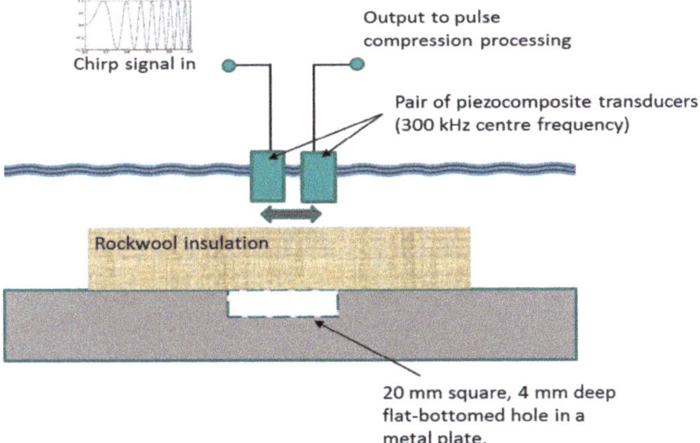

Figure 7. Apparatus used to record pulse compression ultrasonic data from metal surfaces covered in wet insulation layers.

Figure 8 shows the result of scanning the transducer pair over the flat-bottomed hole, where the response at both locations above the top surface and in the presence of the 4 mm deep hole are presented. A pulse compression output waveform is presented at the two locations indicated by the dotted vertical lines. Consider first the one at the left of the figure. When the transducer pair was located over the centre of the 4 mm deep flat-bottomed hole, it can be seen that there were two main peaks in the pulse compression output—an initial peak at 80 μs, which is due to reflection from the top surface of the Rockwool insulation, and a later peak at 120 μs, which is thought to arise from a reflection from the bottom of the hole. To the right of the figure is the pulse compression output when the transducer pair was moved to a position away from the location of the hole. In this case, the reflection from the top of the Rockwool insulation is still visible at 80 μs, but now the subsequent pulse-compression peak has is at 115 μs. The estimated difference of 5 ± 1 μs between the arrival time of the maximum amplitude in each case (120 μs vs. 115 μs) is consistent with a difference in path length of 8 mm (twice the hole depth) between the flat surface and defect regions, where the expected difference in time would be 5.4 μs. The reason for the complicated nature of the arrivals in both waveforms, with double peaks, is thought to arise from inhomogeneity and slight changes in thickness of the Rockwool insulation across the beam diameter, but further complicated by the fact that the width of the transducer beam would be comparable to the width of the hole. Note also that interpretation of the waveform at the hole location is complicated due to the fact that the wavelength at 300 kHz is ~5 mm. Hence, a clear reflection from the underside of the insulation is unlikely to have been resolved in the presence of a signal reflected from the bottom of the hole (which is only 4 mm deep). These measurements also illustrate that transmission though the Rockwool sample could be complicated due to its inhomogeneity, and this fact will need to be taken into consideration in future practical measurements. These experiments do, however, show that transmission through the insulation had been achieved, and that the presence of the hole has been detected.

It is interesting to discuss the minimum depth of hole that could be detected. Pulse compression is a good method for improving depth resolution, by improving SNRs. In our case, the measurements indicated that the depth resolution was ±0.7 mm. It is thus likely that the minimum hole depth detectable would be ~1 mm. This to some extent would depend on the type and thickness of insulation. Thicker or more attenuating insulation layers would not only decrease SNRs but also tend to remove higher frequencies from the signal, affecting bandwidth and hence defect detectability. The longitudinal velocity in the insulation is unlikely to affect this measurement unduly. This will be the subject of future work.

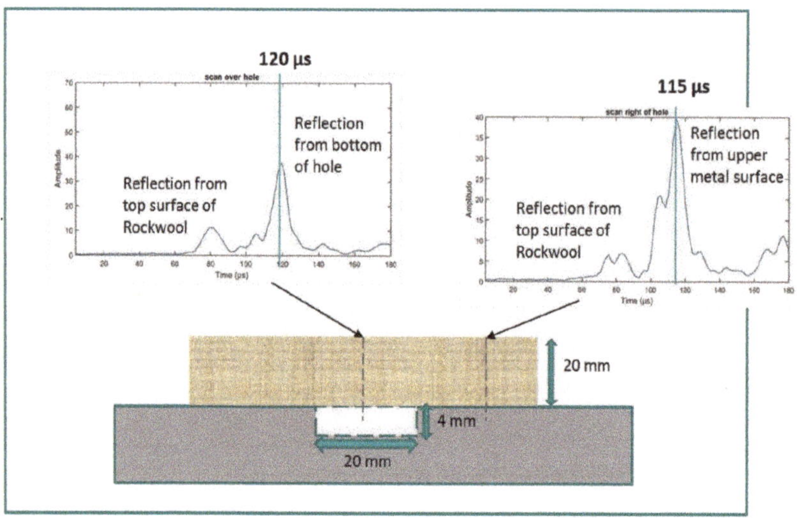

Figure 8. Pulse compression outputs for two locations above a metal surface containing a 4 mm deep flat-bottomed hole. The sample was covered in a 20 mm thick layer of water-saturated Rockwool insulation.

The piezo-composite transducer pair was now scanned over a horizontal area encompassing the artificial defect intended to simulate surface corrosion. The pulse compression output could be used to provide a depth profile of the hole by plotting the time delay of the second pulse compression peak against location in an x–y scan. The result is shown in Figure 9, and the presence of the hole is clearly visible. There are some artefacts within this image, and these could be caused by several factors: for example, the transducers are parallel and their beams overlap in a complicated fashion, and both their individual diameters and their separation distance are not negligible in comparison to the width of the defect. However, it is clear that the defect can be detected using this simple arrangement.

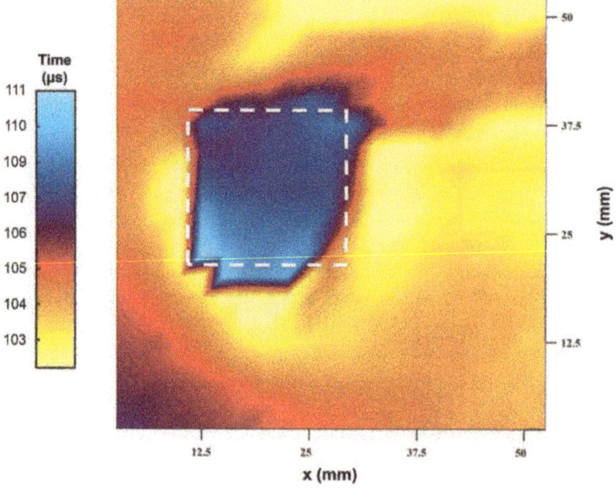

Figure 9. Image of a 20 mm wide hole of 4 mm depth when covered by water-saturated Rockwool insulation. The image was formed using time delay data. The colour scale is arbitrary. The dashed white square outlines the size and location of the artificial flat-bottomed hole.

4.3. Experiments on an Insulated and Cladded Industrial Pipe

A final experiment was conducted on a real insulated steel pipe sample. This sample, shown in the photograph of Figure 10, was designed for use in steam distribution applications. It had a 2 mm thick steel pipe of 40 mm radius coated with a 55 mm thick polyurethane foam insulating layer, which in turn was surrounded by a 2.5 mm thick polymer cladding layer.

Figure 10. Photograph of the industrial insulated pipe sample.

A schematic diagram of this pipe, complete with the dimensions of the different layers is shown in Figure 11. The two transducers operating in pitch–catch mode are shown as the white-coloured cylinders, noting that these were the same transducers as used in the water immersion experiments above. They were arranged in a pitch–catch configuration at the surface of the outer thin polymer protective cladding layer, and propagated ultrasonic signals were collected at a range of different transducer separation angles θ as shown. Here, $\theta = 25°$ represents the angle at which the transducers were positioned so that they were touching laterally (i.e., with their geometric centres being 25.4 mm apart). Note that pulse-echo operation with pulse compression is typically difficult because of the long excitation waveform durations, which overloads receiver amplifiers; hence, we chose only to operate in pitch–catch mode. It was also important to provide a set of measurements at various angles, so as to prove that the ultrasonic signal had been reflected from the steel pipe, and not guided around the cladding from source to receiver—this can be calculated by considering time-of-flight and the estimated speed of sound in the insulation. The red dotted arrows indicate the ultrasonic transmission path in pitch–catch mode. Note that a standard couplant for ultrasonic testing was used, and there was no need to fit a curved 'shoe' to operate at the cladding surface. Further tests (detailed below) were also performed to confirm that the signal was indeed travelling through the insulation, and not via a guided mode along the circumference of the cladding. This may occur in other samples not tested here, and would need to be taken into account in any practical test [27]. Thickness gauging was not possible in the insulated pipe examined, as the pipe wall thickness (2 mm) was too small compared to the wavelength in steel at 300 kHz (~19 mm). The aim here was to demonstrate penetration through the insulation and reflection from the steel surface.

Initial tests showed that no recognizable signal could be detected using a conventional nondestructive testing procedure (using the Panametrics pulser-receiver used previously for transducer characterization). However, the results of using the PuC technique on this sample are shown in Figure 12, using a linear chirp with T = 1 ms, a chirp centre frequency of f_0 = 300 kHz and with B = 200 kHz. It can be seen that PuC resulted in the successful detection of signals reflected from the steel pipe outer surface for various values of θ. The signal amplitude drops with increasing θ, with the expected increased time delay between transmitter and receiver (the two transducer beam axes are not coincident onto the same point on the metal pipe surface but are displaced laterally at increased values of θ). Note the excellent SNR of the output at all angles. The measurements in Figure 12 could have included signals arising from a guided-wave mode within the pipe; these have not been identified in this work, but future experiments will identify the types of wave modes present. Note that, at 2 mm, the pipe thickness was too small to allow multiple reflections to occur for thickness gauging for example. Note that the signals could also possibly have been propagating from transmitter to receiver via guided-wave modes within the cladding layer. Other experiments demonstrated that this was not happening—cuts in the insulation to stop such modes had little effect on the received signal, which was primarily due to reflection from the metal pipe. If this situation was to occur, for example if metal cladding was present, then the two propagation modes could be distinguished by the use of a non-linear chirp for PuC [15,18], but this was not necessary here.

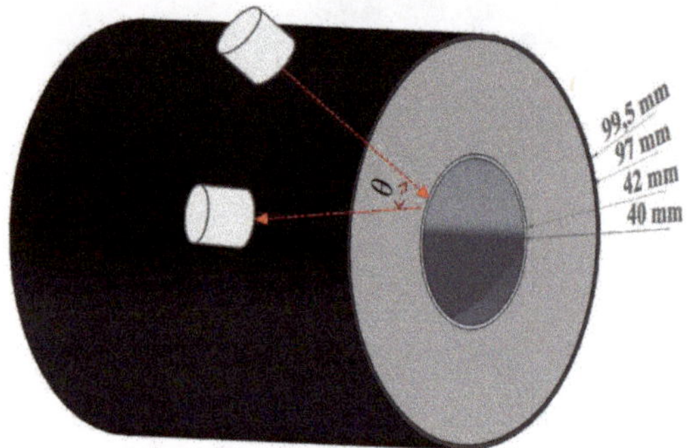

Figure 11. Schematic diagram of the industrial pipe illustrating transducer placement.

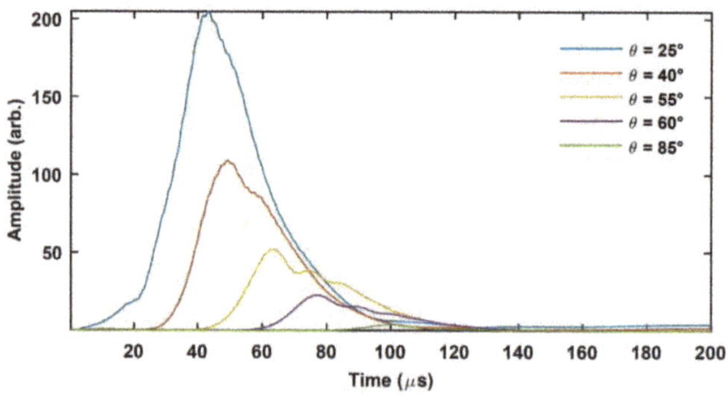

Figure 12. Pulse compression outputs for the sample of Figure 11 at various transducer separation angles θ.

The results of Figure 12 demonstrate that the main aim of this work has been achieved: ultrasonic signals can be detected that reflect from a steel pipe surface, in the presence of insulation and cladding. This leads the way towards developing techniques that could detect corrosion of the metal surface, and hence the characterization of CUI.

5. Discussion and Conclusions

It is evident from the above that pulse compression ultrasound can be used to propagate signals though common forms of insulation. Initial dry contact experiments indicated that transmission was possible in various insulation types, provided pulse compression was used in combination with piezocomposite transducers, the latter providing sufficient bandwidth over the 100–400 kHz frequency range studied here. It was also shown that for Rockwool the acoustic velocity changed dramatically with water content, but this has not been measured for other insulation types. This led to a second set of experiments in water immersion, to further refine the technique in pitch–catch mode, and to determine the possibility of imaging of pitting through a waterlogged Rockwool layer. This demonstrated that imaging of a 4 mm deep depression in a metal surface was possible, even when such a feature was covered in a Rockwool layer saturated with water. A further experiment showed that different surface depth features could be measured using time of flight.

The initial experiments indicated that tests on an actual industrially-relevant insulated pipes could be successful, and this has been shown to be the case for a centre frequency of f_0 = 300 kHz. The results show that penetration through a 55 mm thick insulation layer was possible with an excellent SNR. Further experiments will continue to refine this technique using pulse compression and scanned transducer locations to produce images, either via manual placement or by using arrays, currently under development. While thickness gauging was not possible in the insulated pipe examined, future work which will involve both thicker steel pipes and metallic cladding.

Author Contributions: Conceptualization, D.A.H. and S.L.; methodology, D.A.H., R.L.W., D.R.B. and S.L.; Software, L.A.J.D. and S.L.; Validation, investigation, formal analysis and investigation, R.L.W., L.A., S.L. and D.A.H.; Writing-original draft preparation, D.A.H., S.L., and P.B.; Writing-review and editing, all authors; Supervision, D.A.H., S.L., and M.R.; Project administration, D.A.H.; Funding acquisition, D.A.H. All authors have read and agreed to the published version of the manuscript.

Funding: This research was funded by the UK Engineering and Physical Sciences Research Council (EPSRC), grant number EP/L022125/1, through the UK Research Centre in NDE (RCNDE).

Acknowledgments: Thanks are due to Mike Lowe at Imperial College, UK, for loan of the insulation samples.

Conflicts of Interest: The authors declare no conflict of interest.

References

1. UK Health and Safety Executive (HSE), Corrosion under Insulation of Plant and Pipework v3. SPC/TECH/GEN/18. 2017. Available online: https://www.hse.gov.uk/foi/internalops/hid_circs/technical_general/spc_tech_gen_18.htm (accessed on 20 October 2019).
2. Liss, V.M. Preventing corrosion under insulation. *Chem. Eng.* **1987**, *94*, 97–100.
3. Roberge, P.R. *Corrosion Inspection and Monitoring*; John Wiley & Sons: New York, NY, USA, 2007; ISBN 9780471742487.
4. Dwivedi, D.; Lepková, K.; Becker, T. Carbon steel corrosion: A review of key surface properties and characterization methods. *RSC Adv.* **2017**, *7*, 4580–4610. [CrossRef]
5. Lowe, M.J.S.; Alleyne, D.N.; Cawley, P. Defect detection in pipes using guided waves. *Ultrasonics* **1998**, *36*, 147–154. [CrossRef]
6. Howard, R.; Cegla, F. Detectability of corrosion damage with circumferential guided waves in reflection and transmission. *NDT E Int.* **2017**, *91*, 108–119. [CrossRef]
7. Guan, R.; Lu, Y.; Wang, K.; Su, Z. Fatigue crack detection in pipes with multiple mode nonlinear guided waves. *Struct. Health Monit.* **2019**, *18*, 180–192. [CrossRef]
8. Jones, R.E.; Simonetti, F.; Lowe, M.J.S.; Bradley, I.P. Use of microwaves for the detection of corrosion under insulation: The effect of bends. *AIP Conf. Proc.* **2012**, *1430*, 1665.
9. Cheng, W. Pulsed eddy current testing of carbon steel pipes' wall-thinning through insulation and cladding. *J. Nondestr. Eval.* **2012**, *31*, 215–224. [CrossRef]
10. Sophian, A.; Tian, G.; Fan, M. Pulsed Eddy Current Non-destructive Testing and Evaluation: A Review. *Chin. J. Mech. Eng.* **2017**, *30*, 500–514. [CrossRef]
11. Bailey, J.; Long, N.; Hunze, A. Eddy Current Testing with Giant Magnetoresistance (GMR) Sensors and a Pipe-Encircling Excitation for Evaluation of Corrosion under Insulation. *Sensors* **2017**, *17*, 2229. [CrossRef]
12. Li, Z.; Dixon, S.M.; Cawley, P.; Jarvis, R.; Nagy, P.B. Study of metal magnetic memory (MMM) technique using permanently installed magnetic sensor arrays. *AIP Conf. Proc.* **2017**, *1806*, 110011.
13. Mohamed, I.; Hutchins, D.; Davis, L.; Laureti, S.; Ricci, M. Ultrasonic NDE of thick polyurethane flexible riser stiffener material. *Nondestruct. Test. Eval.* **2017**, *32*, 343–362. [CrossRef]

14. Laureti, S.; Ricci, M.; Mohamed MN, I.B.; Senni, L.; Davis LA, J.; Hutchins, D.A. Detection of rebars in concrete using advanced ultrasonic pulse compression techniques. *Ultrasonics* **2018**, *85*, 31–38. [CrossRef] [PubMed]
15. Hutchins, D.; Burrascano, P.; Davis, L.; Laureti, S.; Ricci, M. Coded waveforms for optimised air-coupled ultrasonic nondestructive evaluation. *Ultrasonics* **2014**, *54*, 1745–1759. [CrossRef] [PubMed]
16. Misaridis, T.; Jensen, J.A. Use of modulated excitation signals in medical ultrasound. Part I: Basic concepts and expected benefits. *IEEE Trans. Ultrason. Ferroelectr. Freq. Control* **2005**, *52*, 177–191. [CrossRef] [PubMed]
17. Misaridis, T.; Jensen, J.A. Use of modulated excitation signals in medical ultrasound. Part II: Design and performance for medical imaging applications. *IEEE Trans. Ultrason. Ferroelectr. Freq. Control* **2005**, *52*, 192–207. [CrossRef] [PubMed]
18. Burrascano, P.; Laureti, S.; Senni, L.; Ricci, M. Pulse compression in nondestructive testing applications: Reduction of near sidelobes exploiting reactance transformation. *IEEE Trans. Circuits Syst.* **2019**, *66*, 1886–1896. [CrossRef]
19. Burrascano, P.; Callegari, S.; Montisci, A.; Ricci, M.; Versaci, M. (Eds.) *Ultrasonic Non-Destructive Evaluation Systems: Industrial Application Issues*; Springer: Cham, Switzerland, 2014.
20. Thermal Insulation Building and Environmental Acoustics. ASTM Volume 4.06. Standard C335/C335M–17. 2107. Available online: https://www.astm.org/ (accessed on 20 December 2019).
21. Turin, G. An introduction to matched filters. *IRE Trans. Inform. Theory* **1960**, *6*, 311–329. [CrossRef]
22. Pallav, P.; Gan, T.H.; Hutchins, D.A. Elliptical-Tukey chirp signal for high-resolution, air-coupled ultrasonic imaging. *IEEE Trans. Ultrason. Ferroelectr. Freq. Control* **2007**, *54*, 1530–1540. [CrossRef]
23. Harput, S.; Arif, M.; McLaughlan, J.; Cowell, D.M.; Freear, S. The effect of amplitude modulation on subharmonic imaging with chirp excitation. *IEEE Trans. Ultrason. Ferroelectr. Freq. Control* **2013**, *60*, 2532–2544. [CrossRef]
24. Smith, P.R.; Cowell, D.M.; Freear, S. Width-modulated square-wave pulses for ultrasound applications. *IEEE Trans. Ultrason. Ferroelectr. Freq. Control* **2013**, *60*, 2244–2256. [CrossRef]
25. Huang, S.W.; Li, P.C. Arbitrary waveform coded excitation using bipolar square wave pulsers in medical ultrasound. *IEEE Trans. Ultrason. Ferroelectr. Freq. Control* **2006**, *53*, 106–116. [CrossRef] [PubMed]
26. Laureti, S.; Khalid Rizwan, M.; Malekmohammadi, H.; Burrascano, P.; Natali, M.; Torre, L.; Rallini, M.; Pyri, I.; Hutchins, D.A.; Ricci, M. Delamination Detection in Polymeric Ablative Materials Using Pulse-Compression Thermography and Air-Coupled Ultrasound. *Sensors* **2019**, *19*, 2198. [CrossRef] [PubMed]
27. De Marchi, L.; Marzani, A.; Caporale, S.; Speciale, N. Ultrasonic Guided-Waves Characterization with Warped Frequency Transforms. *IEEE Trans. Ultrason. Ferroelectr. Freq. Control* **2009**, *56*, 2232–2240. [CrossRef]

Letter

New Proposal for Inverse Algorithm Enhancing Noise Robust Eddy-Current Non-Destructive Evaluation

Milan Smetana, Lukas Behun, Daniela Gombarska and Ladislav Janousek *

Department of Electromagnetic and Biomedical Engineering, Faculty of Electrical Engineering and Information Technology, University of Zilina, Univerzitna 1, 010 26 Zilina, Slovakia; milan.smetana@feit.uniza.sk (M.S.); behun@esys.sk (L.B.); daniela.gombarska@feit.uniza.sk (D.G.)

* Correspondence: ladislav.janousek@feit.uniza.sk; Tel.: +421-41-513-2100

Received: 30 July 2020; Accepted: 25 September 2020; Published: 28 September 2020 

Abstract: Solution of inverse problem in eddy-current non-destructive evaluation of material defects is concerned in this study. A new inverse algorithm incorporating three methods is proposed. The wavelet transform of sensed eddy-current responses complemented by the principal component analysis and followed by the neural network classification are employed for this purpose. The goal is to increase the noise robustness of the evaluation. The proposed inverse algorithm is tested using real eddy-current response data gained from artificial electro-discharge machined notches made in austenitic stainless-steel biomaterial. Eddy-current responses due to the material defects are acquired using a newly developed eddy-current probe that senses separately three spatial components of the perturbed electromagnetic field. The presented results clearly show that the error in evaluation of material defect depth using the proposed algorithm is less than 10% even when the signal-to-noise ratio is as high as 10 dB.

Keywords: non-destructive evaluation; eddy current; 3D sensing; inverse problem; wavelet transform; principal component analysis; neural network

1. Introduction

Electromagnetic non-destructive evaluation of conductive materials is currently used in various strategy sectors, such as aviation, nuclear, petrochemical, biomedicine, and many other industries. There are several methods that utilize the interaction of the electromagnetic field with a conductive structure for the purpose. One of the frequently used methods is the eddy current testing (ECT). It originates from the electromagnetic induction phenomena and the principle of ECT underlies in the interaction of induced eddy currents with structure of an examined body. This method is a powerful tool for non-destructive evaluation of material discontinuities. Its main advantages include high sensitivity, high inspection speed, and versatility. The main limitation is the possibility for investigating of surface and close subsurface defects only. This limitation results directly from the physical principle of the method as the induced eddy currents are strongly attenuated along a material depth. In the non-destructive evaluation, two approaches are defined: forward and inverse problem solution. Forward problem is a direct approach of acquiring responses from a defect by experimental measurements or numerical simulations. The solution of inverse problem involves the identification of the defect under investigation from the responses; for example, identification of its dimensions, geometry, shape, and orientation. In general, solving of the inverse problem is always relatively difficult. Many research teams, mostly from the academic environment, deal with this issue using the ECT method. ECT is a relative method and the inverse problem is thus ill-posed [1]. Characterization of a detected defect from ECT response signals is quite a challenge. Various mathematical procedures are being sought and used with the aim of partial improvements of preciseness and reliability. Scientific works are focused mainly on reducing uncertainty in estimating parameters of defect geometry, especially by applying adaptive Monte Carlo method [2], metamodels-based Markov–Chain–Monte-Carlo [3], Genetic Algorithm [4], Neural Networks [5], Particle Swarn Optimization [6], and others [7,8]. Many published papers address the effect of a defect structure on response signals and then optimize 3D defects [9]. A separate chapter is represented by non-iterative methods and methods for eliminating artefacts such as lift-off variation [10]. The main goal is to correctly estimate the defect geometry at first. It is then possible to reconstruct the predicted shape of the defect using appropriate mathematical procedures. However, all these procedures are considerably computationally time-consuming. If one compares the calculation times for defects of defined shape (electric-discharge-machined, EDM) vs. corrosion defects (stress-corrosion-cracking, SCC), the computational-time difference can reach up to hundred-times under the same conditions. There is currently no algorithm that can provide such real-time results. In addition to the calculation time criterion, other criteria are present: accuracy and unambiguity of calculation, permissible error, resistance and robustness to noise, and others [11,12]. The noise resistance is a very important criterion since the noise greatly increases the error of the defect parameters estimation. The origin of the noise is mainly caused by various artefacts: lift-off variation, superimposed measurement noise, too low probe resolution, and ambient electromagnetic disturbances, [13,14].

The paper focuses on increasing noise robustness in ECT inverse problem solution. A new inverse algorithm is proposed for the purpose. The algorithm combines three methods: wavelet transform, principal component analysis, and neural network. Moreover, newly developed ECT sensor is employed for the inspection. It is experimentally demonstrated that the error of a detected crack depth evaluation from the ECT response signals is quite low even when the signal to noise ratio is as high as 10 dB.

2. New Inverse Algorithm

The solution of ECT inverse problem and successful reconstruction of detected flaw properties or its character strongly depend on the chosen method/s. All conventionally used methods are dependent on database of known eddy-current responses and corresponding crack geometry including dimensions or on optimization and solution of forward problems using numerical means. The contemporary main aim of the methods' development is to obtain relevant crack geometry and/or to visualize 3D crack profile in real time with high preciseness. The preciseness of reconstruction is further deteriorated by presence of noise in the measured signal. Since the nature of the present noise is a stochastic one and it comes from multiple sources, the use of noise robust inverse algorithm is thus essential. The inverse problem solution proposed in the paper applies wavelet transform (WT) on measured signal complemented by principal component analysis (PCA) and followed by neural network (NN) classification. In the ECT field, the most important informations are depth and length of a detected crack [9,13,14]. Specifically, the crack depth is the most critical parameter from the structural integrity point of view. A crack width is not reconstructed in this paper as this parameter is not important from the structural integrity point of view. Artificial as well as natural cracks are very narrow, and their opening is on the level of tenth of millimeter. Moreover, this parameter (in its natural range) almost does not influence the ECT responses.

It should be noted that the ECT is the indirect method and the ECT responses are integral ones. Several methods have been examined in other to develop noise robust algorithm for a crack dimension estimation from ECT responses. The combination of WT, PCA, and NN shows promising results. The flowchart of the proposed new inverse algorithm is shown in Figure 1. The WT is employed for a crack length estimation. Due to generally known limitations of ECT, a crack depth estimation is more difficult. It is possible to determine a crack depth based on the amplitude of a measured signal—the greater the depth, the higher the amplitude of the measured signal. However, this method of estimation is greatly influenced by the lift-off, noise, skin-effect, etc. Therefore, the depth determination from the signal amplitude is often erroneous and inaccurate. For this reason, artificial intelligence is employed in this paper to tackle this issue. The amount of data is reduced at first using PCA. The depth of a defect is then estimated using trained NN. The methods employed in the proposed new algorithm are explained below in details.

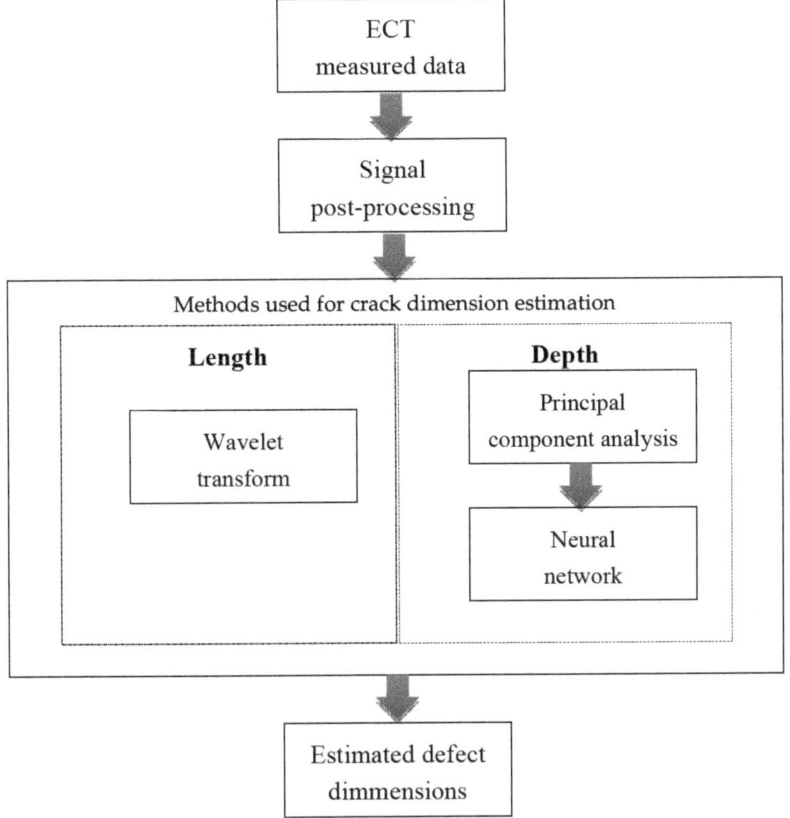

Figure 1. Flowchart of proposed new inverse algorithm.

2.1. Wavelet Transform

Measured ECT signals are of non-stationary character. To identify the frequency components of non-stationary signals various time–frequency analysis methods have been developed including linear and bilinear time–frequency representations, adaptive parametric and non-parametric time–frequency analysis, and more. Linear time–frequency methods decompose signals into a weighted sum of a series of bases localized in both time and frequency domains. The resolution in time–frequency domain, however, is governed by the Heisenberg uncertainty principle: the time localization and frequency resolution cannot reach their highest levels together. To match the complex structure of measured non-stationary signal, the WT is chosen here. The WT as time–scale analysis method gives an effective tool for analyzing self-similar signals. As the basis, the WT employs wavelets and adds a scale variable in the inner product transform to the time variable. It has a good frequency resolution for lower frequencies, which does not further work for higher frequencies. For higher frequency components, WT has a better time localization, but a lower frequency resolution. Thus, for lower frequency components the time localization is worse. To date, various types of wavelet basis have been proposed. However, the question of choice remains an open issue. Choosing a suitable one among all to match the signal structure requires experiences and generally accepted effective methods do not exist. Continuous WT of any energy limited signal $x(t)$ can be defined as:

$$X_W(s,b) = \frac{1}{\sqrt{s}} \int_{-\infty}^{\infty} x(t)\psi^*\left(\frac{t-b}{s}\right)dt \tag{1}$$

where $\psi(t)$ is the basis mother wavelet and the transform is derived by dilating with scale s and translating by the shift parameter b. Energy conservation of transform is maintained by the normalization factor $1/\sqrt{s}$. The Haar wavelet is chosen as the base mother wavelet for processing the measured signals here, because such an application requires the usage of non-symmetrical wavelets.

The WT is used to detect changes in signal amplitude gradient and to estimate scalograms for evaluation of a detected defect dimension, its length in this case. Whilst the length is estimated from the amplitude gradient changes in columns and rows of measured signal matrix, the depth estimation requires an adaptation.

An important step in the use of WT is the appropriate selection of the mother wavelet, which depends on the specific use of WT. The selection of the wavelet is conditioned by monitoring the changes in the gradient. Asymmetric wavelets are suitable for detecting changes in the gradient of the signal under investigation, so the Haar wavelet is chosen. Using Haar wavelet changes in the gradient in the measured signal are detected and detecting the changes the information about the defect and its dimensions is obtained. The mother wavelet is used to calculate the WT coefficients from the matrix of input data.

The convolution process is performed by individual columns of the input matrix to determine the length of the defect. To illustrate the resulting scalogram, the coefficient values for one column are shown in Figure 2. The vector for the value of the scale s = 150 is subsequently extracted from the scalogram. Such a scale represents information about slowly changing signal details, i.e., low frequencies. Subsequently, a matrix is created in which these vectors of WT coefficients are written. Figure 3 displays an example of such matrix converted into a grayscale image. The bright spots specify the edges of the defect for the case of length determination.

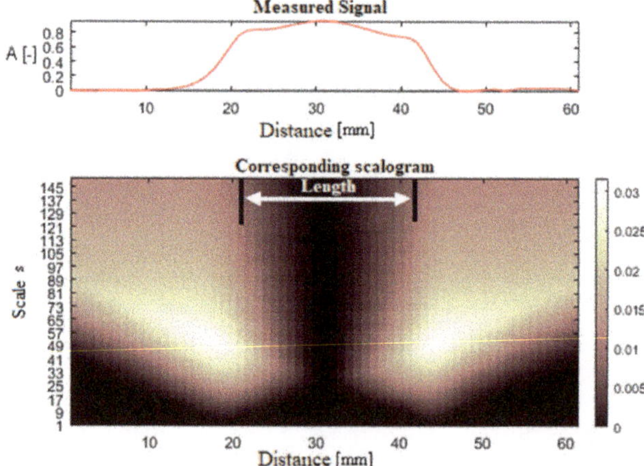

Figure 2. Scalograms for one column of the input data matrix-sample scalogram.

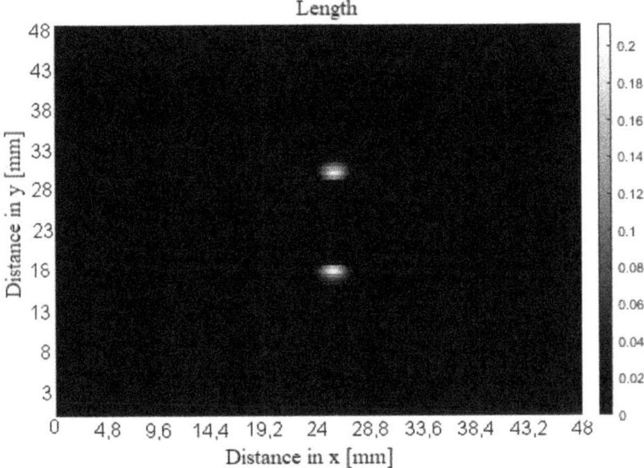

Figure 3. Matrix containing individual column vectors of WT coefficients.

2.2. *Principal Component Analysis*

Usage of the signal amplitude does not work well for the estimation of a detected defect depth; it means the defect dimension in direction towards material thickness. PCA algorithm is adopted for the data processing prior the estimation of a detected defect depth. The first step is to normalize input data by scaling the specific properties $x_j^{(i)}$ by their mean values s_j:

$$s_j = \frac{1}{m}\sum_{i=1}^{m} x_j^{(i)} \tag{2}$$

where j is the index of specific property. The required reduction of data dimension by one is achieved by means of covariance matrix and eigenvectors, where the covariance matrix is:

$$\sum = \frac{1}{m}\sum_{i=1}^{m}\left(x^{(i)}\right)\left(x^{(i)}\right)^T \tag{3}$$

The eigenvector of the covariance matrix is computed by means of singular values decomposition (SVD). The PCA data reduction for the neural network input decreases data from 10^5 to 10^3. This approach significantly reduces computational demands of classification problem. The output data from PCA are fed to the neural network for further classification.

2.3. *Neural Network*

A NN employed for a crack depth estimation uses back-propagation algorithm with forward propagation of input data, backward propagation of error, and consequent changes in input neuron weight values. The real measured data for each defect with known geometry are applied as the training sets and the applied training function is Bayesian regularization based on Levenberg–Marquardt optimization, [3,9,10]. The neural network consists of an input, a hidden and an output layer. The number of input neurons depends on the length of the eigenvectors. The hidden layer consists of ten neurons, and the output layer, since the output is the defect depth value, is formed by one neuron. The activation functions between the input and hidden layer is sigmoid and between hidden and output layer is linear function. The simple neural network model is shown in Figure 4. In the present work, the input layer has 1000 input neurons, which correspond to measured samples in one row. The output layer value represents an estimated parameter-a depth dimension of a detected defect. To achieve faster convergence compared to standard back propagation neural network, the Levenberg–Marquardt algorithm is used. The Levenberg–Marquardt algorithm uses approximation to the Hessian matrix instead of computing it directly. In the method, the Marquardt adjustment parameter μ is introduced. The parameter μ is decreased after each successful step in order to reduce the performance function at each iteration. Besides that, to avoid overfitting during neural network training, the Bayesian approach is used. Typically, the training aims to reduce the sum of squared errors. The Bayesian framework also considers the sum of squares of the network weight and the objective function. It also considers the weights of the network to be random variables. According to the Bayes' rule, the probability distribution can be written as:

$$P(w|D,\alpha,\beta,\eta) = \frac{P(D|w,\beta,\eta)P(w|\alpha,\eta)}{P(D|\alpha,\beta,\eta)} \tag{4}$$

where D corresponds to the input–output data set, η denotes the network model and architecture, w are the network weights, and α, β are objective function parameters. $P(w|\alpha,\eta)$ is the prior distribution derived of the knowledge about the weights before any data are collected, $P(D|w,\beta,\eta)$ is the probability of the data occurring given the weights w. $P(D|\alpha,\beta,\eta)$ is the normalized factor which guarantees that the probability is 1. The network is trained using experimental dataset of 70 similar EDM notches.

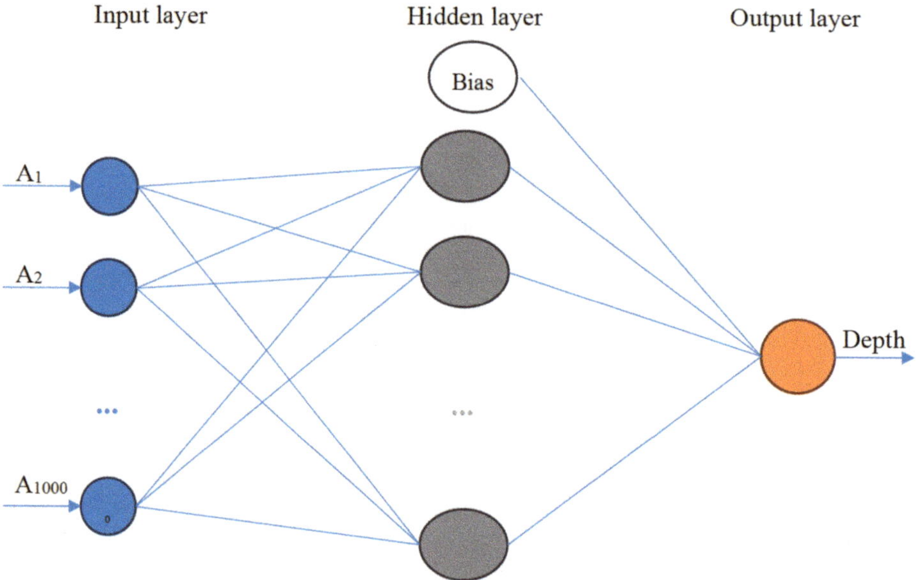

Figure 4. Arrangement of the neural network.

3. Experimental Setup

A modular ECT probe developed and constructed by authors is employed to excite eddy currents in an inspected specimen and to acquire the eddy current responses due to material discontinuities. Schematic arrangement of the ECT probe over a specimen is shown in Figure 5 (dimensions are given in mm).

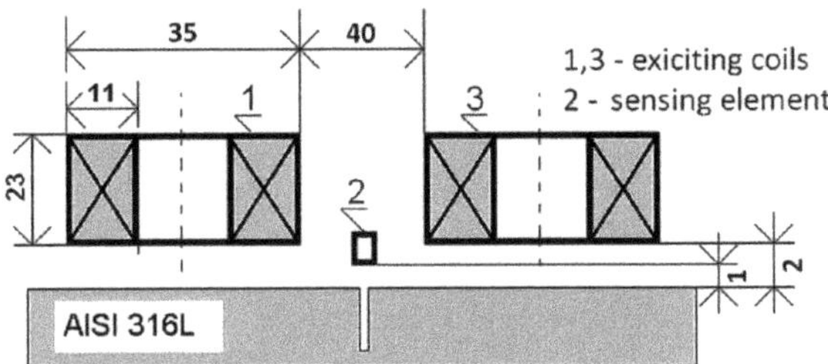

Figure 5. Arrangement of the eddy current testing (ECT) probe over a specimen (dimensions are in mm).

The ECT probe consists of two identical exciting circular coils positioned apart from each other and oriented normally regarding the surface of an inspected plate specimen. The coils are connected in series, but magnetically opposite to induce uniformly distributed eddy currents in the plate. The exciting coils are supplied from a harmonic source with a frequency of 1 kHz and the current density 1 A/mm^2. A detection system (sensing element) of the probe is composed of a fluxgate magnetometer. Three spatial components of the perturbed magnetic flux density field are acquired during a crack inspection. The detection system is located in a centre between the exciting coils to gain high sensitivity as the direct coupling between the exciting coils and the sensing element is minimal at this position.

A non-magnetic conductive plate specimen with a thickness of 10 mm made from the stainless steel AISI316L is inspected in this study, as shown in Figure 6 (dimensions are given in mm). The material has the conductivity of σ = 1.4 MS/m and the relative permeability of μ_r = 1.

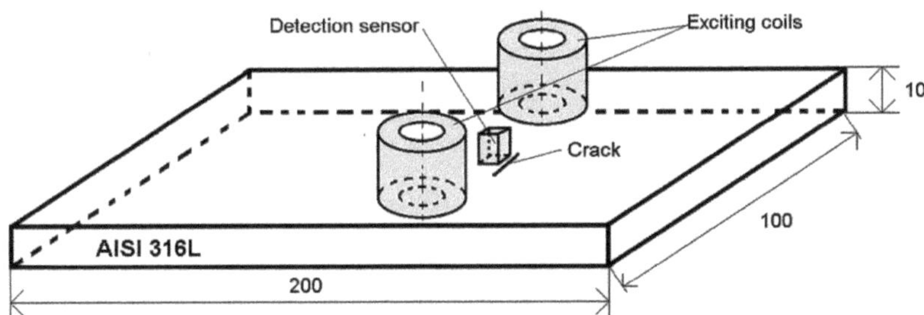

Figure 6. Spatial configuration of the plate specimen and the ECT probe (dimensions are in mm).

Four electro-discharge machined (EDM) notches of cuboid shape are introduced into the plate specimen. The dimensions of the EDM notches (cracks) are summarized in Table 1. The dimensions are denoted as l_c—crack length; w_c—crack width (opening); and d_c—crack depth. One can see that the EDM notches differ in their lengths and depths.

Table 1. Dimensions of the electro-discharge machined (EDM) notches.

Crack No.	l_c [mm]	w_c [mm]	d_c [mm]
1	9	0.20	3
2	15	0.25	5
3	21	0.25	7
4	27	0.25	9

A precise 3-axial mechanical positioning system, so called XYZ stage, is employed to precisely position the ECT probe over a surface of the inspected body and to provide prescribed movement of the probe over the surface. Clearance between the probe bottom and the plate surface, so called lift-off, is kept contact during the whole inspection on a value of 1 mm. Two-dimensional scanning is performed over the cracked surface from the near side. The scanned area over each cracked region is 48×48 mm^2. The probe moves smoothly along the crack length while the crack centre corresponds with a centre of the scanned area. The number of samples in one row is 1000 and the area of scanning is divided into 100 rows. The real and imaginary parts of all three spatial components of the perturbed magnetic flux density vector are sensed and recorded during the inspection of each crack. The XYZ stage control and the data acquisition are done under the LabVIEW environment.

4. Results and Discussions

Four EDM notches introduced in the AISI 3016L plate specimen are inspected using the ECT probe according to the explanation provided in the previous section. Example of the sensed eddy current responses from one crack are presented in Figure 7. The Y-axis of the plots shows sensed voltage difference at the terminals of the fluxgate sensor and it is directly proportional to the measured perturbed magnetic flux density value. All three spatial components of the magnetic flux density vector are shown in corresponding rows of the figure: (Figure 7a) the X component of the perturbed magnetic flux density vector; (Figure 7b) the Y component of the perturbed magnetic flux density vector; and (Figure 7c) the Z component of the perturbed magnetic flux density vector), while the absolute value of the magnetic flux density is shown in the last row (Figure 7d). The sensed signals are shown in the left column. An additional white noise is artificially generated via waveform generator with different signal to noise ratio (SNR) values, i.e., 3 dB, 6 dB, and 10 dB, and added to the measured signal. The right column in Figure 7 shows corresponding signals deteriorated by the noise with SNR of 10 dB.

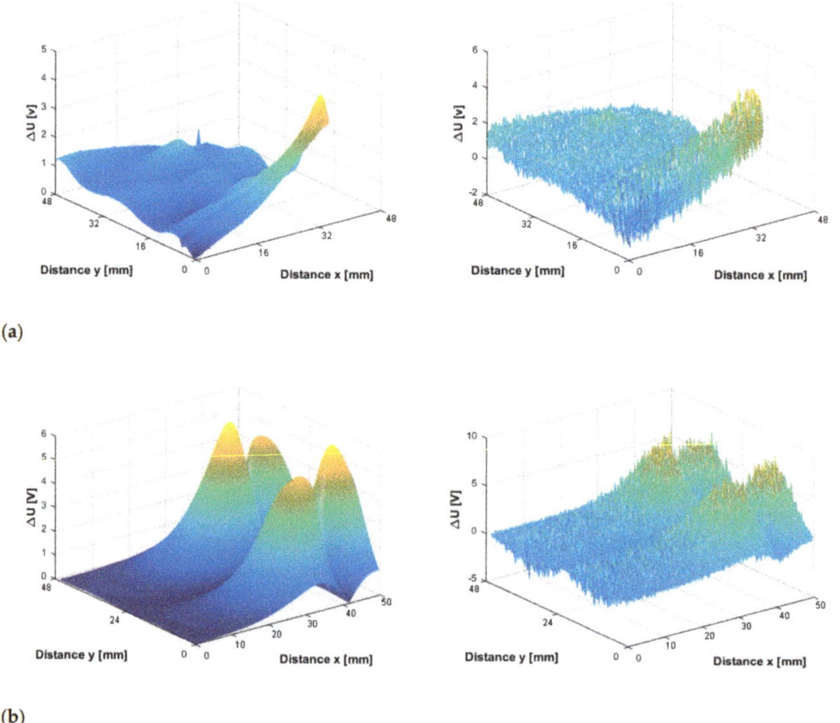

Figure 7. *Cont.*

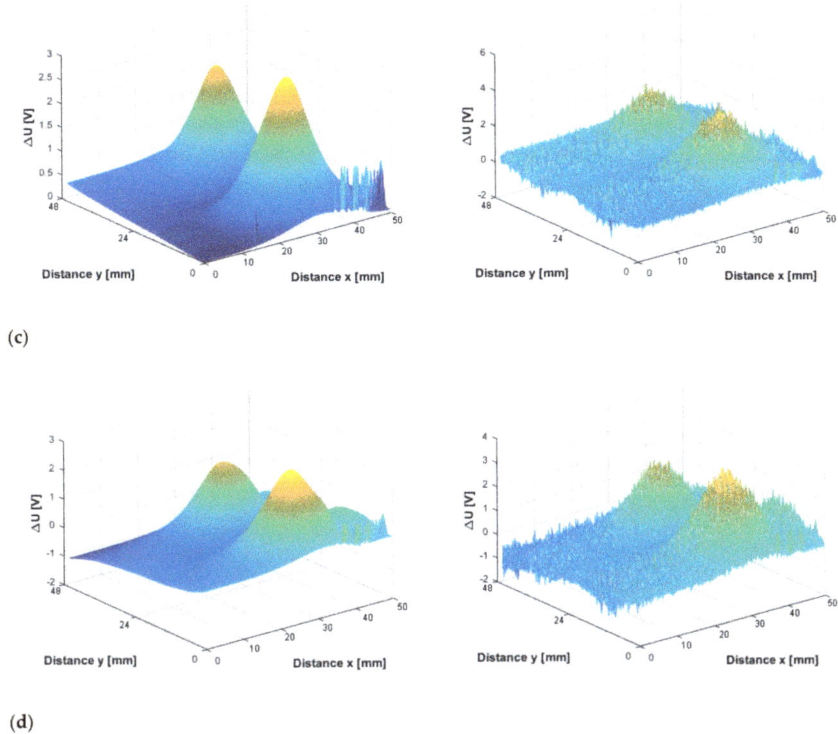

Figure 7. Eddy current responses due to an EDM notch measured during 2-dimensional scanning over a cracked region.

Crack dimensions are reconstructed based on the proposed inverse algorithm presented in the Section 2. The sensed responses as well as the ones with the added white noise of different SNR levels—3, 6, and 10 dB are used as input signals for the reconstruction algorithm, separately.

The results of the crack length l_c reconstruction are reported in Table 2 for different noise levels. Data presented in column 0 dB are the ones reconstructed from the measured signal directly without any artificial noise added. Graphical representation of the results together with the relative error of reconstruction are shown in Figure 8. One can observe that in case of noisy signals the crack length is overestimated what is safe side from the non-destructive evaluation point of view. The estimation error is quite high in case of the crack No. 1, i.e., short and shallow crack. It is caused by large dimensions of the ECT probe employed for inspection and its sensitivity capabilities. The length of three other cracks, i.e., the longer (and deeper) ones, is estimated with quite good precision, especially the one of crack No. 4. It is important to highlight that the results prove that the proposed inversion algorithm is quite robust against the noise.

Table 2. Values of the reconstructed crack length l_c [mm] from the ECT responses with different Signal to Noise Ratio (SNR).

Crack No.	Real Crack Length	SNR [dB]			
		0	3	6	10
1	9.0	10.0	11.5	11.0	9.5
2	15.0	16.0	17.0	16.0	16.0
3	21.0	21.5	24.0	23.5	23.0
4	27.0	25.5	28.5	28.0	27.5

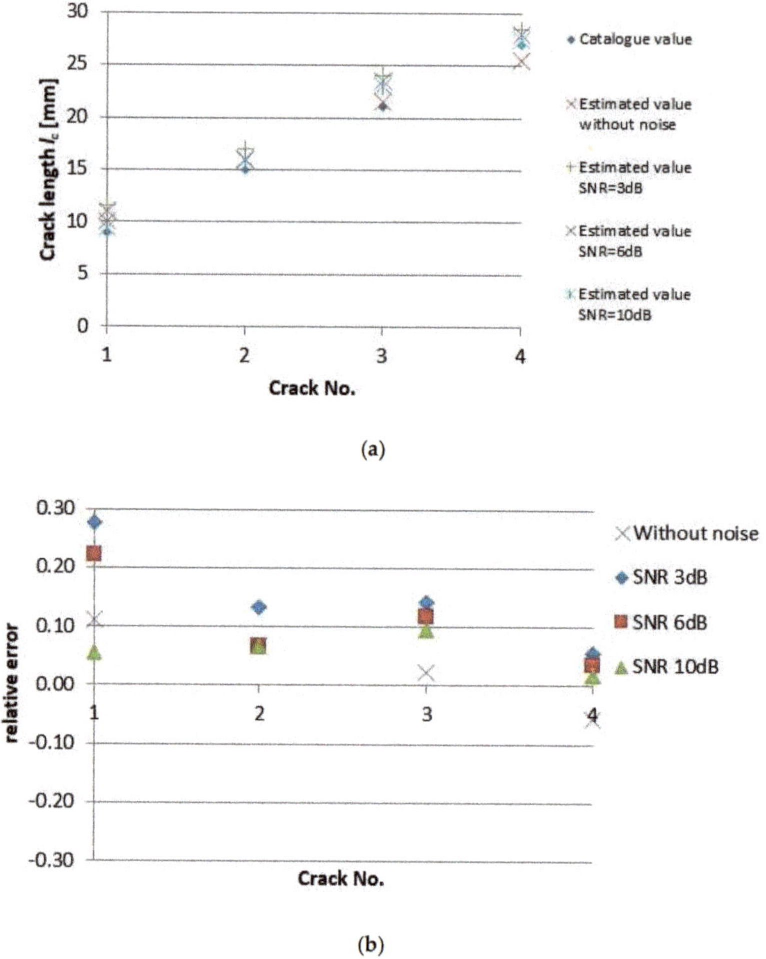

Figure 8. Results of the crack length l_c reconstruction from the ECT responses with different SNR: (**a**) estimated absolute values; (**b**) relative error of estimation.

Precise estimation of a detected crack depth is more critical issue in non-destructive evaluation comparing to a crack length. The crack depth d_c reconstruction results are reported in Table 3 and in Figure 9. Figure 9 displays absolute values of the estimated depth of respective cracks as well as the relative error of reconstruction for each crack and each SNR. As it can be seen, the preciseness of the depth estimation is quite high, and the noise does not have almost any influence on the estimation results. The depth of shallower cracks is slightly overestimated, while the estimated values are a little bit lower than the real ones for the deeper cracks. This corresponds to the depth resolution of the probe itself and the eddy current attenuation along material thickness.

The experimental investigation and presented results clearly proved that the proposed new inverse algorithm presented in the paper is quite robust against the noise.

Table 3. Values of the reconstructed crack depth d_c from the ECT responses with different SNR.

Crack No.	Real Crack Depth	SNR [dB]			
		0	3	6	10
1	3.0	3.1	3.1	3.1	3.1
2	5.0	5.3	5.5	5.5	5.5
3	7.0	7.2	6.4	6.4	6.4
4	9.0	8.8	8.2	8.2	8.2

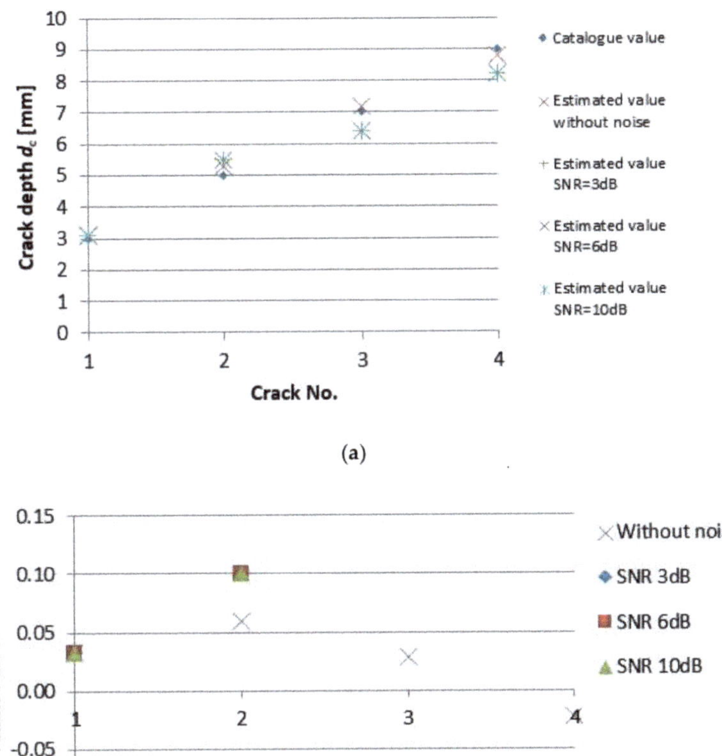

(a)

(b)

Figure 9. Results of the crack depth d_c reconstruction from the ECT responses with different SNR: (**a**) estimated absolute values; (**b**) relative error of estimation.

5. Conclusions

A new algorithm for noise-robust inverse analyses of eddy-current responses in non-destructive evaluation was presented in the paper. The algorithm employs wavelet transformation to estimate a length of a detected crack from the sensed eddy-current responses. The procedure continues with the principal component analysis for data reduction and finally the crack depth is estimated using the neural network. An eddy-current probe newly developed by authors is used for the non-destructive inspection.

The probe drives eddy-currents with uniform distribution in an inspected object. The eddy-current responses are sensed using a magnetic sensor. All three spatial components of perturbed magnetic flux density vector are acquired during the inspection. Two-dimensional scanning of the probe over a cracked surface is performed using XYZ stage from the near side. A plate specimen with four electro-discharge machined notches was inspected in this study. The sensed responses were further deteriorated by noise of different levels. Dimensions of the notches were estimated from the sensed responses using the proposed algorithm. An artificial noise of different levels was added to the measured responses in order to evaluate noise-robustness of the developed inverse algorithm. It was presented that the estimated crack depth differs by less than 10% from the real ones even when the signal to noise ratio level is 10 dB. The obtained results demonstrated that the noise level in the investigated range almost do not have any impact on the preciseness of the estimation.

Author Contributions: Conceptualization, M.S., D.G., and L.J. Methodology, M.S., L.B., and D.G. Software, L.B. Validation, M.S. and L.B. Formal analysis, M.S. and L.B. Investigation, L.B. Resources L.B. and L.J. Data curation, L.B. Writing—original draft preparation, M.S. and D.G. Writing—review and editing, L.J. Visualization L.J., and D.G. Supervision, L.J. All authors have read and agreed to the published version of the manuscript.

Funding: This research received no external funding.

Acknowledgments: This work was supported by project ITMS: 26210120021, co-funded from EU sources and European Regional Development Fund.

Conflicts of Interest: The authors declare no conflict of interest.

References

1. Yusa, N.; Huang, H.; Miya, K. Numerical evaluation of the ill-posedness of eddy current problems to size real cracks. *NDT E Int.* **2006**, *40*, 185–191. [CrossRef]
2. Fan, M.; Wu, G.; Cao, B.; Gyan, T.S.; Li, Z.; Tian, G. Uncertainty metric in model-based eddy current inversion using the adaptive Monte Carlo method. *Measurement* **2019**, *137*, 323–331. [CrossRef]
3. Cai, C.; Miorelli, R.; Lambert, M.; Rodet, T.; Lesselier, D.; Lhuillier, P.E. Metamodel-based Markov-Chain-Monte-Carlo parameter inversion applied in eddy current flaw characterization. *NDT E Int.* **2018**, *99*, 13–22. [CrossRef]
4. Biju, N.; Ganesan, N.; Krishnamurthy, C.V.; Balasubramaniam, K. Defect sizing simulation studies for the tone-burst eddy current thermography using genetic algorithm based inversion. *J. Nondestruct. Eval.* **2012**, *31*, 342–348. [CrossRef]
5. Zhu, P.; Cheng, Y.; Banerjee, P.; Tamburrino, A.; Deng, Y. A novel machine learning model for eddy current testing with uncertainty. *NDT E Int.* **2019**, *101*, 104–112. [CrossRef]
6. Duca, A.; Rebican, M.; Duca, L.; Janousek, L.; Altinoz, T. Advanced PSO algorithms and local search strategies for NDT-ECT inverse problems. In Proceedings of the 2014 International Symposium on Fundamentals of Electrical Engineering (ISFEE), Bucharest, Romania, 28–29 November 2014; p. 5, ISBN 978-1-4799-6821-3.
7. Grimberg, R. Electromagnetic non-destructive evaluation: Present and future. *J. Mech. Eng.* **2011**, *57*, 204–217. [CrossRef]
8. Deng, Y.; Liu, X. Electromagnetic imaging methods for nondestructive evaluation applications. *Sensors* **2011**, *11*, 11774–11808. [CrossRef] [PubMed]
9. Haddar, H.; Jiang, Z.; Riahi, M.K. A robust inversion method for quantitative 3D shape reconstruction from coaxial eddy current measurements. *J. Sci. Comput.* **2017**, *70*, 29–59. [CrossRef]
10. Ramos, H.G.; Ribeiro, A.L. Image post-processing and inversion for eddy current crack detection problems. In Proceedings of the 2016 IEEE Metrology for Aerospace (MetroAeroSpace), Florence, Italy, 22–23 June 2016; p. 6, ISBN 978-1-4673-8292-2.
11. Ribeiro, L.A.; Pasadas, D.; Ramos, H.G.; Rocha, T. Regularization of the inversion process in eddy current characterization of superficial defects. In Proceedings of the 20th International Workshop on Electromagnetic Non-destructive Evaluation, Sendai, Japan, 21–23 September 2015; Yusa, N., Uchimoto, T., Kikichi, H., Eds.; IOS Press: Amsterdam, The Netherlands, 2016; pp. 48–54, ISBN 978-1-61499-638-5.
12. Ahmed, S.; Miorelli, R.; Salucci, M.; Massa, A. Real-time flaw characterization through learning-by-examples techniques: A comparative study applied to ECT. In Proceedings of the 21st International Workshop

on Electromagnetic Non-destructive Evaluation, Lisbon, Portugal, 25–28 September 2016; Ramos, H.G., Ribeiro, A.L., Eds.; IOS Press: Amsterdam, The Netherlands, 2017; pp. 228–235, ISBN 978-1-61499-766-5.
13. Behun, L.; Smetana, M. Decreasing uncertainty in width estimation of EDM cracking from eddy-current Signals. In Proceedings of the 11th International Conference ELEKTRO 2016, Strbske Pleso, Slovakia, 16–18 May 2016; pp. 474–477, ISBN 978-1-4673-8698-2.
14. Behun, L.; Smetana, M.; Capova, K. Estimation of defect geometry in eddy current non-destructive evaluation of conductive biomaterials. *Acta Tech. CSAV* **2018**, *63*, 43–52.

Article

Study on the Influence of Measuring AE Sensor Type on the Effectiveness of OLTC Defect Classification

Daria Wotzka * and **Andrzej Cichoń**

Faculty of Electrical Engineering Automatic Control and Informatics, Opole University of Technology, 45-758 Opole, Poland; a.cichon@po.edu.pl

* Correspondence: d.wotzka@po.edu.pl

Received: 6 May 2020; Accepted: 29 May 2020; Published: 30 May 2020

Abstract: The principal objective of this study is to improve the diagnostics of power transformers, which are the key element of supplying electricity to consumers. On Load Tap Changer (OLTC), which is the object of research, the results of which are presented in this article, is one of the most important elements of these devices. The applied diagnostic method is the acoustic emission (AE) method, which has the main advantage over others, that it is considered as a non-destructive testing method. At present, there are many measuring devices and sensors used in the AE method, there are also some international standards, according to which, measurements should be performed. In the presented work, AE signals were measured in laboratory conditions with various OLTC defects being simulated. Five types of sensors were used for the measurement. The recorded signals were analyzed in the time and frequency domain and using discrete wavelet transformation. Based on the results obtained, sets of indicators were determined, which were used as features for an autonomous classification of the type of defect. Several types of learning algorithms from the group of supervised machine learning were considered in the research. The performance of individual classifiers was determined by several quality evaluation measures. As a result of the analyses, the type and characteristics of the most optimal algorithm to be used in the process of classification of the OLTC fault type were indicated, depending on the type of sensor with which AE signals were recorded.

Keywords: OLTC; AE sensor; acoustic emission; feature extraction; supervised classification; machine learning

---

1. Introduction

The reliability of the power system operation, to a great extent, depends on the proper operation of power transformers. These are devices constituting one of the main elements of the power transmission and distribution network. Their failures occur relatively rarely but result in huge costs. One of the most frequent causes of transformer failures is a faulty operation of on-load tap changers (OLTC). To ensure the continuity of the transformer's electrical circuit and to maintain appropriate winding resistance parameters, the appropriate condition of OLTC contacts is essential. Due to the destructive action of the electric arc, the contacts are subject to wear processes. This phenomenon is particularly important in a power switch, where the switching process takes place at the flow of the transformer load current. Excessive wear of both fixed and movable contacts can lead to an increase in the contact resistance, which increases the temperature of the contact at the current flow and its further degeneration [1–4].

An important structural parameter of the contacts is their switching capacity, which depends on the way the contacts are made, the time of their closing and opening, and the medium in which the electric arc is extinguished. Due to the medium used to extinguish the electric arc and the method of breaking the electric circuit, OLTCs are divided into three groups: oily, with vacuum chambers, and thyristors. In this work, we consider OLTCs operating in electro-insulating oil.

There are many methods to determine the technical condition of power switches, which are based on vibration measurement, arcing or motor current/voltage signal analysis, dynamic resistance measurement, and acoustic emission (AE) signal measurement [2,5–7]. In [8], the authors investigated the possibility of using fractal analysis for damage detection in OLTC, while, in [9], a description of the dynamic resistance measurement (DRM) DV test method is given. In [10], a joint vibration and arcing measurement were applied for interpretation of events occurring during OLTC operation.

AE signals are often and successfully used to detect damage and assess the technical condition of equipment in the industry [11]. The advantage of the AE method is the possibility to carry out the technical condition tests of equipment or materials in a non-invasive way. The registered signals are processed in the time and frequency domain, with the use of various transformations, e.g., Fourier, discrete and continuous wavelet, Hilbert, and Gabor [2].

Today, scientists and engineers have relatively easy access to computers of high computing power. Moreover, there are numerous highly developed methods of artificial intelligence that are used in various fields. They are also successfully implemented for the diagnosis of electrical equipment. Classification tools, which are used when there exists a priori knowledge of the type of failure, can be applied in a process of supervised learning using a machine learning algorithm (MLA). If no knowledge of the type of defect is available, clustering methods can be applied using unsupervised learning. In the first case, genetic algorithms, hidden Markov models, chaos theory, group method of data handling (GMDH), fuzzy algorithms, and artificial neural networks (ANN) are usually used, in the second type, e.g., self-organizing maps (SOM) are applied. Classification or clustering is made based on a set of features, which are first obtained from registered measurement data. For example, the authors of the paper [12] proposed an expert system for the detection of various types of OLTC and circuit breakers. They proposed a feature extraction method based on a decomposition of acoustic signals in time and frequency domain. In their work, they used the reference database. However, this article does not give any detailed information about the algorithms specifically used, nor does it provide numerical values, based on which comparative analysis could be performed. In [13], the authors used an SOM neural network in the process of crack monitoring based on processed AE signals. In [14], authors apply supervised classification techniques and various feature extraction methods recognition of the aging state of a polyethylene-insulated cable for high voltage direct current (HVDC) usage. In [15] authors propose an efficient approach for classification of AE signals related to corrosion. They have applied random forests (RF) and k-nearest neighbor (KNN) algorithm. In [16], a hybrid method for the selection of features using the support vector machines (SVM) and KNN is proposed. Similar analyses in application to gear are described, e.g., in [17], where the authors obtained a maximum of 95.93% for testing accuracy when using psychoacoustic, and lower values for standard analyses. Similarly, not perfect accuracies were achieved in [18], where the authors apply the deep graph convolutional network for the diagnosis of faults of roller bearings. Rolling bearing fault diagnosis based on resonance-based sparse signal decomposition (RSSD) and waveform based on vibroacoustic signals are considered in [19], where the authors determine frequency spectrum and use these characteristic frequencies for fault diagnosis. The authors of [20] and [21] also use waveform transforms in their research works. In [21], a discrete wavelet transform and fast Fourier transform are applied for feature extraction and then a feed-forward with ANN was applied for recognizing the medium of the discharge source. In [22], authors applied ANN for the localization and identification of the AE source in various parts of the electric power transformer. In [23], the authors consider the influence of different feature sets on the results of SVM classification. They investigate rotating machinery using vibration signal for faults diagnosis. They analyze data in time, frequency and time–frequency domains, applying statistic features, empirical-mode decomposition, energy and Lempel–Ziv complexity features. They have achieved 78%-95% accuracy for the recognition of six types of failures.

As shown above, there are many measurement and analytical possibilities to be used for diagnostics with the use of the AE method. However, there is a lack of knowledge on determining which algorithm of supervised machine learning is the most optimal to use depending on the type of measuring sensor. Therefore, our work attempts to answer this question.

The main contribution of this work lies in the determination of the differences and selection of a set of optimal parameters (features) for the classification of technical condition of oily OLTC using the AE method, based on signals registered with various sensors. Furthermore, the novelty is in the analysis of the suitability of different sensors for OLTC fault detection and determination of the most suitable MLA for the particular sensor.

2. Materials and Methods

2.1. Measurement Setup

The tests performed in the laboratory conditions were carried out using the OLTC model with a separate selector and power switch type VEL-110-27 from ELIN VEL-110. Inside the tank, there is a complete OLTC system consisting of a selector and a power switch located in the insulation sleeve. To limit the height of the system, the length of the tap selector has been reduced to six on/off switches. The tested OLTC model was placed in a tank filled with insulating oil. The system is equipped with a drive unit that enables the automatic switching of individual tapes. The change-over can be carried out between any two adjacent tapes. During the test, the entire tank was filled with insulating oil. In the presented measuring system, the tests were performed in two operating states: with nominal current flow and without load. The source used was three phase and supplied all three phases of the switch. The contacts in all three phases of the switch were connected to each other. Using a single-phase current forcing device, the 50 Hz AC flow was regulated from 0 to 250 A, in all phases symmetrically. Such a system guaranteed the possibility of generating an electric arc in all three phases of the power switch. The current value was generated at a low voltage of 50 V. The forcing device was located at a large distance from the measuring system, which limited the generation of possible interference from the transformer inductance. The contacts installed in the tested switch were the same as in the actual construction. Therefore, the results of the measurements of the discharges in the tested model of OLTC can be compared with the results obtained in the real object. In the laboratory conditions, comparative tests of particular types of measuring transducers were carried out. AE signals were measured with all transducers, which made it possible to perform a comparative analysis and to select the type of transducer for which changes in the AE signal structure generated by various defects in OLTC were most visible. The results of previous works, presenting the results of measurements and analyses carried out for the OLTC under consideration, can be found, e.g., in [24,25].

2.1.1. Sensors Applied for Measurements

The AE signal may be distorted and reverberated along with its propagation; therefore, the number and type of AE sensor, its installation, and placement must be carefully investigated [26,27]. In [28], authors investigate the influence of the distance between AE transmitter and receiver on the gathered analysis results. In this work, five types of acoustic-electric transducers are used to measure AE signals generated by OLTC:

- Broadband contact transmitter WD AH 17 (by Physical Acoustics Corporation) [29], marked as WD;
- Contact transducer D9241A (by Physical Acoustics Corporation) [30], marked as DS;
- Narrowband contact transducer R-15a (by Physical Acoustics Corporation) [31], marked as R15;
- Hydrophone 8103 (by Brüel & Kjær) [32], marked as MiG;
- Hydrophone TC4038 (by Reson) [33] marked as MiC.

The broadband transducer type WD AH 17 (WD) was attached to the outer wall of the ladle with a permanent magnet. The transducer used has a high sensitivity (55 dB ± 1.5 dB with V/ms^{-1}) and a six-way bandwidth from 100 kHz to 1 MHz in the ± 10 dB range. This transmitter is equipped with a differential measuring system for the AE signals. The use of this system allows for eliminating interference signals, which appear under the influence of the electromagnetic field affecting the transmitter and measuring cable [34]. It is of particular importance while making measurements at the current flow during the OLTC switching process. This contact transducer is immune to the appearing interferences because of the differential system used to measure AE signals. Its frequency band allows for performing analysis of AE signals generated by OLTC in no-load condition, lower frequency bands as well as during current flow when components from higher frequency bands appear in AE signal.

The transducer type D9241A (DS) was also attached to the outer wall of the tank with a permanent magnet. This AE sensor is an epoxy sealed, enclosed device, which was developed for application in electrically noisy environments. Its ceramic face ensures proper insulation from the transformer tank. For noise reduction, the transducer is integrated with a differential BNC connector. Its typical operating frequency is in the range from 20 to 180 kHz range. Its sensitivity is around 82 dB V/(m/s).

The third type of contact transducer was the R15a—a narrow band device, which, similar to the other sensors, was mounted to the outer wall of the tank. The sensor cavity is built from a solid stainless steel rod, which makes it rugged and reliable. It also has a ceramic face for the improvement of the electrical isolation from the object under test. Its operating frequency is in the range from 50 to 400 kHz and its sensitivity of 63 dB with V/μbar. It enables application in a wide temperature range, from −65 to 175 °C.

The advantage of using a broadband contact transducer is the possibility of easy attachment to the tested object and analysis of the reactive controlled AE signals in a wide frequency band. However, the AE signal recorded with the use of contact transducers is distorted due to signal attenuation. It is created when the AE signal passes through barriers located inside the tested object [35]. When the AE signal is recorded in OLTC, the attenuation is caused by the necessity to pass the measured signal through the isolation sleeve, thick oil layer, and metal tank walls.

For the comparison of AE signals occurring inside the tank with the signals recorded on its external wall, another measuring system using a TC4038 (MiC) hydrophone was used. The hydrophone was immersed in insulating oil between the wall of the ladle and the examined power switch. The hydrophone used has a wide bandwidth of 10 to 800 kHz within ± 5 dB and a high sensitivity of 228 dB ± 3 dB with V/μPa. During measurements, 30 dB amplification was used. The cut-off frequency of the high pass filter was 10 kHz, while the low pass filter was 1 MHz.

For comparison purposes, the second hydrophone of type 8103 (MiG) was also applied. It enables AE signal registration within a wide frequency range, from 0.1 to 180 kHz, where it has a flat frequency response and with a sensitivity of 30 μV/Pa. This sensor is equipped with low noise, double-shielded integral cable, which enables appropriate electromagnetic shielding. The operating temperature is in the range from −40 to +80 °C. This type of sensor has a high level of corrosion resistivity; therefore, it may be applied in transformer oil. The use of the hydrophone allows for the precise analysis of time–frequency structures. This kind of transducer is, however, very sensitive to electromagnetic interferences arising from the flow of current through OLTC contacts. This makes the interpretation of the recorded signals much more difficult. The use of the hydrophone is mainly related to the difficulty of placing the transmitter inside the OLTC being diagnosed.

The mounting positions for contact and immersion sensors were determined in preliminary works, where locations of the highest signal to noise ratio (SNR) were selected.

2.1.2. Measurement Settings

The AE signal received was amplified in a 2/4/6 type preamplifier system, which has a possibility of step-by-step amplification adjustment: 20, 40, 60 dB. During measurements in the preamplifier, 20 dB gain was used and, in some cases, the gain value was increased to 40 dB. From the preamplifier, the AE measurement signal was fed to the amplification system, subjecting it to 20 dB of gain.

The common part of all applied measuring systems is the element of measurement signal acquisition. The time sequences of AE series received were recorded using Acquitek CH 3160 measuring card. Because the measuring card has four measurement channels, the experiments were performed in two series: three sensors were attached in the first one and two in the second one.

The electrical discharges generated during the switching process generate AE signals at frequencies up to 350 kHz. Taking into account the Shannon-Kottlekevich sampling claim, to avoid the effect of mutual overlap of periodically repeated spectra, the sampling frequency should be at least twice as high as the maximum frequency of the analyzed AE signal.

In the measurement practice, a sampling frequency several times higher than that of Nyquist is generally used. In our measurements, the sampling frequency of 1 MHz, with a 12-bit distribution of the AC converter was selected.

In the measurements of AE signals generated by the power switch, 300,000 samples were registered, which allows for the recording of the full switching cycle in 300 ms. Laboratory tests of the selector required increasing the number of samples to 2,000,000 and recording the signals in 2 s. During the OLTC tests in operation, the number of AE signal samples was selected depending on the length of the OLTC drive operation time.

2.1.3. OLTC Defects Considered

As mentioned, for most types of OLTC, the working environment is the insulating oil. During the swiching process, a sound pressure wave is produced, which propagates and reaches the metal tank, where it may be registered by contact transducers. This signal contains information characterizing the operation of the power switch and selector [36]. The result of the operation of oil OLTC is the progressive degradation of its components, including the main contacts of the power switch. Therefore, one of the methods enabling its diagnosis of their mechanical condition is the diagnosis of the degree of contact wear. This most frequent defect was modeled in the research and then the influence of its occurrence on the recorded AE signals was examined. The modeling of the defect consisted of a manual change of the thickness of OLTC-head fixed contacts. This was done successively by: the separation of fixed contacts from the place of their fixing, placing a special pad with a thickness of d mm under each of them, and then their reassembly. As a result, the contact working surface was d mm closer to the moving main contacts. The specially prepared pads were made of steel sheet. Their shape was selected in such a way as to ensure, on the one hand, good support of the contact and not to allow for assembly clearance, and, on the other hand, not to introduce distortions, curves or other unintentional dislocation of the contact in relation to the original position. After the pads have been placed under each fixed contact, the contacts themselves have been placed in their proper places and then bolted with mounting screws. After the preparation of the model, the functionality of the model was checked to detect potential problems in the operation of the modified device. After a thorough check and confirmation of the correctness of assembly, measurements were taken. The proposed laboratory system allows for the adjustment of contact thickness d from 1 to 3 mm, which is related to the actual degree of wear. Moreover, it is possible to simulate the non-uniformity of their operation. Therefore, five classes were considered, marked as:

- Class C1: No damage - power switch operation with new contacts.
- Class C2: Operation of the power switch with 1 mm thick contacts.
- Class C3: Operation of the power switch with 2 mm thick contacts.
- Class C4: Operation of the power switch with 3 mm thick contacts.
- Class C5: Non-simultaneous operation of the power switch contacts.

2.1.4. Measurement Disturbances

When recording AE signals, the possibility of interference must be considered [37,38]. External interference refers to the environmental conditions prevailing in the place of measurement. An example of this type of disturbance, which accompany measurements in real conditions, is the presence of exhaust air discharges occurring in overhead elements of station equipment, electric power lines, culvert isolators, etc. The measuring cable is particularly exposed to electromagnetic interference. Often, due to the location of the tested unit at the substation, the measurements require the use of a cable which average length is several meters. To counteract the influence of electromagnetic interference on the measuring cable, shielding is used. However, a more effective solution is to use a fiber optic link, which ensures the complete elimination of this type of interference. Moreover, it increases the safety of the measurements, as the electrical separation between the measuring transmitter and the rest of the system is applied.

When analyzing the influence of interferences, it is also important to stress the importance of external electromagnetic type interferences, which include own noise of preamplifiers, amplifiers, and data recording systems [39–42]. In the case of a small maximum value of the signal measured with the noise of the measuring equipment, there are analytical difficulties related to the extraction of relevant information from the recorded signal. These shots are created on passive and active elements of the measurement track. Among the disturbances generated by the measuring equipment, the following can be distinguished: thermal (thermal), shot (Schottky), flickering, generation-recombination, and explosion noise.

The most important is heat noise (Johnson's), which is the main source of interference in electronic circuits. They occur on every resistive element regardless of the technology. The mechanism of their formation consists of the interaction of thermal vibrations of the semiconductor's crystal network with free electrons at temperatures above absolute zero. This movement causes uncontrolled changes in voltage or current values.

The described disturbances especially occur during measurements in real conditions at the substation. During laboratory tests, the influence of interferences is much smaller. Additionally, it should be noted that the interference amplitude is several times lower than the measurement signal generated by the power switch. The described disturbances may be more significant in other types of switches where the amplitude of the measurement signal is not as high as in the examined case. Taking into account the construction of the tested OLTC, the high amplitude of the measurement signal and the measurement conditions, it can be implied that the influence of the described disturbances is negligible.

2.2. Data Analysis and Classification Methods

In Figure 1, the methodology of the data analysis and classification procedure is depicted. The procedure is divided into four blocks: A, B, C, D. It starts (A) with gathering signals in a series of measurements. Next, (B) the recorded AE signals are subjected to digital processing analyses, in which the signal is reduced by pre-samples and tail. Similar steps, pre-trigger removing and tail cutting, are taken as a standard procedure, e.g., in [15]. In [15], the authors applied energy criterion for waveform cutting and additionally resampling for lowering the resulted signal. They also proposed an *alter-class matrix* method, which allows for noise and outliers' introduction. In our case, the signals, registered by various sensors, have different characteristics, and our aim is not to reduce the information included in the registered signals.

In the subsequent step (C), feature extraction takes place. During this phase, a set of features is determined, in particular: the power spectrum density (PSD) using Welch method is calculated for various window sizes, which constituted the (1) *Welch PSD* feature set; based on this, parameters are determined for the *Frequency* feature set, which contains such measures as mean and median frequency, spectra centroid, spectra skewness and spectra kurtosis; furthermore, the energy of detail coefficients, calculated using Haar wavelet decomposition is determined as (2) *Haar Wavelet* feature set. This wavelet method was selected within preliminary works. The base wavelet has a fundamental impact on the results obtained from the discrete waveform transformation, which then affects the calculated energy values of the details. The wavelet selection procedure included calculating the value of the coefficient of variation, which is a classic measure of the variation of results. The procedure was repeated for 29 wavelengths, including Haar, Symlet, and Daubechies of various ranks in the range from 2 to 15. On this basis, the most suitable wavelet was selected: Haar wavelet packet energy percentage, with 6 wavelet packet energy features [43–45]. Also in [46], authors analyze AE signals features of which were determined with the use of wavelet transform.

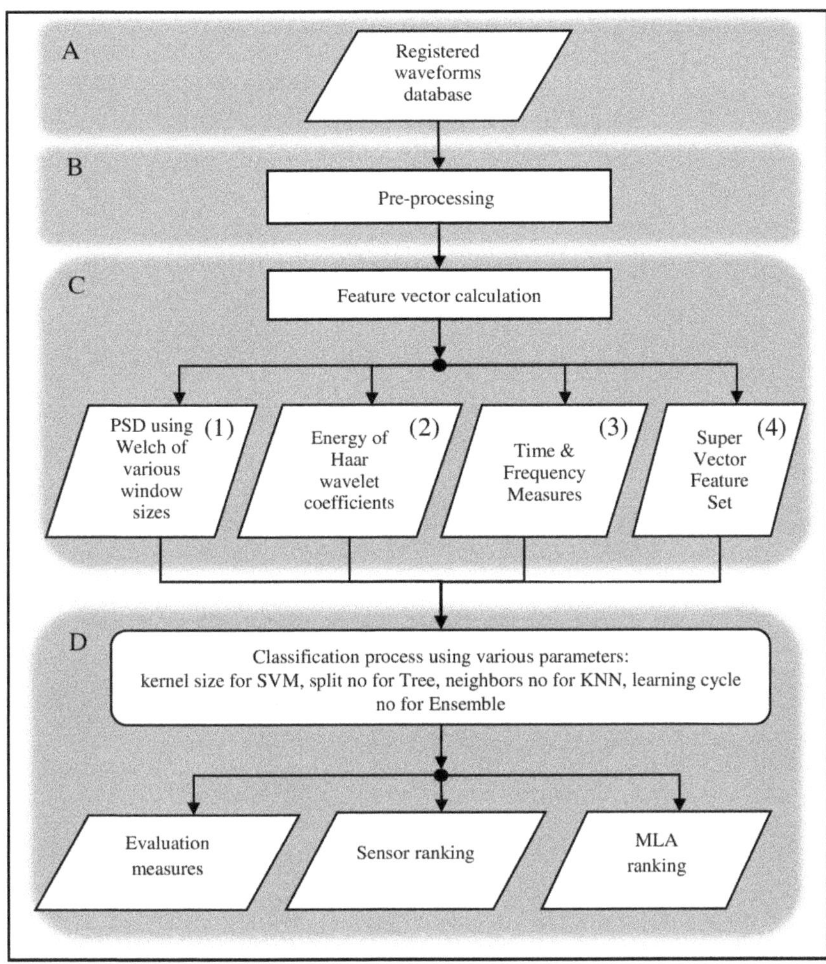

Figure 1. Methodology of the data analysis and classification procedure.

Furthermore, the Hilbert transform was applied to the absolute value of the registered signal amplitude, and the envelope was determined by this means. The envelope was subjected to *Time Feature* set determination, in which the following measures are included: shape and peak coefficients, maximum value, arithmetic, geometric and harmonic mean, median, kurtosis, slope, standard deviation, and variance. The *Time and Frequency Measures* were combined as subsequent (3) features set. Additionally, all the measures were combined into one (4) *Super Vector* features set and applied for classification. Similar methods of feature extraction are performed by other groups of scientists. For example, *Time features and Frequency features* were determined according to reference [47,48] and [15]. In [15], the authors classify AE signals processed with the use of wavelet transformations and using random forest and KNN algorithms to locate corrosion in devices applied in the chemical industry. After the features are determined, they are applied for the (D) classification process, which results in a ranking of the sensors and MLA. This step is described in detail further in this section.

In Figure 2, a graphical representation of the major methodology steps (see Figure 1B,C) is presented. The examplary AE waveform was registered using the R15 sensor during the simulation of defect D1 (without damage). The original waveform is depicted in Figure 2a. The envelope is depicted in Figure 2b. The PSD Welch signal, calculated for the size of the window of 1024, is depicted in Figure 2c, and the energy for six Haar wavelet coefficients is depicted in Figure 2d.

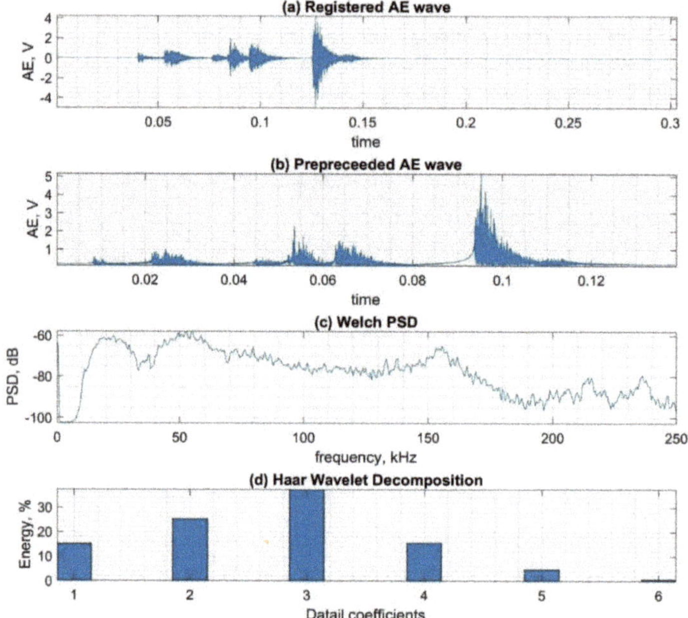

Figure 2. Process of preparing data for classification, considering sample signal for class C1 sensor type R15. (**a**) Registered waveform, (**b**) pre-processed signal, (**c**) power spectrum density calculated using the Welch method, (**d**) Energy of wavelet decomposition coefficients calculated using Haar wavelet.

In Figure 3, the example waveforms and corresponding Welch PSD of class C1 signals registered with all considered sensors are presented. Significant differences in the waveforms and power spectra may be recognized.

The environment applied for analysis and classification was Matlab. It has a set of built-in artificial intelligence algorithms for classification, including supervised MLA and neural networks. The application of this environment allows one to use the resources of modern computing methods without the need to implement these algorithms from scratch. In this situation, it was possible to test a large group of algorithms, available in this tool, along with the modification of their parameters. Four groups of algorithms were tested: support vector machines (SVM), ensemble, k-nearest-neighbors (KNN), and decision trees. In the group of SVM algorithms, four types of kernel functions were studied: linear, quadratic, cubic, and Gaussian. In the group of KNN algorithms, three types of distance measurement functions (distance metric) were studied: Euclidian, cubic Minkowski, and cosine. In the group of ensemble algorithms, three types of algorithms were tested: bugged tree, subspace discriminant, and subspace KNN. To facilitate subsequent analysis, the following designations have been introduced for the tested MLA: SVL (support vector machine with the linear kernel), SVC (support vector machine with the cubic kernel), SVQ (support vector machine with the quadratic kernel), SVG (support vector machine with Gaussian kernel), ESD (ensemble subspace discriminant), EBT (ensemble bugged tree), ESK (ensemble subspace KNN), KNE (k-nearest-neighbors with Euclidian distance metric), KCO (k-nearest-neighbors with cosine distance metric), TRE (decision tree), and KCM (k-nearest-neighbors with cubic Minkowski distance metric).

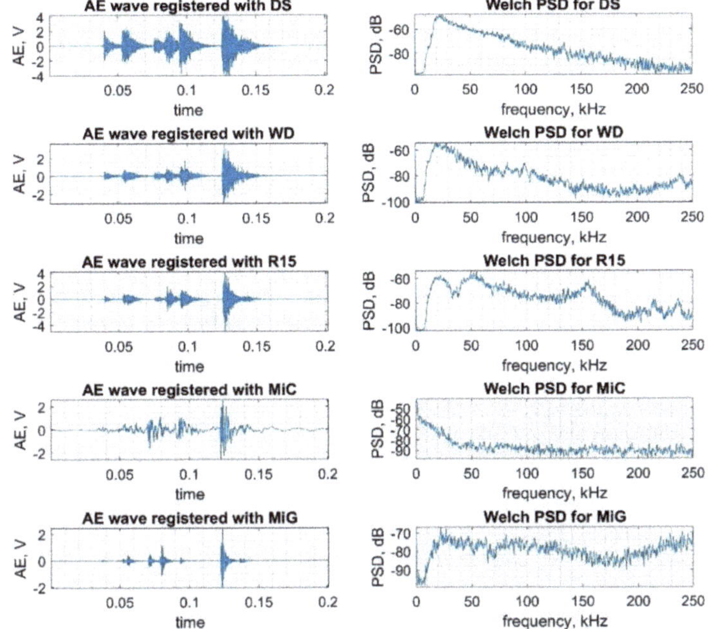

Figure 3. Example waveforms and corresponding Welch power spectrum density (PSD) of class C1 signals registered with the considered sensors.

The proposed methods of classification resulted from the fact that a relatively small amount of measurement data is available (see Table 1). The signals were gathered under laboratory conditions where the particular defects were modeled; thus, the expert knowledge about the defect was given a priori. Such experiments, using OLTC, are relatively difficult to perform, and the number of units available for measurements is small. Therefore, the application of, e.g., neural networks or deep networks would require "data-augmentation" [6]. Such methods are commonly used in various industry areas, e.g., in processing and classifying images. To ensure a proper learning process of the MLA on a relatively small number of training data, the cross-validation method with a coefficient of 5 was applied during the classification process.

Table 1. Number of samples used for each sensor related to the class considered.

Class no	Number of Samples Used for Each Sensor				
	DS	WD	R15	MiC	MiG
C1	29	29	29	15	15
C2	30	30	30	27	27
C3	30	30	29	19	19
C4	30	30	30	30	30
C5	29	30	30	29	29

For the above-mentioned procedure, carefully selected signals have been used, i.e., there are no erroneous signals in the learning vector or such with strong interferences. It is important for the supervised learning algorithm not to teach it with erroneous signals or with such low SNR.

The interferences and distortions of low amplitude that occur in OLTC are still contained in the AE signals we considered in the classification process. High-quality signals with high SNR value are assumed for the proposed MLA. Application in cases of strong AE interferences can be considered in future works.

In [49], the authors examined the influence of the number of features on the classification results obtained using KNN, RF, and SVM. The evaluation was based on the analysis of accuracy, sensitivity, specificity, precision, and F1 score measures. Their analysis clearly shows that, regardless of the type of algorithm, the values of quality measures increase as the number of features increases, with a maximum of 650 features tested. The highest increases in accuracy were observed for the first few dozen feature elements. For a higher number, these increases were not much larger. In our case, we consider the following sizes of feature sets: *Welch PSD* (129 features), *Haar Wavelet* (6 features), *Time and Frequency Measures* (16 features), and *Super Vector* (151 features). As we will show later in this paper, no significant differences were observed between the *Welch PSD* and *Super Vector* feature sets.

As a measure of quality, validation accuracy was used in the analysis of results presented in detail in the following section. However, other quality measures were also considered in the analyses: test accuracy, sensitivity, specificity, precision, F1 score, and Matthews correlation coefficient (Matthews CC). Additionally, the CPU time was calculated. The results, included in the next section, allow us to state that, in the considered problem, there exists no CPU time issue, since the CPU times are fractions of seconds, except for a few MLA parameters, for which the analyses have taken a few minutes, but did not affect (improve) the quality of the classification. On this basis, it was possible to identify specific MLA parameters for which the best results were achieved. These results were found in the validation process based on the mentioned quality measures.

In further sections of the article, detailed results for the *Super Vector* feature set, for which the best values of evaluation parameters were obtained, are presented. For the other feature sets, the results depicting the summary statements, from which the ranking can be determined, are shown.

3. Results and Discussion

3.1. Analysis of the Influence of MLA Parameters on the Results Obtained for the Super Vector feature Set

To select the optimal values of classification algorithms, an analysis of the influence of particular parameter values on classification efficiency was carried out. The parameters were changed depending on the type of algorithm as follows: SVM—*the size of kernel scale* in the range from 0.1 to 50, decision tree—*no of tree splits* in the range from 1 to 200, KNN—*no of neighbors* in the range from 1 to 100, ensemble—*no of learning cycles* in the range from 1 to 100. The tests were performed for each of the methods separately for signals recorded by individual sensors.

In Figure 4, the calculation results, which depict the validation accuracy values gathered for particular algorithms using the *Super Vector* features set for the classification process are presented. The colors of particular lines correspond to data related to signals registered by the individual sensors.

Depending on the type of kernel function used for classification and the type of sensor with which AE signals were recorded, differences were observed as follows.

For algorithms type SVM:

- With a first degree polynomial (SVL), the highest efficiency is achieved for the smallest kernel scale. It then decreases with the increase of the scale size, but not for every type of signal. For signals recorded with DS and MiC sensors it drops rapidly, even below 0.6; for R15 to 0.9; and, for WD, the accuracy is the highest, regardless of the size of the analyzed parameter. In contrast, for MiG sensors, the value is initially smaller, then increases, so that, for values above 40, it falls again, but with values close to 0.9.
- This situation is different for the higher orders of the polynomial (SVQ, SVC) used and with the Gauss function-core (SVG). However, in these cases, the effectiveness is initially worse and increases for increasing kernel sizes.

For decision trees and ensemble algorithms:

- Efficacy is not dependent on the number of tree splits or from the number of learning cycles, especially for the ESK algorithm. However, significant differences in these values are visible depending on the type of sensor used.
- For Tree, EBT, and ESD the values are worse for the smallest value of the parameter and then increase and stabilize at a constant level.
- The worst results were obtained for the ESK algorithm and here also the biggest disproportions are visible for the signals recorded by individual sensors.

For algorithms in the KNN family:

- For each type of applied kernel function, there is a negative impact on the level of effectiveness of the increase in the number of neighbors. Each time, the best results were obtained for the WD and the MiG sensors.

Fluctuations in the accuracy value for the successive values of the parameters under consideration result from the fact that the learning process uses the method of cross-validation, which each time divides the sequence into sets of learning, validation, and test, which causes the results to be not deterministic, but avoids over-learning, which is important in our case because we have only a limited number of measurement samples (see Table 1).

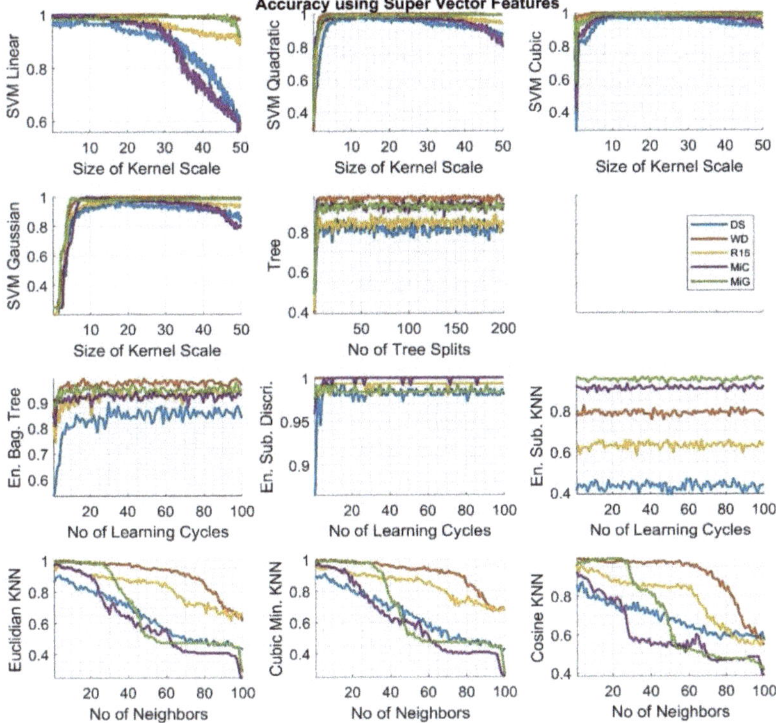

Figure 4. Calculation results—the validation accuracy value for particular algorithms using the *Super Vector* feature set. The legend for the colors corresponding to the individual sensors can be found on the right side of the image.

Graphs in Figure 5 depict the duration of the learning and prediction process (CPU time), which varied depending on the type of algorithm and its parameters used. However, no significant differences in the classification of signals recorded by individual sensors were observed. In the case of Euclidian, cubic Minkowski, and cosine KNN algorithms, these times are maximum one second for the smallest area of analysis (number of neighbors equal to one) and a fraction of a second for the remaining sizes of analyzed areas of adjacent classes.

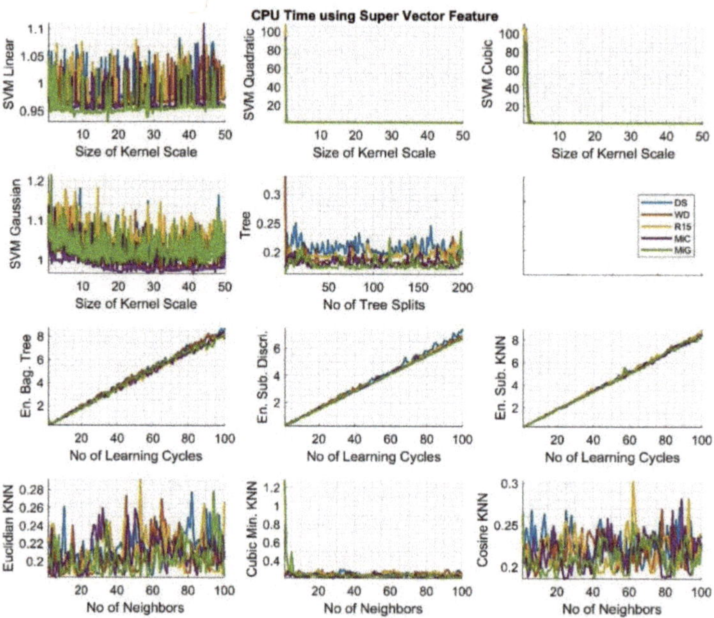

Figure 5. Calculation results—the CPU time for particular algorithms using the *Super Vector* feature set. The legend for the colors corresponding to the individual sensors can be found on the right side of the image.

The situation is similar for decision trees, which, regardless of the number of splits, do not exceed a fraction of a second. In the case of algorithms from the ensemble group, the CPU time increases linearly relating to the increasing number of learning cycles, but, even for 100 iterations, the time is only eight seconds. In the case of algorithms from the SVM group, the duration does not significantly depend on the number of kernel scale sizes, and, excluding the smallest kernel, it is about one second. Only for SVM with the kernel function of the second- and third-order polynomial type, for the smallest size of the kernel scale, the CPU time exceed 100 s.

To determine the specific values of parameters, which are the most suitable for the signals, a statistical analysis of the obtained classification results was performed based on validation accuracy measure, the results of which are presented in the form of boxplots in Figure 6. Based on boxplot diagrams, it is very clear that the best outcome is achieved when classifying signals recorded with the WD sensor. However, for other sensors, the situation is no longer obvious and changes depending on the type of algorithm used. The results of this analysis also indicate that different results are obtained depending on the type of MLA algorithm, for example, the smallest deviations and the highest values are obtained for the ESD algorithm.

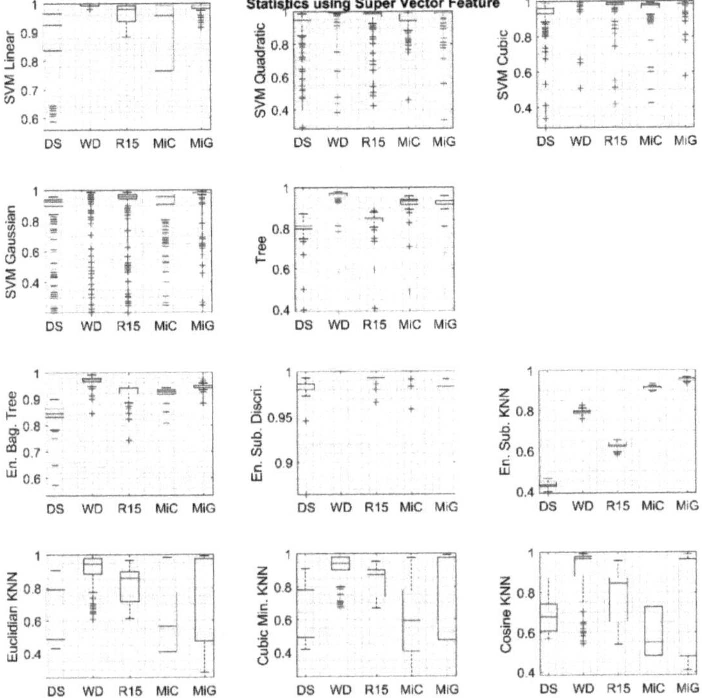

Figure 6. Boxplots calculated over the gathered classification results using the *Super Vector* feature set. Each column in the figures corresponds to a different sensor.

The analysis of the data shown in the figures above has allowed for determining the following:

- The optimal kernel scale value for SVM algorithms was selected at level 10.
- The optimal value of tree splits for the decision tree algorithm was chosen at level 20.
- The optimal number of learning cycles for ensemble group algorithms was chosen at level 40.
- The optimal number of neighbors for KNN algorithms is chosen at level 1, which is the most precise.

Figure 7 presents the results of calculations for each highest accuracy value obtained for signals recorded by individual sensors. It can be observed that, in most cases, a good or very good match is obtained. In the case of the ESK, the results were the worst. The presented data are related to the *Super Vector* feature set applied in the classification process.

3.2. Discussion on the Optimal Sensor and MLA with Dependence from the Applied Feature Set

To qualitatively assess the obtained results, histograms were calculated using 15 bins, based on which it was determined which of the sensors applied for the recording of the AE signals is best suited for classification and which of the MLA most often achieved highest efficiency, regardless of the type of sensor used. The results related to classification using the *Super Vector* feature are presented in Figure 8. The results related to *Time and Frequency* features, *Haar wavelet* features, and *Welch PSD* features are presented in Figures 9–11, respectively.

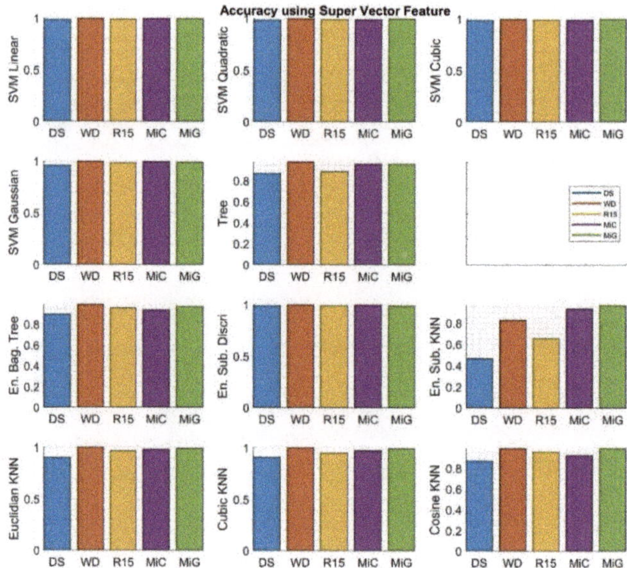

Figure 7. Very best validation accuracy values calculated for each of the algorithms and sensors depicted as a bar plot related to the *Super Vector* feature set.

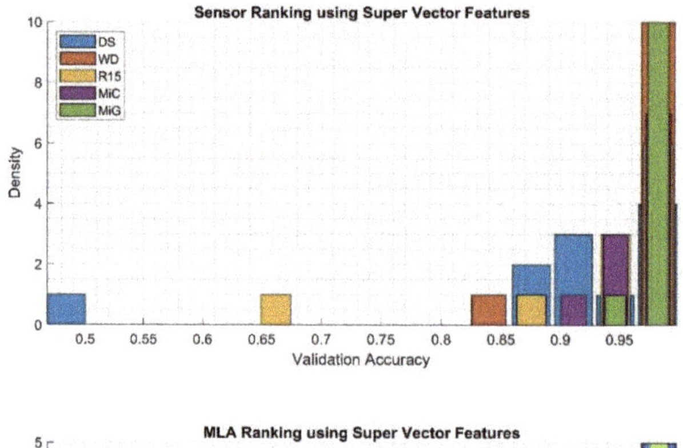

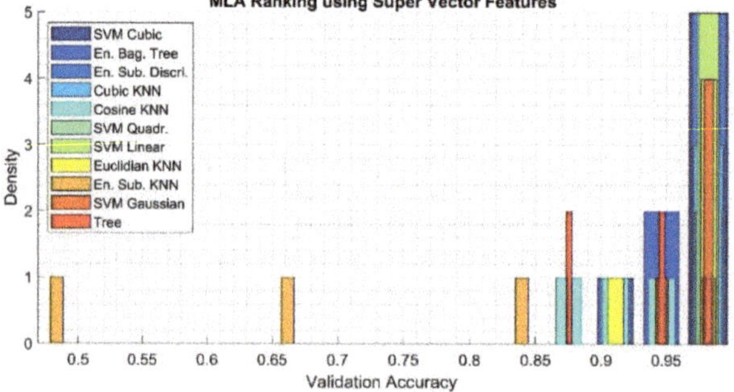

Figure 8. Results of analysis depicting quantities of the best sensor (upper) and the best machine learning algorithm (MLA) (bottom) applied for classification using the *Super Vector* feature determined using validation accuracy value.

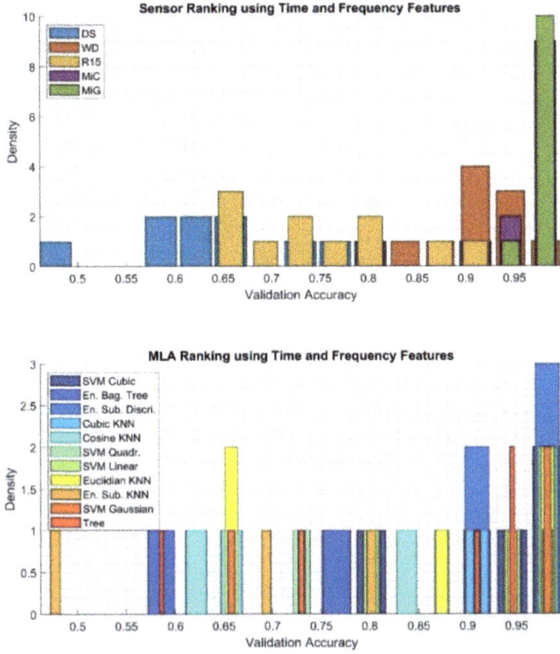

Figure 9. Results of analysis depicting quantities of the best sensor (upper) and the best MLA (bottom) applied for classification using the *Time and Frequency* feature set determined using validation accuracy value.

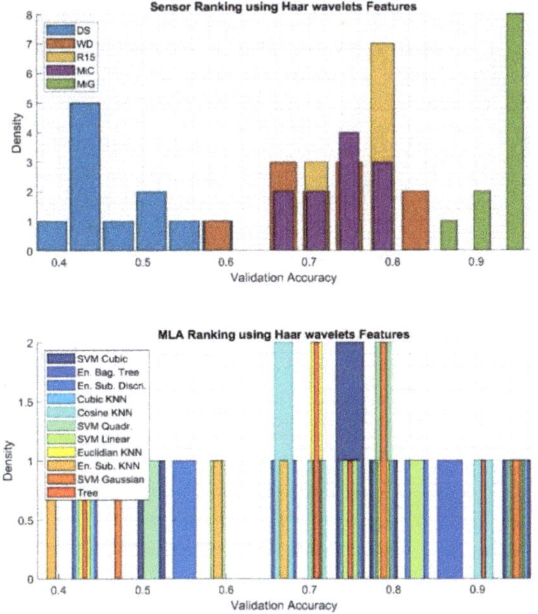

Figure 10. Results of analysis depicting quantities of the best sensor (upper) and the best MLA (bottom) applied for classification using the *Haar wavelets* features determined using validation accuracy value.

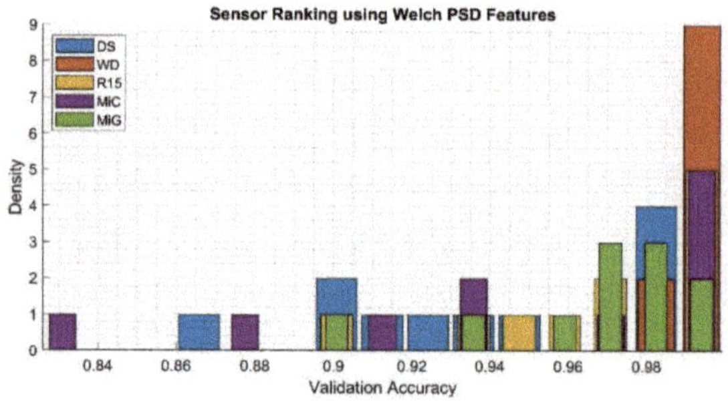

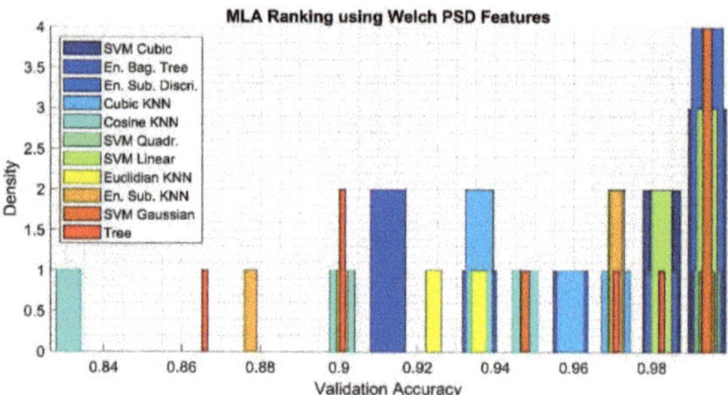

Figure 11. Results of analysis depicting quantities of the best sensor (upper) and the best MLA (bottom) applied for classification using the *Welch PSD* features determined using validation accuracy value.

The very best results were obtained in the classification process with the *Super Vector* feature set, which is presented in Figure 8. In this case, the highest validation accuracy values were obtained for signals recorded with the WD and MiG sensors (density equal to 10). Similar results, with density equal to 7 and 6, were obtained for signals recorded by sensors MiC and R15. Values below 0.6 were obtained for the DS sensor. Any type of algorithm from the SVM family, ESD, or KNE can be successfully used for this type of data set due to similar validation accuracy values. It is not recommended to use an ESK algorithm for which the lowest values of the quality measure under consideration were obtained.

Based on data presented in Figure 9 top, which is related to the results gathered using the *Time and Frequency* feature set, one can recognize that AE signals registered using MiG and MiC sensors achieved, most often, the highest validation accuracies: 10 and 9, respectively. Much worse results were obtained for the classification of signals registered using the DS sensor. While the signals recorded by the R15 sensor were average to good. Based on data presented in Figure 9 bottoms, which depict the results of the evaluation of the applied algorithm, it can be stated that all algorithms achieved similar results. The very best results were gathered for the ESD, while the worst were obtained using the ESK algorithm.

Based on data presented in Figure 10 top, which is related to results gathered using the *Haar wavelets* feature set, it can be seen that by far the best results are achieved for signals registered with the MiG sensor. Moderate validation accuracy values, in the range 0.6-0.8, were gathered for signals registered by WD, R15, and MiC, while the worst were gathered for signals registered by the DS sensor. In terms of determining the optimal algorithm, the evaluation is not trivial because all algorithms obtain flat density distributions. In most cases, the worst results were obtained for the ESK algorithm.

Based on the data presented in Figure 11 top, which is related to the results gathered using the *Welch PSD* feature set, one can recognize that AE signals registered using the WD sensor were most often classified achieving the highest validation accuracy (density equal to 9). Twice as infrequently are such good results obtained for the classification of signals recorded with R15 and MiC sensors, while only in two cases were such good classification results obtained for signals recorded with MiG sensors. Slightly worse results were obtained for the classification of signals with the DS sensor. It should be noted that the worst classification results were obtained in the case of signals recorded with MiC (validation accuracy less 0.84). Based on data presented in Figure 11 bottom, which depicts the results of analysis concerning the evaluation of the applied algorithm and kernel functions, it can be stated that ESD and SVG achieved four times the best results, while SVL, SVC, and SVQ were as good in classification one time fewer. The worst results were obtained using the KCO algorithm.

In Table 2, summarized results gathered from comparative analysis obtained for all four considered feature sets using six various measures of classification quality are depicted. The presented data include the best and the worst cases: the maximal and minimal calculated values of the measure, the algorithm for which it was determined, and the sensor, with which the AE signals were registered. As has already been stated several times in this paper, based on analyses of validation accuracy, here, in an overall comparison of the results obtained, it was confirmed that the highest values were obtained for the *Super Vector* feature set, ESD algorithm, and the WD sensor.

Table 2. Summary of the best and worst values of considered measures, depicting the related MLA and sensor, related to the various feature sets.

Measure Name	Super Vector F.	Welch PSD F.	Haar Wavelet F.	Time–Frequency F.
	Best Value of the Measure, MLA, Sensor			
Accuracy	1, ESD, WD	1, ESD, WD	0.958, SVL, MiG	0.991, ESD, MiC
Sensitivity	1, ESD, WD	1, ESD, WD	0.961, SVL, MiG	0.993, ESD, MiC
Specificity	1, ESD, WD	1, ESD, WD	0.989, SVL, MiG	0.998, ESD, MiC
Precision	1, ESD, WD	1, ESD, WD	0.957, SVL, MiG	0.993, SVL, MiC
F1 score	1, ESD, WD	1, ESD, WD	0.958, SVL, MiG	0.992, SVL, MiC
Matthews CC	1, ESD, WD	1, ESD, WD	0.948, SVL, MiG	0.990, SVL, MiC
	Worst value of the measure, MLA, Sensor			
Accuracy	0.201, SVG, WD	0.203, SVG, WD	0.202, SVG, DS	0.201, SVG, WD
Sensitivity	0.200, SVG, DS	0.200, SVG, DS	0.200, SVG, DS	0.200, SVG, DS
Specificity	0.800, SVG, DS	0.800, SVG, DS	0.800, SVG, DS	0.800, SVG, DS
Precision	0.286, SVQ, WD	0.298, SVQ, DS	0.258, EBT, DS	0.359, SVC, DS
F1 score	0.252, SVQ, DS	0.286, SVQ, DS	0.254, EBT, DS	0.270, SVC, R15
Matthews CC	0.109, SVQ, DS	0.135, SVQ, DS	0.089, EBT, DS	0.162, SVC, DS

The same results were obtained for the *Welch PSD* feature set. In contrast, using *Haar wavelets*, the MiG sensor turns out to be the best for measuring AE signals and the SVL algorithm for classification. When using the *Time–Frequency* feature set, the most appropriate sensor is the MiC, while, in the classification task, similar results are obtained for the ESD and SVL algorithms. The analysis of the lowest values led to the unequivocal statement that the worst-fitting algorithm was SVG only for the following parameters: accuracy, sensitivity, specificity, and precision. On the other hand, the F1 score and Matthews correlation coefficient parameters indicate different algorithms (SVQ, SVC, and EBT) depending on the applied feature set as the training sequence. The worst results for the vast majority were obtained for signals recorded with the DS sensor. In individual cases, the signals recorded with WD and R15 sensors were also poorly classified.

4. Conclusions

Based on gathered analysis results, and, in particular, the ranking presented in Table 2, it is possible to consider the differences calculated for the individual sensors. Additionally, one can investigate the differences calculated for individual classification methods.

Summarizing the obtained results, it can be stated that each time choosing the optimal set of parameters of the learning algorithm, including the kernel function, it is possible to classify the defect with a very high level of accuracy using the power density spectrum analysis calculated with the Welch method (*Welch PSD*). Identical results were obtained for the largest set of features under consideration, the *Super Vector* feature set. In both cases, the best results were obtained for AE signals recorded with the WD sensor and applying the ESD MLA. For the other types of analysis, the values of the assessment quality measures exceed 94%. When wavelet analysis is used, the best results were obtained for the MiG sensor. When considering *Time–Frequency Measures*, the MiC sensor is best used.

The DS has a bandwidth of only up to 180 kHz and this is probably the reason why the worst classification values were obtained for the signals recorded with it. In comparison to this, the WD sensor's bandwidth is the widest—up to 1MHz. For the R15 sensor, with a bandwidth up to 400 kHz, average results were obtained in comparison with other sensors. In the case of the MiG hydrophone, the band is also up to 180 kHz, and yet, with the use of wavelets, a very good match was achieved, especially for SVL MLA. Moreover, for the MiC sensor, which has a bandwidth of up to 800 kHz, good classification results were obtained using ESD and SVL.

The obtained results may be used in automated expert systems for diagnostic purposes, where the knowledge base will include fingerprints derived from the digital processing of recorded signals. The system may contain different types of MLA algorithms that will recognize the OLTC defect depending on the type of sensor used for measurement.

Author Contributions: Conceptualization, D.W., A.C.; methodology, D.W., A.C.; software, D.W.; validation, D.W.; data curation, A.C.; writing—original draft preparation, D.W.; writing—review and editing, D.W.; visualization, D.W.; project administration, D.W., A.C.; All authors have read and agreed to the published version of the manuscript.

Funding: This research received no external funding.

Conflicts of Interest: The authors declare no conflict of interest.

References

1. Wang, Y.Y.; Zhou, J.J.; Chen, W.J.; Du, L.; Chen, R.G.; Zhang, L.; Song, W.G. Study of an assessment method for the reliability of tap-changers in power transformer based on fault-tree analysis. In Proceedings of the 2008 International Conference on High Voltage Engineering and Application, ICHVE 2008, Chongqing, China, 9–12 November 2008; pp. 604–608. [CrossRef]
2. Feizifar, B.; Usta, O. A new arc-based model and condition monitoring algorithm for on-load tap-changers. *Electr. Power Syst. Res.* **2019**, *167*, 58–70. [CrossRef]
3. Jongen, R.; Gulski, E.; Siodla, K.; Parciak, J.; Erbrink, J. Diagnosis of degradation effects of on-load tap changer in power transformers. In Proceedings of the ICHVE 2014—2014 International Conference on High Voltage Engineering and Application, Poznan, Poland, 8–11 September 2014. [CrossRef]
4. Jongen, R.; Seitz, P.P.; Smit, J.; Strehl, T.; Leich, R.; Gulski, E. On-load tap changer diagnosis with dynamic resistance measurements. In Proceedings of the Proceedings of 2012 IEEE International Conference on Condition Monitoring and Diagnosis, CMD 2012, Bali, Indonesia, 23–27 September 2012; pp. 485–488. [CrossRef]
5. Duan, R.; Wang, F. Mechanical condition monitoring of on-load tap-changers using chaos theory & fuzzy C-means algorithm. In Proceedings of the IEEE Power and Energy Society General Meeting, Denver, CO, USA, 26–30 July 2015. [CrossRef]
6. Duan, R.; Wang, F. Fault Diagnosis of on-load tap-changer in converter transformer based on time-frequency vibration analysis. *IEEE Trans. Ind. Electron.* **2016**, *63*, 3815–3823. [CrossRef]
7. Duan, R.; Zuo, Y.; Zheng, H.; Yao, R.; Deng, M.; Sun, Z. Condition monitoring of on-load tap-changers in converter transformers based on vibration method. In Proceedings of the 2nd IEEE Conference on Energy Internet and Energy System Integration, Beijing, China, 20–22 October 2018. [CrossRef]
8. Secic, A.; Kuzle, I. On the novel approach to the on Load Tap Changer (OLTC) diagnostics based on the observation of fractal properties of recorded vibration fingerprints. In Proceedings of the 17th IEEE International Conference on Smart Technologies, Ohrid, Macedonia, 6–8 July 2017; pp. 720–725. [CrossRef]
9. Osmanbasic, E.; Skelo, G. Tap changer condition assessment using dynamic resistance measurement. *Procedia Eng.* **2017**, *202*, 52–64. [CrossRef]

10. Seo, J.; Ma, H.; Saha, T.K. A joint vibration and arcing measurement system for online condition monitoring of onload tap changer of the power transformer. *IEEE Trans. Power Deliv.* **2017**, *32*, 1031–1038. [CrossRef]
11. Li, X. A brief review: Acoustic emission method for tool wear monitoring during turning. *Int. J. Mach. Tools Manuf.* **2002**, *42*, 157–165. [CrossRef]
12. Hussain, A.; Lee, S.J.; Choi, M.S.; Brikci, F. An expert system for acoustic diagnosis of power circuit breakers and on-load tap changers. *Expert Syst. Appl.* **2015**, *42*, 9426–9433. [CrossRef]
13. Yen, C.L.; Lu, M.C.; Chen, J.L. Applying the self-organization feature map (SOM) algorithm to AE-based tool wear monitoring in micro-cutting. *Mech. Syst. Signal. Process.* **2013**, *34*, 353–366. [CrossRef]
14. Morette, N.; Ditchi, T.; Oussar, Y. Feature extraction and ageing state recognition using partial discharges in cables under HVDC. *Electr. Power Syst. Res.* **2020**, *178*, 6428–6438. [CrossRef]
15. Morizet, N.; Godin, N.; Tang, J.; Maillet, E.; Fregonese, M.; Normand, B. Classification of acoustic emission signals using wavelets and Random Forests: Application to localized corrosion. *Mech. Syst. Signal. Process.* **2016**, *70–71*, 1026–1037. [CrossRef]
16. Wang, Y.; Feng, L. Hybrid feature selection using component co-occurrence based feature relevance measurement. *Expert Syst. Appl.* **2018**, *102*, 83–99. [CrossRef]
17. Kane, P.V.; Andhare, A.B. Critical evaluation and comparison of psychoacoustics, acoustics and vibration features for gear fault correlation and classification. *Meas. J. Int. Meas. Confed.* **2020**, *154*, 107495. [CrossRef]
18. Zhang, D.; Stewart, E.; Entezami, M.; Roberts, C.; Yu, D. Intelligent acoustic-based fault diagnosis of roller bearings using a deep graph convolutional network. *Meas. J. Int. Meas. Confed.* **2020**, *156*, 107585. [CrossRef]
19. Chen, B.; Shen, B.; Chen, F.; Tian, H.; Xiao, W.; Zhang, F.; Zhao, C. Fault diagnosis method based on integration of RSSD and wavelet transform to rolling bearing. *Meas. J. Int. Meas. Confed.* **2019**, *131*, 400–411. [CrossRef]
20. Kyprianou, A.; Lewin, P.L.; Efthimiou, V.; Stavrou, A.; Georghiou, G.E. Wavelet packet denoising for online partial discharge detection in cables and its application to experimental field results. *Meas. Sci. Technol.* **2006**, *17*, 2367–2379. [CrossRef]
21. Sameh, W.; Gad, A.H.; Eldebeikey, S.M. An intelligent classifier of electrical discharges in oil immersed power transformers. In Proceedings of the 2019 21st International Middle East Power Systems Conference, Cairo, Egypt, 17–19 December 2019; pp. 866–871. [CrossRef]
22. Meitei, S.N.; Borah, K.; Chatterjee, S. Modelling of acoustic wave propagation due to partial discharge and its detection and localization in an oil-filled distribution transformer. *Frequenz* **2020**, *74*, 73–81. [CrossRef]
23. Ge, J.; Yin, G.; Wang, Y.; Xu, D.; Wei, F. Rolling-bearing fault-diagnosis method based on multimeasurement hybrid-feature evaluation. *Information* **2019**, *10*, 359. [CrossRef]
24. Boczar, T.; Borucki, S. Time-frequency analysis of the AE signals generated by pds on bushing and stand-off insulators. *Arch. Acoust.* **2006**, *31*, 325–333.
25. Cichoń, A.; Borucki, S.; Berger, P. Selecting sensor for on load tap changer contacts degree of wear diagnostics. *Acta Phys. Pol. A* **2013**, *124*, 395–398. [CrossRef]
26. Gao, C.; Yu, L.; Xu, Y.; Wang, W.; Wang, S.; Wang, P. Partial discharge localization inside transformer windings via fiber-optic acoustic sensor array. *IEEE Trans. Power Deliv.* **2019**, *34*, 1251–1260. [CrossRef]
27. Jiang, J.; Wang, K.; Wu, X.; Ma, G.; Zhang, C. Characteristics of the propagation of partial discharge ultrasonic signals on a transformer wall based on Sagnac interference. *Plasma Sci. Technol.* **2020**, *22*, 024002. [CrossRef]
28. Khalid, K.N.; Rohani, M.N.K.H.; Ismail, B.; Isa, M.; Rosmi, A.S.; Wooi, C.L.; Yii, C.C. Influence of PD source and AE sensor distance towards arrival time of propagation wave in power transformer. In Proceedings of the First International Conference on Emerging Electrical Energy, Electronics and Computing Technologies, Melaka, Malaysia, 30–31 October 2019; Volume 1432, p. 012006. [CrossRef]
29. MISTRAS Group. *Product Brochure—WD Sensor*. Available online: https://www.physicalacoustics.com/by-product/sensors/WD-100-900-kHz-Wideband-Differential-AE-Sensor (accessed on 29 May 2020).
30. MISTRAS Group. *Product Brochure—D9241A Sensor*. Available online: https://pdf.directindustry.com/pdf/physical-acoustics/d9241a-sensor/27111-396871.html (accessed on 29 May 2020).
31. MISTRAS Group. *Product Brochure—R15 α Sensor*. Available online: https://pdf.directindustry.com/pdf/physical-acoustics/r15-alpha/27111-396891.html (accessed on 29 May 2020).
32. Brüel & Kjær. *Product Brochure—Hydrophones—Types 8103, 8104, 8105 and 8106*. Available online: https://www.bksv.com/-/media/literature/Product-Data/bp0317.ashx (accessed on 29 May 2020).

33. Teledyne Reson. *Product Brochure—Hydrophone TC4038*. Available online: http://www.teledynemarine.com/reson-tc-4038?BrandID=17 (accessed on 29 May 2020).
34. Boczar, T.; Cichoń, A. Comparative analysis of acoustic emission signals generated by electrical discharges measured by the hydrophone and the wideband contact transducer. *J. Phys. IV Proc. France* **2005**, *129*, 93–96. [CrossRef]
35. Boczar, T.; Borucki, S.; Cichoń, A.; Lorenc, M. The influence of the propagation path length on the results of the time-frequency analysis of the acoustic emission generated by partial discharges in insulation oil. *Int. J. Phys. IV Proc. France* **2006**, *137*, 35–41. [CrossRef]
36. Cichon, A.; Berger, P. Possibility of using acoustic emission method for testing load tap changers during normal operation of the transformer. In Proceedings of the ICHVE 2014—2014 International Conference on High Voltage Engineering and Application, Poznań, Poland, 8–11 September 2014. [CrossRef]
37. Boczar, T.; Cichoń, A.; Borucki, S.; Szczyrba, T. Analysis of acoustic signals interfering diagnostic measurements of electric power transformer tap changers. *Acta Phys. Pol. A* **2013**, *124*, 387–390. [CrossRef]
38. Xie, J.; Liu, Y.; Liu, L.; Tang, L.; Wang, G.; Li, X. A Partial discharge Signal Denoising Method Based on Adaptive Weighted Framing Fast Sparse Representation. *Proc. Chin. Soc. Electr. Eng.* **2019**, *39*, 6428–6438. [CrossRef]
39. Kozioł, M.; Nagi, Ł.; Kunicki, M.; Urbaniec, I. Radiation in the optical and UHF range emitted by partial discharges. *Energies* **2019**, *12*, 4334. [CrossRef]
40. Kunicki, M. Variability of the UHF signals generated by partial discharges in mineral oil. *Sensors (Switzerland)* **2019**, *19*, 1392. [CrossRef] [PubMed]
41. Nagi, L.; Kunicki, M. Ionizing radiation generated by the electrical discharges from medium and high voltage in the air. In Proceedings of the 2017 IEEE International Conference on Environment and Electrical Engineering and 2017 IEEE Industrial and Commercial Power Systems Europe (EEEIC/I&CPS Europe), Milan, Italy, 6–9 June 2017; pp. 1–5. [CrossRef]
42. Nagi, Ł.; Kozioł, M.; Zygarlicki, J. Optical radiation from an electric arc at different frequencies. *Energies* **2020**, *13*, 1676. [CrossRef]
43. Cichon, A.; Fracz, P.; Zmarzly, D. Characteristic of acoustic signals generated by operation of on load tap changers. *Acta Phys. Pol. A* **2011**, *120*, 585–588. [CrossRef]
44. Cichoń, A.; Borucki, S.; Boczar, T.; Zmarzły, D. The possibilities of using acoustic signals generated by the on load tap changer. In Proceedings of the 41st International Congress and Exposition on Noise Control Engineering 2012, New York, NY, USA, 19–22 August 2012.
45. Cichon, A.; Borucki, T.; Boczar, T.; Zmarzly, D. Characteristic of acoustic emission signals generated by electric arc in on load tap changer. In Proceedings of the International Symposium on Electrical Insulating Materials, Kyoto, Japan, 6–10 September 2011. [CrossRef]
46. Suzuki, H.; Kinjo, T.; Hayashi, Y.; Takemoto, M.; Ono, K.; Hayashi, Y. Wavelet transform of acoustic emission signals. *J. Acoust. Emiss.* **1996**, *14*, 69–84.
47. Sause, M.G.R. *Situ Monitoring of Fiber-Reinforced Composites Theory, Basic Concepts, Methods, and Applications*; Springer Series in Materials Science; Springer: New York, NY, USA, 2016.
48. Karandaeva, O.I.; Yakimov, I.A.; Filimonova, A.A.; Gartlib, E.A.; Yachikov, I.M. Stating diagnosis of current state of electric furnace transformer on the basis of analysis of partial discharges. *Machines* **2019**, *7*, 77. [CrossRef]
49. Zheng, K.; Wang, X. Feature selection method with joint maximal information entropy between features and class. *Pattern Recognit.* **2018**, *77*, 20–29. [CrossRef]

Article

Distribution Transformer Parameters Detection Based on Low-Frequency Noise, Machine Learning Methods, and Evolutionary Algorithm

Daniel Jancarczyk [1,*], Marcin Bernaś [1] and Tomasz Boczar [2]

[1] Department of Computer Science and Automatics, University of Bielsko-Biala, 43-309 Bielsko-Biala, Poland; mbernas@ath.bielsko.pl

[2] Institute of Electric Power Engineering and Renewable Energy, Opole University of Technology, 45-758 Opole, Poland; t.boczar@po.opole.pl

* Correspondence: djancarczyk@ath.bielsko.pl

Received: 6 July 2020; Accepted: 1 August 2020; Published: 4 August 2020

Abstract: The paper proposes a method of automatic detection of parameters of a distribution transformer (model, type, and power) from a distance, based on its low-frequency noise spectra. The spectra are registered by sensors and processed by a method based on evolutionary algorithms and machine learning. The method, as input data, uses the frequency spectra of sound pressure levels generated during operation by transformers in the real environment. The model also uses the background characteristic to take under consideration the changing working conditions of the transformers. The method searches for frequency intervals and its resolution using both a classic genetic algorithm and particle swarm optimization. The interval selection was verified using five state-of-the-art machine learning algorithms. The research was conducted on 16 different distribution transformers. As a result, a method was proposed that allows the detection of a specific transformer model, its type, and its power with an accuracy greater than 84%, 99%, and 87%, respectively. The proposed optimization process using the genetic algorithm increased the accuracy by up to 5%, at the same time reducing the input data set significantly (from 80% up to 98%). The machine learning algorithms were selected, which were proven efficient for this task.

Keywords: low-frequency sensor; power transformer; machine learning; low-frequency noise; genetic algorithm

1. Introduction

The transformer is a passive electrical device that transfers electrical energy from one electrical circuit to another, or multiple circuits. The transformer is composed of the main parts: The primary and secondary winding wound around the same core and the air or oil cooling system. Noise emitted by the transformer is a vibrio-acoustics problem. The acoustic vibrations of a transformer can be generated by the following main phenomena [1,2]:

- Coil vibrations depending on the current amplitude and winding clamping compression;
- Core vibration depending on magnetostriction or loosening of core clamping;
- Air circulation caused by fans; and
- Work of the pumps circulating the insulation oil.

The process of the generation of noise from the vibration of a transformer is shown in Figure 1.

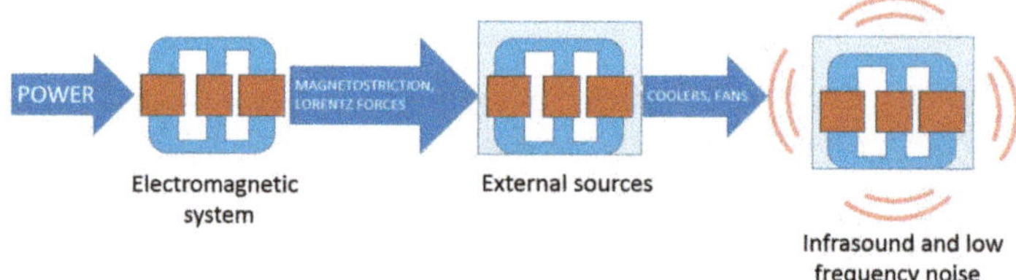

Figure 1. The process of the generation of noise from the vibration of a transformer.

In transformers, the load noise is predominantly produced by axial and radial vibration of the windings. Load noise can also be caused by vibrations in the transformer tank walls and magnetic shields due to the electromagnetic forces produced by the load currents. These electromagnetic forces are proportional to the square of the load currents. The frequency of load noise is usually twice the power frequency. In some cases, the natural mechanical frequency of winding clamping systems may tend to resonate with electromagnetic forces, thereby severely intensifying the load noise. Transformer cores are constructed by stacking layers of thin iron laminations, separated from its neighbors by a thin non-conducting layer of insulation. When the core becomes magnetized, the magnetic field acts between the adjacent plates, stretching and squeezing the adhesive and insulation between them. A transformer is magnetically excited by an alternating voltage and current so that it becomes extended and contracted twice during a full cycle of magnetization. This change in dimension is independent of the direction of magnetic flux, occurring at twice the line frequency. The main source of heat generation in transformers is caused by copper loss in the windings and core. This heat is often removed by cooling fans, which blow air over radiators or coolers. Noise produced by cooling fans usually contribute more to the total noise for transformers of a smaller rating and for low-induction transformers. Cooling equipment noise typically dominates the very low- and very high-frequency ends of the sound spectrum, whereas the core noise dominates in the intermediate range of frequencies between 100 and 600 Hz [3–7].

For a person with normal hearing, the human hearing range starts low at about 20 Hz. On the other side of the human hearing range, the highest possible frequency heard without discomfort is 20 kHz. Ultrasound is sound waves with frequencies higher than the upper audible limit of human hearing (above 20 kHz). In this description, we focus on the lower end of the frequency spectrum. We are interested in infrasound and low-frequency noise; see Figure 2. The range of analyzed frequencies is up to 200 Hz, taking under consideration body resonance infrasound.

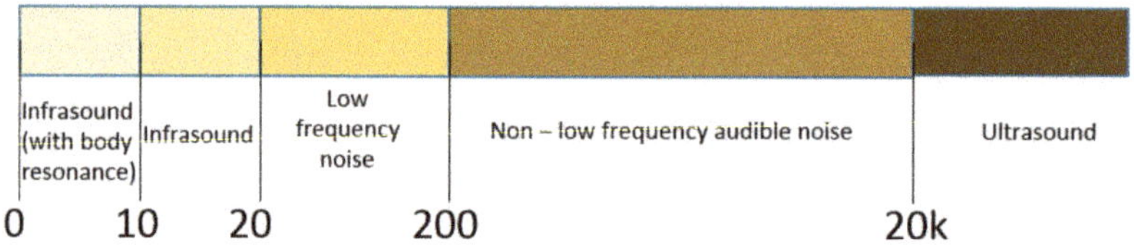

Figure 2. The frequency spectrum of sound and its nomenclature.

The acoustic emission depends on the anomalies, age, and rated power of the machine. For this reason, the analysis of the generated sound, especially within the spectrum of low-frequency noise, can be useful in determining the rated parameters of the transformer and its diagnostic parameters. It is worth noting that the measurement of low-frequency signals is carried out non-invasively and during normal operation of the distribution transformer.

Our research was extended to remove drawbacks of the previous method [8], and to develop an automatic method for determining the technical parameters and diagnostics of distribution transformers based on the analysis of the characteristics of their low-frequency signals. Using low-frequency signal allows measurement of the device from a distance (50 m). This research presents a method for detecting the base parameters of distribution transformers (type and rated power).

2. State-Of-The-Art

Transformers can vary, from a miniature high-frequency audio transformer to a large power transformer, but the operating parameters are the same. These can be divided into eight groups and are posted on the nameplate of any transformer of significant size: Nominal apparent power (VA rating), cooling, transformer rating frequency, voltage, phase, connections, and taps [9]. Based on the low-frequency characteristics of the transformer, we are able to distinguish between the transformer's operating status, its cooling type, and its apparent power.

In our first study [10], we tried to ascertain the emission level of low-frequency signals generated by distribution transformers at rated conditions. The medium-voltage devices (indoor and overhead type and variety apparent powers) were under study.

The results of the study showed that distribution transformers are a source of infrasound and low-frequency signals. The results demonstrated the similarity in the shape of waveforms of averaged amplitude spectra and time-frequency changes. The waveforms are characterized by relatively dynamic (exponentially) decreasing values of registered sound pressure, which occurs with an increase in the frequency in the range from 10 to 100 Hz [10,11].

Research of a similar nature was conducted for the problem of noise from electrical infrastructure. Piana et al. [12] claimed that the low-frequency disturbance occurs for the tested transformer at two harmonics of for each of two frequencies. Other scientists proposed prognostic and system health management (PHM) for power transformer fault diagnosis. Potential uses for PHM is a condition-based maintenance. This system presents opportunities for the detection of mechanical failures, or for system life cycle management. Li et al. [13] presented the study of a power transformer fault diagnosis using a machine learning-based method with a neural network model. The proposed method uses dissolved gas analysis (DGA) as input data. A frequently used method in the diagnosis of oil-filled power transformers is a partial discharge (PD) detection using an acoustic emission (AE) technique. Many cases of power transformer breakdowns are related to insulation system failures, which might have been caused by the high activity of partial discharges [14]. Kunicki et al. [15] proposed a method for detecting defects of power transformers. This method is based on machine learning classification of selected faults. In this case, input data is AE measurement. The classification process consists of two parts: The first part checks whether the source of the emitted signal is partial discharge or another AE source while the second part allows the identification of the specific AE source type.

The aforementioned research is focused on the life cycle of a distribution transformer measured in an isolated environment, where the external noise influence is minimalized. The classification is based on various machine learning algorithms (ML), which finds a pattern in data based on expert feedback. Therefore, in [12–15], the supervised learning [16] was used to train a model for various applications. However, in this research, the analyzed data were gathered from a significant distance, thus unsupervised learning was used [17] to find anomaly in the background noise, the source of which was unknown. In the case of maintenance operation issues [14], reinforcement learning [18] was used to find optimum actions for a given operation status of a transformer. In this paper, the proposed method utilizes both supervised and unsupervised learning. The supervised learning is used to build a classification model based on evolution strategy and state-of-the-art ML algorithms: k nearest neighbors (kNN) [19], naive Bayes classification (Bayes) [20], support vector machine (SVM) [21,22], random forests [23,24], and neural networks [25–28]. The unsupervised learning is applied to tackle the background noise.

The accuracy of the constructed model (using the ML algorithm) depends not only on the algorithm but also the type of input data and its representation. Thus, many models use a preprocessing method, which transfers data into a new variable space [17]. The transformation can be a simple operation as scale transformation (e.g., to the dB scale) or a more complex one, when the nature of data changes [29]. The principal component analysis [29–32] and canonical correlation analysis [33,34] are commonly used methods for TS.

The neural networks (especially deep ones), due to the variety of their structures, are both classification and pattern recognition tools [25,26]. Using multiple hidden layers allows the creation of linear and non-linear models. A drawback of this approach is that the training procedure requires a large amount of data and computation power to obtain a high model accuracy. Moreover, finding the optimum set of weights in the case of a multiple hidden layer structure is an NP-complete problem [35]. Thus, a substitute for classic multiple layer perceptron networks was proposed (deep neural network), in which the network is divided into layers with specific functions [28,32]. Significant results using deep neural networks have led them to be the most commonly employed classifiers in machine learning [35,36].

The proposed method is based on two sources of data (transformer sound and background sound gathered in various locations). Thus, these two sources of the sound are considered as independent ones. However, each registered series of sound according to [17,37] is characterized by significant sequential correlations and should be represented as temporal features. There are several methods fitting for this type of data: Hidden Markov [38], sliding window [39], Kalman filter [40], random fields [41], recurrent neural networks [42], and the Welch method. An extended analysis of the methods can be found in [43,44]. In this study, which is an extension of [8], the same Welch method [45] was used to retain the consistency of results. Furthermore, this method was used with success for noise analysis in [11,12].

The previous study [8] proposed a method to automatically detect a known working transformer in close vicinity (50 m). Low-frequency noise generated by transformers (two indoor and two overhead ones) was registered by a dedicated sensor from a distance of 50 m and then classified using the proposed machine learning method. It is worth noting that the research was performed in a real environment. The method used an exhaustive search and Bayesian optimization to find a frequency interval that gave the best detection results. The drawback of the previous method [8] was that only one interval could be searched at a time, while the results showed that depending on the method, various intervals were selected near the following frequencies: 2, 50, and 100 Hz. Despite the drawbacks of the method, a 99% accuracy was obtained for the state of the transformers (on/off) using the random forest, KNN, and naïve Bayes methods.

3. Proposed Method

In the research for this paper, the frequency was extended to 200 Hz as it was proven that most information was stored close to 100 Hz and a higher frequency was not examined. Furthermore, the research was performed for one interval at a time while several harmonics were found (1, 50, and 100 Hz). To provide sounder results, the transformer database was significantly extended (from 4 to 16) and several of its parameters were researched. Based on the initial observation and previous research [8], a method was proposed, which finds the optimum frequency representation for distribution transformer features: Model, type, and apparent power. The overview of the method, with a background profile analysis, is presented in Figure 3.

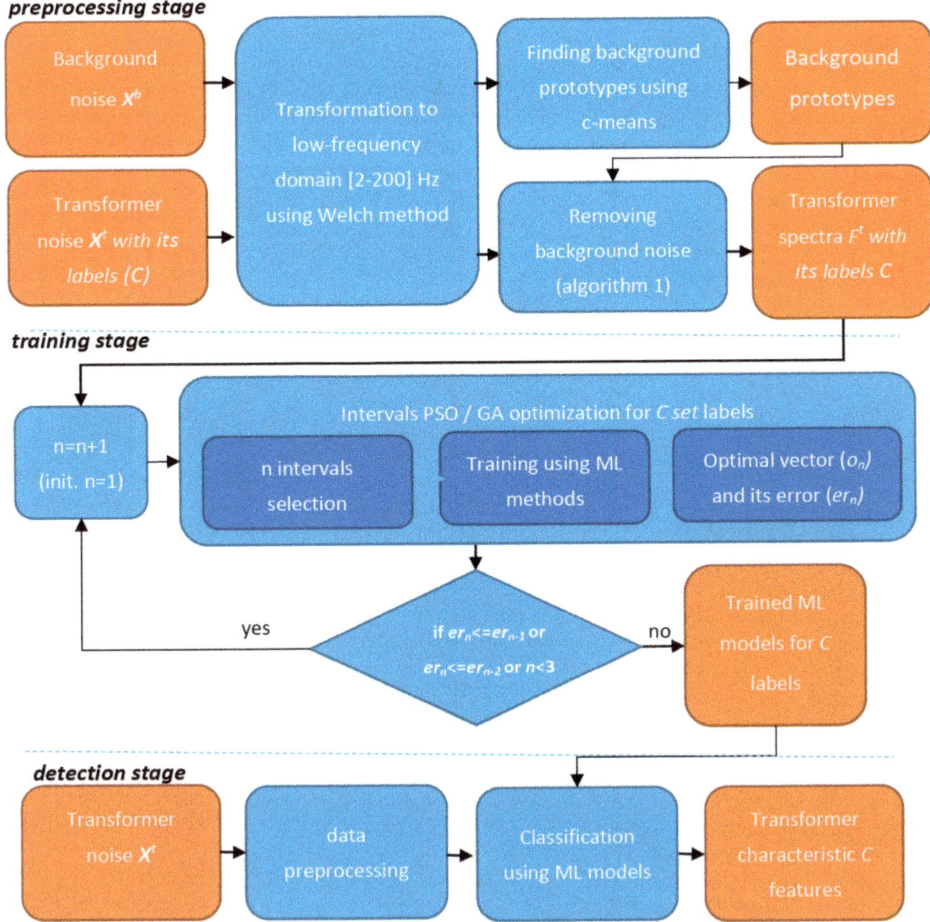

Figure 3. Proposed method for the detection of transformer parameters.

3.1. Preprocessing Stage

The input data of the method is a sound registered as a time series (X) by a dedicated sensor node. The sensor measures the transformer noise with background noise (X^t) or only background noise (X^b) if the transformer is not present in the vicinity or it is turned off. Each registered time series X is a sequence of sound pressure values x_i, where i defines its order.

The samples, registered for each distribution transformer (X^t), are converted using the Welch method [8,42] to obtain its frequency representation (F_x^t). The obtained spectra $F_x^t = [f_1, f_2, \ldots, f_m]$ represent the low-frequency spectrum in the range from 2 to 200 Hz with the maximum considered resolution df equal to 0.125 Hz (thus $m = 1585$). The df value lower than 0.125 Hz would require a significant recording time (over 10 s a sample) and as it was proven in this research that the higher resolution did not increase the classification accuracy. The spectra samples are represented as $\boldsymbol{F}^t$ set ($\boldsymbol{F}^t = \{F_x^t, x = 1, \ldots, tmax\}$, where $tmax$ is a number of samples). Each vector F_x^t is described by basic transformer parameters ($C = \{c_1, c_2, c_3\}$), i.e., transformer model (c_1), its type (c_2), and apparent power value (c_3). The characteristic spectra registered in the vicinity of two transformers are presented in Figure 4.

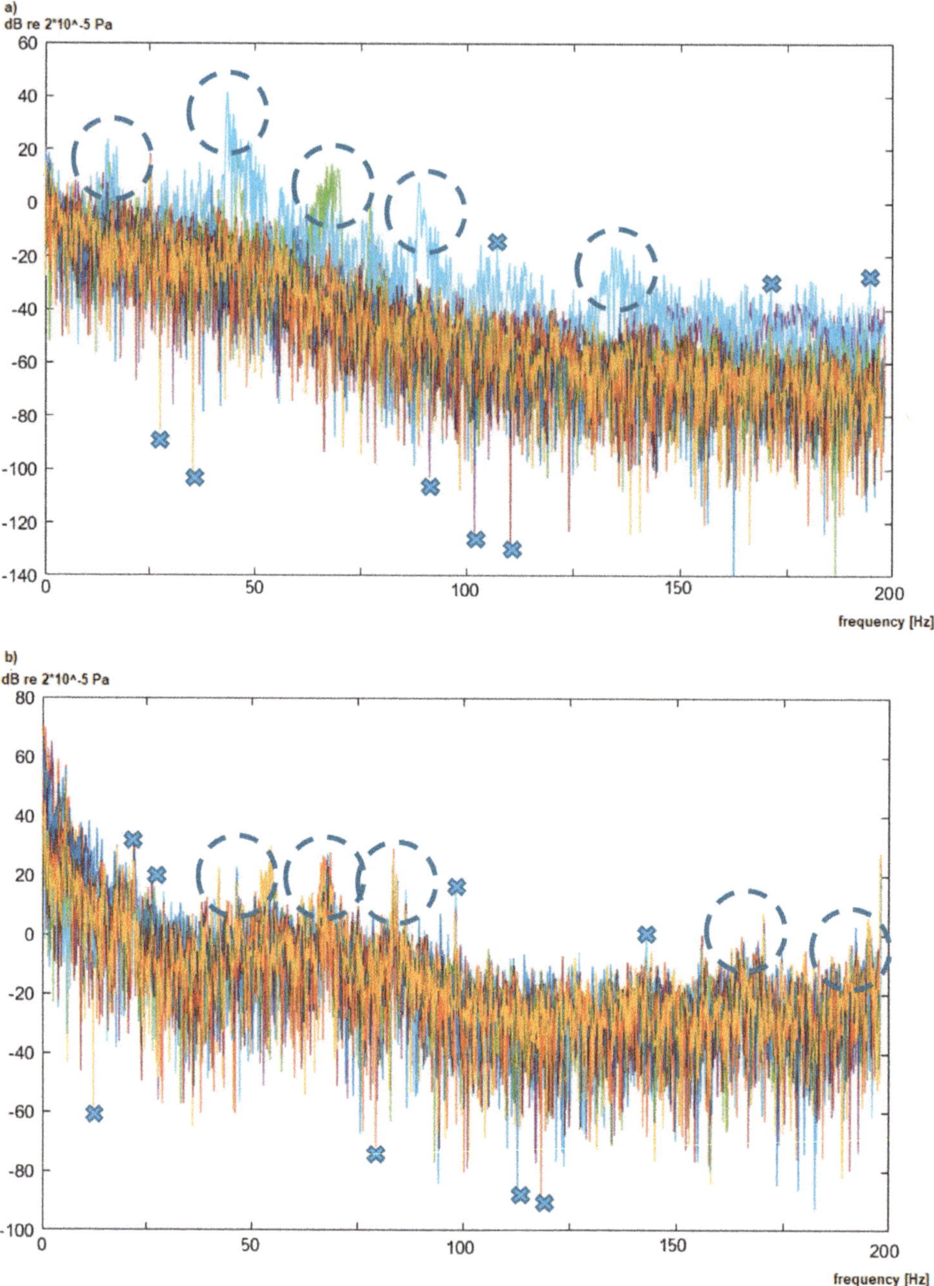

Figure 4. The example of spectra registered in the vicinity of: (**a**) transformer air-type, and (**b**) interior type.

Each plot in Figure 4a,b presents measurements of the noise characteristic in the vicinity of the same transformer. Measured values for the specific frequency (marked with dashed circles), single peaks (marked with crosses), as well as changes of the amplitude within the whole spectrum can vary significantly. This is caused by the measured distance of 50 m, where other sources of noise are registered as well. Moreover, the transformer load, as presented in [6], can also influence the value of noise, especially in the 50 Hz area. Therefore, an additional source of data was used to identify the anomalies, which originate from analyzed transformers. The background noise in multiple localizations was registered (denoted as F_x^b, $x = 1, \ldots, bmax$), where the noise is not biased by a transformer noise. The registered background characterizes interiors, fields, forests, and the home environment. An example of the background is presented in Figure 5.

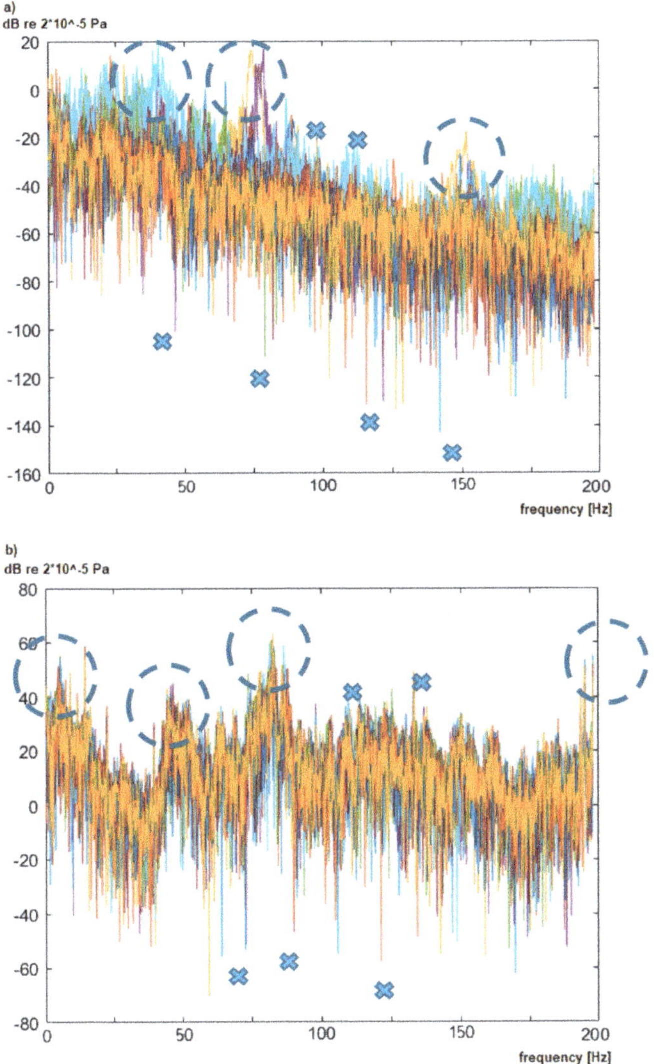

Figure 5. Noise background for various positions: (**a**) outdoor background, (**b**) indoor background.

The analyzed background noise (without transformers in vicinity) shows that its characteristic are not constant and also change in time. It is characterized by the same type of variations; however, it is registered more frequently for specific frequencies (e.g., 75 Hz). In case of indoor background, registered in Figure 5b, additional noise was registered for the 50 Hz harmonics and near 10 Hz. This noise is generated by electric devices, which are part of the production areas. It is nearly impossible to gather background characteristic for each transformer, because they are part of the energy infrastructure and they cannot be turned off easily. Therefore, the gathered backgrounds (Figure 5) are characteristic for a specific area, not for a specific transformer. Moreover, as it is shown in Figures 4 and 5, the background changes with time due to temporal occurrences like the influence of constructions, large objects, or vehicles.

Initial work was conducted to find a universal background using the following estimates: Mean, median average, and maximum and minimum value; however, this approach generated worse results. Therefore, based on the analysis, we propose to define several background characteristics and adopt them for each spectrum based on the similarity measure. The fuzzy c-means algorithm [46] was used to find characteristic backgrounds and then, based on the similarity to the analyzed spectra, the appropriate one was selected. To select the background $\boldsymbol{F^b}$ for transformer spectra F_x^t, the following procedure (Algorithm 1) is given:

Algorithm 1

1. V = cmean($\boldsymbol{F^b}$,*vmax*), where: cmean-fuzzy c-mean function [42], vmax-number of prototypes,
2. mt = mean(F_x^t);
3. mxw = 0; minVal = INF; minid = −1;
4. for j = 1:vmax
5. mx = mean(V(j)));
6. val1 = mean(((F_x^t−mt)−V(j)−mx).^2.^0.5);
7. val2 = mean(F_x^t)−mean(V(j));
8. if (val1 < minVal && val2 > 0)
9. minVal = val1;
10. minid = j;
11. mxw = mx;
12. end
13. end
14. result = F_x^t−V(minid).

Algorithm 1 generates a set of characteristic backgrounds denoted as V based on background data (line 1). Then, the most similar background noise V(j) is searched for in the analyzed F_x^t spectrum (lines 4–10). The V spectra are compared based on similarities for the same frequencies (*val1*) as well as its average value (*val2*). Figure 6 presents the similarity calculation process. At first, the average value for each potential background is calculated as presented in Figure 6a. The difference between average values (denoted as val2) is defined if the background noise is stronger than the registered transformer noise. Only those prototypes V(j) are considered, in which the average noise is lower than the analyzed spectrum, F_x^t. Next, the potential background spectra shapes are compared according to formula in line 6. The result of the subtraction is presented in Figure 6b. The obtained mean value of this substation is treated as a similarity measure. In the example presented in Figure 6, the V(1) prototype will not be selected because its mean noise level is higher (val2 = −50) and its similarity measure (val1) is lower than in the case of V(2). Thus, the V(2) prototype is selected as the F_x^t background. The background is subtracted from the spectrum (line 14). The operation is performed for every spectrum $\boldsymbol{F^t} = \{F_x^t,\ x = 1,\ \ldots,\ tmax\}$. Then, the input set F^t with the subtracted background is used to find the optimum frequency interval representation in the next step.

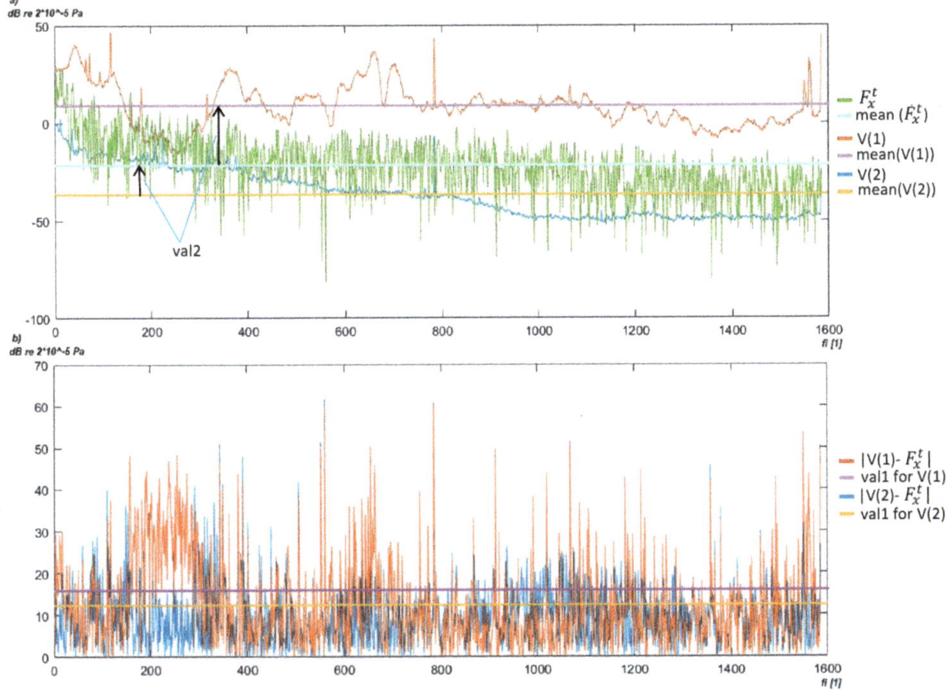

Figure 6. Example of the c-means background cluster fitted to a transformer: (**a**) calculate val2 measure, (**b**) calculate val1 measure.

3.2. Training Stage Using ML and Evolutionary Algorithms

The search for optimum data representation (frequency space and its resolution) is performed using the evolutionary algorithm. Two state-of-the-art evolution methods, genetic algorithm (ga) and particle swarm optimization (PSO), were applied for each of the C features separately.

The first method, due to the discrete search space, was based on integer programming. In this implementation, special creation, crossover, and mutation functions cause variables to be integers [47]. The implemented genetic algorithm attempts to minimize a penalty function, which is combined with binary tournament selection to select individuals for subsequent generations [48].

The PSO algorithm is based on the implementation proposed by Kennedy and Eberhart [49], using modifications suggested in Mezura-Montes and Coello Coello [50] and in Pedersen [51]. In contradiction to the ga algorithm, it creates the initial particles, and assigns initial velocities to them. Then, it evaluates the objective function at each particle location and determines the best (lowest) function value and the best location. Finally, it chooses new velocities, based on the current velocity, the particles' individual best locations, and the best locations of their neighbors. Initially, the particles are uniformly distributed within bounds.

Both methods use a similar population/swarm size parameter equal to 200 as well as a total error function equal to 1×10^{-4}.

The searched frequency space is described as an ordered set of frequency intervals described by a t_i tuple. The initial number of intervals n is equal to 1 and its value increases with each iteration. Each tuple $t_i = [l_i, h_i, s_i]$, $i = 1, \ldots, n$, where: $l_i = [2, 200]$, $h_i = [2, 200]$, $l_i < h_i$, and $h_i < l_{i+1}$ define respectively the lower (l_i) and upper (h_i) frequency bound and $s_i \in N$—defines the sample resolution for the given tuple. The described constraint allows limitation of the searching space to a considered frequency interval [2, 200] and ensures that intervals will not overlap.

Using evolution algorithms, a population of potential vectors $o_n = \{t_i, i = 1, \ldots, n\}$ is selected. With each generation, the o_n is selected, which has the lowest error ratio calculated as the misclassification accuracy using cross-validation results. Additionally, the secondary aim of optimization is to find the minimal set of frequency, which did not influence the result significantly. Thus, the following fitness function (*ff*) was proposed:

$$ff\left(F^t, o_n\right) = \min\left(cv\left(ML\left(F^t, o_n\right)\right) + \frac{\sum_{i=1}^{n}\left\lceil\frac{h_i - l_i + 1}{s_i}\right\rceil}{1000000}\right. \tag{1}$$

where ML is a function performing training for data and cv-return cross-validation error in the range [0, 1].

Using this approach, we ensure that the result with the highest accuracy will be selected, while the results with less samples will be favored in case of a comparable accuracy level. Taking under consideration the maximum number of samples, the value fitness function will be changed by 0.0015.

Several optimization algorithms as well as ML methods were analyzed to find an optimum solution. In this section, all used algorithms will be described.

The M model is a result of the selected ML method and data provided as a F^t vector and interval set o_n. The following methods were researched to find the optimum classifier:

- kNN model, where class is determined by the *k* closest vectors in a defined frequency space. As a measure of distance, the classic Euclidean distance was used with an initial *k* value equal to 5;
- Bayes approach, where a family of probabilistic classifiers was applied according to [20], thus no additional parameters were needed. It was assumed that each frequency characteristic is independent (Bayes theorem);
- Multivariable support vector machines, where binary learners were used to train the characteristics of each transformer and its parameters. In contrast to the radial basis function 45 applied in [8], in this research, multi-linear SVM was used, which finds a hyperplane that is a linear function of each input feature and the rest of the features. The implementation was adopted from [18,19,52];
- Random forest, where 10 trees create a forest, was used make the method more robust [23]. The result of a class is determined by voting. The parameter value was selected empirically;
- Neural network, where the multilayer perceptron network (MLP) was selected as the architecture. The MLP was selected to reduce the computation complexity of each iteration for the evolution algorithm. Several architectures were researched. Finally, 2 hidden layers and 20 neurons per layer were used to take the non-linear characteristic of the analyzed data into consideration [26].

3.3. Detection Stage

The detection stage is used to verify a proposed model and can also be applied for detecting transformer parameters C in the considered area. The registered noise is preprocessed in a similar way as data in the preprocessing stage. The sample is transformed to the frequency domain using the Walsh method and algorithm 1 is performed. However, in this case, the generated prototypes in the first stage are used. The preprocessed data are then processed in parallel by three ML models generated for each of three C features. As the result, the basic transformer parameters are determined.

4. Results and Discussion

Measurements were made using specialized equipment from Brüel & Kjær (company name details: Brüel & Kjær Sound & Vibration Measurement A/S, DK-2850 Nærum, Denmark). The system consists of a $\frac{1}{2}$-inch free-field microphone type 4190, a preamplifier type 2669 L, and a digital signal meter with registration function LAN-XI type 3050-A-60 from Brüel & Kjær. The connected system and its block diagram are shown in Figure 7.

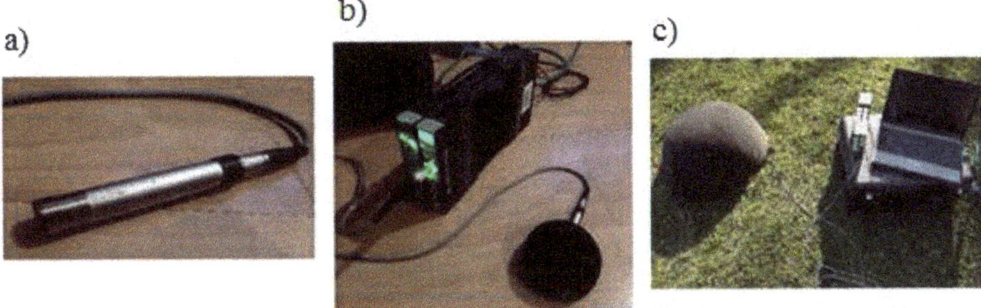

Figure 7. Used measurement system from Brüel & Kjær: (**a**) microphone, (**b**) digital meter, and (**c**) complete system.

The microphone was designed for very precise measurements in the free field. Its lower cut-off frequency is 1.2 Hz and the transmission characteristic is linear within ±3 dB from 1.2 Hz to 20 kHz. This microphone is characterized by high sensitivity (50 mV/Pa) and a dynamic range from 15 to 148 dB. The lower range of the measured frequencies can be defined by the user using high-pass filters (0.7 or 7 Hz).

The system was managed from the personal computer using PULSE LabShop application version 15.1.0. This is dedicated software that defines all operating parameters, records the measured signals, and preprocesses and visualizes them after measurements in offline mode.

The measurements were taken in a continual process over several hours, with a sampling frequency of 51.2 kHz (sampling for the whole human hearing range). Additionally, all measurements were carried out far from major roads and motorways, which are considered to be sources of low-frequency noise. The minimal sample length was defined by the Welch window size and it was defined as 10 s.

The research was conducted using 16 distribution transformers and 6 backgrounds for various areas and types (indoor and overhead, dry-type, and oil-type transformers and their apparent power in the range of 100–2500 kVA). All tested transformers reduce the voltage from 15 to 0.4 kV with a mains frequency 50 Hz. A detailed description of the tested transformers is provided in Table 1.

Table 1. List of the transformers under study.

Transformer Number	Manufacturer	Apparent Power	Transformer Type	Cooling Type
Transformer 1	Schneider Electric	2500 kVA	indoor type	dry-type
Transformer 2	Schneider Electric	2000 kVA	indoor type	dry-type
Transformer 3	Schneider Electric	1600 kVA	indoor type	dry-type
Transformer 4	Schneider Electric	1250 kVA	overhead type	oil-type
Transformer 5	Schneider Electric	630 kVA	overhead type	oil-type
Transformer 6	ABB	400 kVA	overhead type	oil-type
Transformer 7	ABB	400 kVA	overhead type	oil-type
Transformer 8	ABB	250 kVA	overhead type	oil-type
Transformer 9	ABB	250 kVA	overhead type	oil-type
Transformer 10	ABB	250 kVA	overhead type	oil-type
Transformer 11	ABB	250 kVA	overhead type	oil-type
Transformer 12	ABB	250 kVA	overhead type	oil-type
Transformer 13	ABB	160 kVA	overhead type	oil-type
Transformer 14	ABB	100 kVA	overhead type	oil-type
Transformer 15	ABB	100 kVA	overhead type	oil-type
Transformer 16	ABB	63 kVA	overhead type	oil-type

Initial work was performed to find an optimum background representation. It was achieved by finding a set of background prototypes giving the best results in terms of classification. The classification considered the following basic transformer parameters: Specific transformer model (c_1), transformer type (indoor/outdoor) (c_2), and its apparent power (c_3). The results are presented in Table 2. The values were obtained using the kNN classifier.

Table 2. The result of finding background representation.

	Cross-Validation Error [%]		
Error Rate for Number of Vectors (vmax)	Specific Transformer (c_1),	Transformer Type (c_2)	Transformer Power (c_3)
No background	23.3	0.4	18
1	23.1	0.4	16.1
2	22.8	0.4	15.9
3	23	0.5	17.22
4	24	0.6	18

The results show that the background selection influences c_3 more than c_1 and c_2. This is caused by the fact that background noise has a bigger impact on the magnitude of low-frequency noise registered and can influence the classification. The initial results confirmed that the characteristics of the background noise for air transformers and indoor transformers are significantly different and allow a decrease of the error by more than 1% even at the preprocessing stage. The indoor transformers are usually installed in urbanized areas, thus more sources of noise are present. Then, the proposed method was applied for the kNN classifier with both the pso and ga training methods. Additionally, the variant with and without background subtraction (application of Algorithm 1) was tested. The result of method training is presented in Figure 8.

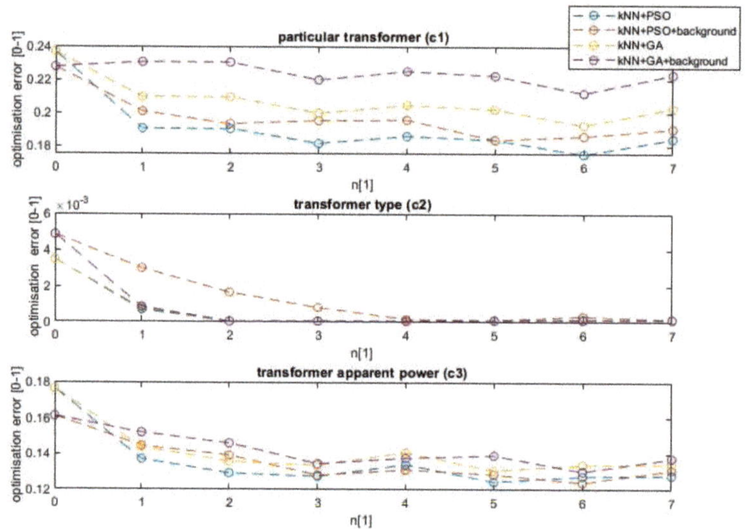

Figure 8. The training results for the optimization algorithm number of intervals n and feature c_1, c_2, and c_3.

The value n equal to 0 represents the value without interval optimization. The process of finding the optimum interval using the evolution strategy, from the first interval (n equal to 1) reduces the method error on average by 4% in the case of detecting a model of a transformer, five times in the case of the transformer type, and by 5% in the case of the transformer apparent power value. The further increase of the number of intervals steadily decreases the error up to n equal to 3. The best results were achieved for n equal to 6. Above this threshold, the method is over trained and its effectiveness starts to fall. The results confirmed that the used PSO method outperforms the tested ga algorithm by 2% on average for all analyzed features. Additionally, the background data decreases the error in the case of feature c_3. Further analysis showed that the optimization of the data using the background gives a slightly worse result (0.5%) than raw data in the case of c_1 feature. The difference is caused by the characteristic background of a specific area and the tendency to classify the transformer model based on the area background and not transformer noise. Similar training was performed for the remaining ML algorithms. The comparative result for the proposed method is presented in Figure 9 for the PSO algorithm and ML methods as shown in the legend.

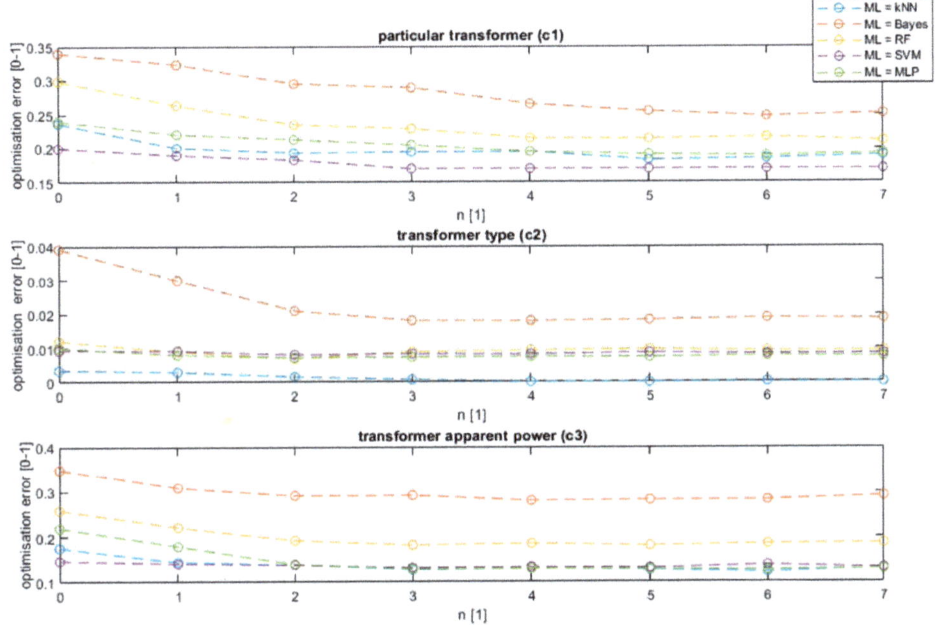

Figure 9. Training for all ML algorithms and analyzed features.

The proposed algorithm of interval selection with background subtraction decreased the error in the case of all analyzed ML models. The method was effective in the case of all features; however, the biggest optimization (in percent scale) was noticed in the case of transformer model detection (c_1) and transformer apparent power detection (c_3). In the case of the type of transformer detection, the detection error was close to zero, thus a significant improvement was noticed only in the case of the Bayes classifier. The decreased error can be noticed at the first interval (n = 1), which supports the results obtained in previous research [8], but further decreases can be observed with the interval number increase, obtaining the optimum value at n = 5. This value allows optimal detection of all harmonics of 50 Hz generated by a transformer, and the infrasound interval, which proved to be vital frequencies for transformer classification. The biggest decrease in the error rate can be noticed in case of simple classifiers, such as kNN, Bayes, or random forest. On the other hand, in the case of a complex non-linear SVM model, the decrease is equal to 2% on average. The proposed model not only allows an increase of the detection accuracy but also allows a decrease of the data usage by selecting intervals and the resolution at which the data are vital for classification. The research shows that various ML models require various data sizes; nevertheless, some frequencies' intervals are characteristic for all classifiers and were repeated in all models. Their ranges are presented in Figure 10.

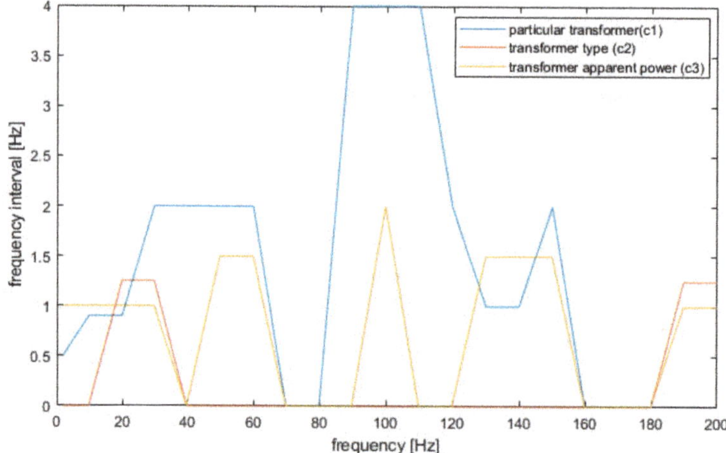

Figure 10. The data usefulness for classification and detection.

The analysis definitely shows that in the case of transformer model detection, nearly all frequencies are vital to improve detection. Only frequencies near 70 and 170 Hz contain less information. The results confirm the observations shown in Figure 5, where the external background noise was strong for those intervals. It is worth noting that detection of a specific transformer is most biased by additional background noise (like other electric devices). In the case of type discrimination (c_2 feature), the vital information can be obtained for frequencies until 25 Hz and near 200 Hz. These frequencies are sufficient to precisely define a transformer type and thus are the most robust to background noise. Finally, in the case of the power of the transformer, the vital information is stored near frequencies of 25, 50, 100, 150, and 200 Hz. This result is consistent with previous results for the frequency range 2–100 Hz [8] and the results found in [1–3,10,11]. Finally, in most cases, the frequency interval of 1 Hz is sufficient for precise classification for each feature, which strongly decreases the data usage. It is worth noting that more precision is required near 20 Hz (0.5 Hz). The final results for classification and data reduction using the proposed method are presented in Table 3.

Table 3. Results of the classification for a specific algorithm.

Feature	Algorithm	Accuracy (Based)	Accuracy (Proposed Method)	Data Reduction (Proposed Method)
Transformer model (c_1)	kNN	77%	81%	87%
	Bayes	70%	79%	70%
	Random Forest	68%	75%	91%
	SVM	83%	84%	11%
	MLP	74%	77%	81%
Transformer type (c_2)	kNN	98.9%	99.99%	98%
	Bayes	97%	99.99%	98.5%
	Random Forest	98.8%	99.8%	97.7%
	SVM	99.8%	99.99%	97.8%
	MLP	99.2%	99.98%	97.1%
Transformer Power (c_3)	kNN	78%	87%	93%
	Bayes	73%	78%	93%
	Random Forest	74%	79.1%	95%
	SVM	85%	87.7%	50%
	MLP	79%	86.8%	92%

The research presented in [12–15] proposes methods that precisely detect transformer flaws and discharge with an accuracy above 90%; nevertheless, the sensors have to be placed in close vicinity of a transformer. The proposed method, using low-frequency sound, allows the detection of basic parameters of various transformers from a distance of 50 m with an accuracy of at least 80%. It is worth noting that this type of sound has a strong influence on health [1,2] so this type of simple detection from a distance verifies the parameters of a transformer from a distance. The presented results show that various data sets need to be used depending on the task. The data reduction was achieved by decreasing the frequency resolution up to 1 Hz. An additional reduction is characteristic for a specific feature. In the case of indoor/outdoor transformer detection, only two frequency ranges were vital, thus the achieved reduction is high (98%). In the case of the specific transformer model, it was crucial to analyze a wide frequency range to find dissimilarities, thus the reduction was the smallest and depended on the interval resolution reduction. It is worth noting that the SVM ml MODEL requires bigger data precision (only 50% reduction); however, it obtained the best accuracy. The other classifiers obtained lower results; however, the data reduction was more significant, e.g., kNN obtained an 87% reduction at the cost of an accuracy decrease by 3%. In the case of the transformer type, all classifiers obtained a high 99% accuracy; however, the kNN outperformed the other classifiers in accuracy and data reduction. Finally, the power of a transformer can be precisely detected using kNN, SVM, and the MLP network. In the case of the SVM classifier, the accuracy was higher by 0.7%; however, this was at the cost of a lower data reduction. On the other hand, the MLP and kNN classifier obtained a comparable accuracy and data reduction.

5. Conclusions

The paper proposed a method to detect distribution transformer parameters from a distance, without the need of installing a multiple sensor array on the transformer. The proposed method uses genetic algorithms to find the optimum frequency representation for detecting a transformer model, its type, and rated power. In every case, the proposed method allowed an increase of the accuracy by 5% on average while finding the optimum intervals and their resolution and decreasing the input data set from 50% up to 98% depending on the task. The research confirmed that in the case of generated power, its harmonics are based on 50 Hz; however, an important interval can also be found near 20 Hz, which are infrasound signals. The specific transformer model can generate the noise in all of the analyzed spectrum; thus, in the case of all machine learning algorithms, the reduction of the interval caused a decrease in the detection accuracy. Finally, the type of the transformer due to its different characteristics (power level and interval construction) obtained a near 100% accuracy while a significant data reduction (over 90%) was achieved. The model, with the SVM classifier, can be applied for solutions requiring maximum accuracy, while models based on kNN and MLP can be applied in edge sensors, due to the significant data reduction.

Further research will be focused on anomaly detection during the operation of transformers based on various characteristics. This method will allow analysis of the technical condition of the transformer based on the measurement of its low-frequency signals made online (without switching off).

Author Contributions: Conceptualization, D.J.; Methodology, M.B.; Software, M.B., D.J., T.B.; Validation, T.B., D.J., M.B.; Formal Analysis, T.B.; Investigation, D.J., M.B., T.B.; Writing-Original Draft Preparation, M.B. and D.J.; Writing—Review and Editing, M.B., T.B. and D.J. All authors have read and agreed to the published version of the manuscript.

Funding: This research received no external funding.

Conflicts of Interest: The authors declare no conflict of interest.

References

1. Ying, L.; Wang, D.; Wang, J.; Wang, G.; Wu, X.; Liu, J. Power Transformer Spatial Acoustic Radiation Characteristics Analysis under Multiple Operating Conditions. *Energies* **2018**, *11*, 74. [CrossRef]
2. Bartoletti, C.; Desiderio, M.; Di Carlo, D.; Fazio, G.; Muzi, F.; Sacerdoti, G.; Salvatori, F. Vibro-Acoustic Techniques to Diagnose Power Transformers. *IEEE Trans. Power Deliv.* **2004**, *19*, 221–229. [CrossRef]

3. Zou, L.; Guo, Y.; Liu, H.; Zhang, L.; Zhao, T. A Method of Abnormal States Detection Based on Adaptive Extraction of Transformer Vibro-Acoustic Signals. *Energies* **2017**, *10*, 2076. [CrossRef]
4. Bouayed, K.; Mebarek, L.; Lanfranchi, V.; Chazot, J.-D.; Marechal, R.; Hamdi, M.-A. Noise and vibration of a power transformer under an electrical excitation. *Appl. Acoust.* **2017**, *128*, 64–70. [CrossRef]
5. Masti, R.S.; Desmet, W.; Heylen, W. On the influence of core laminations upon power transformer noise. In Proceedings of the International Conference on Noise and Vibration Engineering (ISMA), Leuven, Belgium, 20–22 September 2004; pp. 3851–3861.
6. Girgis, R.S.; Bernesjo, M.; Anger, J. Comprehensive analysis of load noise of power transformers. In Proceedings of the 2009 IEEE Power & Energy Society General Meeting, Calgary, AB, Canada, 26–30 July 2009; pp. 1–7.
7. Zawieska, W.M. The active control issues related to the noise generated by power transformers. *Mechanics* **2005**, *24*, 155–161.
8. Jancarczyk, D.; Bernaś, M.; Boczar, T. Classification of Low-frequency Signals Emitted by Power Transformers Using Sensors and Machine Learning Methods. *Sensors* **2019**, *19*, 4909. [CrossRef]
9. Orosz, T. Evolution and modern approaches of the power transformer cost optimization methods. *Period. Polytech. Electr. Eng. Comput. Sci.* **2019**, *63*, 37–50. [CrossRef]
10. Jancarczyk, D.; Bernas, M.; Sidzina, M.; Janusz, J. Comparative Analysis of Infrasound Noise Emitted by Power Transformers. In *Scientific Papers Nr 59*; Faculty of Electrical and Control Engineering Gdansk University of Technology: Gdańsk, Poland, 2018. (In Polish)
11. Jancarczyk, D. Research and Analysis of Infrasound Noise Emitted by Power Transformers. *Electr. Eng. Pozn. Univ. Technol. Acad. J.* **2018**, *95*, 153–161. (In Polish)
12. Piana, E.A.; Roozen, N.B. On the Control of Low-Frequency Audible Noise from Electrical Substations: A Case Study. *Appl. Sci.* **2020**, *10*, 637. [CrossRef]
13. Li, A.; Yang, X.; Dong, H.; Xie, Z.; Yang, C. Machine Learning-Based Sensor Data Modeling Methods for Power Transformer PHM. *Sensors* **2018**, *18*, 4430. [CrossRef]
14. Sikorski, W. Development of Acoustic Emission Sensor Optimized for Partial Discharge Monitoring in Power Transformers. *Sensors* **2019**, *19*, 1865. [CrossRef] [PubMed]
15. Kunicki, M.; Wotzka, D. A Classification Method for Select Defects in Power Transformers Based on the Acoustic Signals. *Sensors* **2019**, *19*, 5212. [CrossRef] [PubMed]
16. Barber Bayesian, D. *Reasoning and Machine Learning*; Cambridge University Press: Cambridge, UK, 2012.
17. Mahdavinejad, M.S.; Rezvan, M.; Barekatain, M.; Adibi, P.; Barnaghi, P.; Sheth, A.P. Machine learning for internet of things data analysis: A survey. *Digit. Commun. Netw.* **2018**, *4*, 161–175. [CrossRef]
18. Murphy, K.P. *Machine Learning: A Probabilistic Perspective*; MIT Press: Cambridge, MA, USA, 2012.
19. Jagadish, H.V.; Ooi, B.C.; Tan, K.L.; Yu, C.; Zhang, R. Idistance: An adaptive b^+-tree based indexing method for nearest neighbor search. *ACM Trans. Database Syst. (TODS)* **2005**, *30*, 364–397. [CrossRef]
20. Zhang, H. The optimality of naive bayes. *Am. Assoc. Artif. Intell.* **2004**, *1*, 3.
21. Cortes, C.; Vapnik, V. Support-vector networks. *Mach. Learn.* **1995**, *20*, 273–297. [CrossRef]
22. Scholkopf, B.; Smola, A.J. *Learning with Kernels: Support Vector Machines, Regularization, Optimization, and Beyond*; MIT Press: Cambridge, MA, USA, 2001.
23. Breiman, L. Random forests. *Mach. Learn.* **2001**, *45*, 5–32. [CrossRef]
24. Breiman, L. Bagging predictors. *Mach. Learn.* **1996**, *24*, 123–140. [CrossRef]
25. Glorot, X.; Bengio, Y. Understanding the difficulty of training deep feedforward neural networks. In Proceedings of the 13th International Conference Artificial Intelligence and Statistics (AISTATS), Sardinia, Italy, 13–15 May 2010; pp. 249–256.
26. Eberhart, R.C. *Neural Network PC Tools: A Practical Guide*; Academic Press: Cambridge, MA, USA, 2014.
27. He, K.; Zhang, X.; Ren, S.; Sun, J. Deep residual learning for image recognition. In Proceedings of the IEEE Conference on Computer Vision and Pattern Recognition, Las Vegas, NV, USA, 27–30 June 2016; pp. 770–778.
28. LeCun, Y.; Bengio, Y.; Hinton, G. Deep learning. *Nature* **2015**, *521*, 436–444. [CrossRef]
29. Hotelling, H. Analysis of a complex of statistical variables into principal components. *J. Educ. Psychol.* **1933**, *24*, 417. [CrossRef]
30. Jolliffe, I. *Principal Component Analysis*; Wiley Online Library: Hoboken, NJ, USA, 2002.
31. Abdi, H.; Williams, L.J. Principal component analysis. *Wiley Interdiscip. Rev. Comput. Stat.* **2010**, *2*, 433–459. [CrossRef]

32. Bro, R.; Smilde, A.K. Principal component analysis. *Anal. Methods* **2014**, *6*, 2812–2831. [CrossRef]
33. Hotelling, H. Relations between two sets of variates. *Biometrika* **1936**, *28*, 321–377. [CrossRef]
34. Bach, F.R.; Jordan, M.I. Kernel independent component analysis. *J. Mach. Learn. Res.* **2002**, *3*, 1–48.
35. Blum, W.; Burghes, D.; Green, N.; Kaiser-Messmer, G. Teaching and learning of mathematics and its applications: First results from a comparative empirical study in england and Germany. *Teach. Math. Appl. Int. J. IMA* **1992**, *11*, 112–123. [CrossRef]
36. Schmidhuber, J. Deep learning in neural networks: An overview. *Neural Netw.* **2015**, *61*, 85–117. [CrossRef]
37. Bernas, M.; Placzek, B. Period-aware local modelling and data selection for time series prediction. *Expert Syst. Appl.* **2016**, *59*, 60–77. [CrossRef]
38. Rabiner, L.R. A tutorial on hidden markov models and selected applications in speech recognition. *Proc. IEEE* **1989**, *77*, 257–286. [CrossRef]
39. Sejnowski, T.J.; Rosenberg, C.R. Parallel networks that learn to pronounce english text. *Complex Syst.* **1987**, *1*, 145–168.
40. Kalman, R.E.; Bucy, R.S. New results in linear filtering and prediction theory. *J. Basic Eng.* **1961**, *83*, 95–108. [CrossRef]
41. Lafferty, J.; McCallum, A.; Pereira, F. Conditional random fields: Probabilistic models for segmenting and labeling sequence data. In Proceedings of the Eighteenth International Conference on Machine Learning, ICML, Williamstown, MA, USA, 28 June–1 July 2001; Volume 1, pp. 282–289.
42. Williams, R.J.; Zipser, D. A learning algorithm for continually running fully recurrent neural networks. *Neural Comput.* **1989**, *1*, 270–280. [CrossRef]
43. McCallum, A.; Freitag, D.; Pereira, F. Maximum Entropy Markov Models for Information Extraction and Segmentation. In Proceedings of the 17th International Conference on Machine Learning, ICML, Stanford, CA, USA, 29 June–2 July 2000; Volume 17, pp. 591–598.
44. Baltrušaitis, T.; Ahuja, C.; Morency, L. Multimodal Machine Learning: A Survey and Taxonomy. *IEEE Trans. Pattern Anal. Mach. Intell.* **2018**, *41*, 423–443. [CrossRef] [PubMed]
45. Welch, P. The use of fast Fourier transform for the estimation of power spectra: A method based on time averaging over short, modified periodograms. *IEEE Trans. Audio Electr.* **1967**, *15*, 70–73. [CrossRef]
46. Bezdec, J.C. *Pattern Recognition with Fuzzy Objective Function Algorithms*; Plenum Press: New York, NY, USA, 1981.
47. Deep, K.; Singh, K.P.; Kansal, M.L.; Mohan, C. A real coded genetic algorithm for solving integer and mixed integer optimization problems. *Appl. Math. Comput.* **2009**, *212*, 505–518. [CrossRef]
48. Deb, K. An efficient constraint handling method for genetic algorithms. *Comput. Methods Appl. Mech. Eng.* **2000**, *186*, 311–338. [CrossRef]
49. Kennedy, J.; Eberhart, R. Particle Swarm Optimization. In Proceedings of the IEEE International Conference on Neural Networks, Perth, Australia, 27 November–1 December 1995; pp. 1942–1945.
50. Mezura-Montes, E.; Coello, C.A.C. Constraint-handling in nature-inspired numerical optimization: Past, present and future. *Swarm Evol. Comput.* **2011**, *1*, 173–194. [CrossRef]
51. Pedersen, M.E. *Good Parameters for Particle Swarm Optimization*; Hvass Laboratories: Luxembourg, 2010.
52. Escalera, S.; Pujol, O.; Radeva, P. On the decoding process in ternary error-correcting output codes. *IEEE Trans. Pattern Anal. Mach. Intell.* **2010**, *32*, 120–134. [CrossRef]

Article

Application of Correlation Analysis for Assessment of Infrasound Signals Emission by Wind Turbines

Tomasz Boczar, Dariusz Zmarzły, Michał Kozioł * and Daria Wotzka

Faculty of Electrical Engineering Automatic Control and Informatics, Opole University of Technology, 45-758 Opole, Poland; t.boczar@po.edu.pl (T.B.); d.zmarzly@po.edu.pl (D.Z.); d.wotzka@po.edu.pl (D.W.)

* Correspondence: m.koziol@po.edu.pl

Received: 28 October 2020; Accepted: 30 November 2020; Published: 2 December 2020

Abstract: The study reported in this paper is concerned with areas related to developing methods of measuring, processing and analyzing infrasound noise caused by operation of wind farms. The paper contains the results of the correlation analysis of infrasound signals generated by a wind turbine with a rated capacity of 2 MW recorded by three independent measurement setups comprising identical components and characterized by the same technical parameters. The measurements of infrasound signals utilized a dedicated measurement system called INFRA, which was developed and built by KFB ACOUSTICS Sp. z o.o. In particular, the scope of the paper includes the results of correlation analysis in the time domain, which was carried out using the autocovariance function separately for each of the three measuring setups. Moreover, the courses of the cross-correlation function were calculated separately for each of the potential combinations of infrasound range recorded by the three measuring setups. In the second stage, a correlation analysis of the recorded infrasound signals in the frequency domain was performed, using the coherence function. In the next step, infrasound signals recorded in three setups were subjected to time-frequency transformations. In this part, the waveforms of the scalograms were determined by means of continuous wavelet transform. Wavelet coherence waveforms were calculated in order to determine the level of the correlation of the obtained dependencies in the time-frequency domain. The summary contains the results derived from using correlation analysis methods in the time, frequency and time-frequency domains.

Keywords: infrasound measurement system; wind turbine; infrasound correlation analysis

---

1. Introduction

The task of the adequate recording of infrasound signals generated by sources of emission poses a relatively difficult measurement task in practical application. This is attributable to the possibility of a number of potential sources responsible for generating acoustic signals in the low frequency bandwidth, including infrasound that accompanies normal operation of wind turbines. Such sources include noise caused by vehicle traffic, agricultural machines or passing trains, as well as natural ones caused by waves of the water surface or blowing winds that sweep obstacles on its way. We can bear in mind that most often wind turbines operate as part of wind farms, which include from several to even several thousand individual generators. Moreover, even a few wind farms can be located in a relatively small area in relation to the possible range of the emitted infrasound. Additionally, location-specific conditions occur in which a given wind farm comprises turbines with different technical design, made by various manufacturers, with different dimensions and capacities, which may affect the differentiation of the infrasound emission. There are wind farms in which turbines with various service lives also operate side by side. For instance, the study reported in the article [1] focused on the significant differences in the results of infrasound and acoustic noise measurements in the audible band, which were recorded close to wind farms, depending on the type of the supporting structure applied in a given turbine. The authors presented the results of measurements that were carried out in the vicinity of wind turbines with different supporting structures (in the form of a truss or tubular structure) and of different heights. On the basis of the results of a comparative analysis, it was found that wind turbines with tall towers built with trusses emit much lower noise levels than ones with towers with a tubular design, and this applies to both the audible bandwidth (the level is approximately 10 dB lower) and infrasound range (a few dB). Their level was on average 10 dB higher than the background noise level, both for audible and infrasound noise levels. However, the article [2] presents the results of a comparative analysis of infrasound generated by wind turbines equipped with a synchronous and asynchronous generator, and the article [3] assesses the impact of a number of metrological parameters on the results obtained in this respect.

We can also note that the length of the propagation of infrasound waves in the air ranges from 17 m to 340 m, which directly determines the actual effect of obstacles on wave propagation in an open space. Hence, any objects whose dimensions are smaller than the length of the propagating infrasound wavelengths do not pose an obstacle. This phenomenon leads to the inconsiderable damping of infrasound signals during their propagation in the air that results only from the distance between the source and the receiver. Therefore, infrasound waves have good propagation characteristics, and their interaction at the lowest frequency values is possible even over distances of tens of kilometers. Relatively poor damping, wavelength and frequency, combined with the possibility of standing waves that can be formed in field specific conditions, as well as the possible resonance phenomenon lead to objective difficulties in the unambiguous and adequate location of the source of infrasound generation. An important element is also related with the need to take into account the acoustic background during the measurements of infrasound emitted by wind turbines, the level of which, in many cases, may be close to the useful signals [4–8]. In particular, this applies to the conditions when wind speeds with values above 12–15 m/s occur during infrasound recording. In the case of wind farms comprising many wind turbines, there is usually objective difficulty in measuring the noise background, as it requires stalling all operating installations by the investor or the occurrence of wind less conditions when the wind speed is below the value when turbines can start (usually below 3 m/s).

Another important issue is also associated with the lack of identical, and in many countries a complete lack of normatively specified values with the levels of permissible long-term exposition to infrasound noise in the working environment. Moreover, these levels are constantly variable, and in many regions or provinces, local regulations and laws are enforced, which have been commonly developed under the pressures of local communities. On the other hand, the issues of infrasound noise occurring in the generally accessible environment are practically not subjected to legal regulations. Additionally, there is no single common reference method of measuring and analyzing infrasound signals. In this regard, the measurement methodology specified in the IEC 61400-11 standard is employed, and it was developed with the purpose of measuring acoustic signals emitted by wind turbines in the audible range. Only the regulations contained in Annex A.2 offer the potential to extend the bandwidth to infrasound range, but these provisions do not specify details of the procedure to be applied for infrasound measurements. This standard describes a procedure for determination of the acoustic power level of wind turbines for different wind speeds on the basis of registered changes in the acoustic pressure level. On the other hand, the authors of the article [9] argue that the currently applied techniques and methodology for measuring and analyzing acoustic noise tend to obscure the ratio of low-frequency impulse noise and infrasound in the generated spectrum of signals emitted by wind turbines. It has also been raised that the widespread use of a level A-weighting filter for infrasound analysis, which is normatively used to assess noise levels in the audible range, instead of a level G-weighting filter, forms an inadequate tool and constitutes an unreliable indicator of this assessment [10].

We can note that the problem of the potential adverse effects of infrasounds generated by the operation of wind turbines to the health of people living in their vicinity may result in the lack of consent of local communities to the installation of wind farms, even at relatively large distances from residential areas. As a consequence, a slowdown and in extreme situations even an inhibition of the development of onshore wind energy can follow this. People's fears and, consequently, opposition of local communities, in extreme cases may even lead to the closure of operating wind farms by decision of local authorities.

Therefore, the task of developing adequate techniques of measuring and analyzing infrasound signals generated by wind turbines and assessing their potential effects on human health still poses a significant problem not only from the cognitive and scientific point of view. We can remark that this problem varies in degree depending on the country, and above all is largely determined by a derivative of the state of knowledge and awareness of the local community, which results mainly from the information policy conducted in this area by the state or, respectively, by local authorities on a specific local community scale.

The level of social acceptance of working wind turbines and projected new investments varies from country to country. We need to be aware that this acceptance is guided by a number of derivative unmeasurable and subjective factors. To a large extent, public opinion is formed by the publicity focusing on the potential adverse effect associated with the operation of wind turbines on a broadly understood human health.

The positions presented in various countries and the results of the research work carried out in this area were presented in articles [10–38].

The purpose of the research, whose representative results are reported in this paper, was to determine the scope and applicability of correlation analysis in the time, frequency and time-frequency domains in the analysis of infrasound signals generated by a wind turbine registered in a system consisting of three independent measurement setups comprising identical elements. The use of correlation analysis offers the possibility to verify whether and how infrasound waves propagating around the investigated wind turbine are the same, and if they differ and if they are dependent on the direction of the wind.

2. Wind Turbine under Study

The object of this investigation was a single, three-bladed V110 wind turbine manufactured by Vestas with a rated electrical capacity of 2.0 MW, which has been in service for over 6 years. The diameter of the turbine rotor is 110 m, the sweeping area is 9.503 m^2 and the tower height is 120 m. The speed at which the start-up and electricity production is initiated is equal to 4.0 m/s. However, the nominal wind speed at which the rated power of 2.0 MW is achieved is 12 m/s. For safety reasons, automatic shutdown of the turbine occurs for the wind speeds of 21 m/s. This turbine design has been comprehensively tested by the manufacturer in terms of the generated acoustic signals in the audible range, and the maximum value of the sound intensity emitted during their operation is 107.6 dB. The turbine has a 690 V four-pole asynchronous generator.

The investigated turbine is located in central Poland, in a lowland area, over 1800 m from the nearest built-up areas, and the nearest tarmac road at a distance of around 2 km. Due to the low acoustic nuisance and the relatively large distance, this environment forms a source of a low level of acoustic background in the infrasound range.

3. Characteristics of Measurement System

Dedicated measurement system called INFRA was built by KFB ACOUSTICS (KFB Acoustics Sp. z o.o., Wrocław, Poland) and was subsequently applied to for the measurements of infrasound signals generated during the operation of the wind turbine. The system was designed for the purposes of the scientific project entitled "Numerical and experimental analysis of low-frequency acoustic phenomena generated during the operation of wind turbines" funded by the National Science Center. The system used enables measurements of acoustic signals and wind speed and direction simultaneously by application of three independent and identical measurement setups. Registrations are synchronized in time and measurement data are transmitted wirelessly using Wi-Fi technology. This way, the range of distances for which measurements can be performed can be feasibly increased, and the necessity to use connection cables which are troublesome in field conditions is eliminated. It is possible to record infrasound, comparative in any three directions in relation to the wind turbine under study.

The standard and common TCP/IP (Transmission Control Protocol/Internet Protocol) transmission protocol was applied for the transfer of measurement data. Depending on the type of applied measuring sensors, the INFRA system can be utilized both for recording acoustic signals in the audible range and in the low frequency bandwidth, including infrasound. The system consisted of the following functional elements: A base station, three independent and identical measurement setups, a weather station and a laptop. The software part of the system comprised dedicated INFRA v. 1.2 software, which offers archiving and pre-processing of recorded on-line measurement data, as well as provides the operator with the option of setting measurement parameters. The intuitive graphical interface enables the user to view real-time waveforms of infrasound signals recorded by the three independent setups, as well as instantaneous wind speed and direction values. The export of recorded data occurs in the *.mat file format and was followed by its further processing in the Matlab environment.

The measuring station, which is presented in Figure 1, included the following functional elements: Wi-Fi router Ubiquiti BULLET M2 (Ubiquiti Networks, New York, NY, USA), omnidirectional ProEter10 CyberBajt antenna (Cyberbajt, Zamość, Poland) and a battery power supply system. Throughout the measurements, the antenna was directly coupled with the router via a dedicated connector, thus a 2.4 GHz Wi-Fi access point was created. The router comprises an Atheros MIPS 24KC, 400 MHz microprocessor and the maximum RF output power (TX Power) is 30 dBm. It is contained in a special case resistant to moisture, high temperature differences, dust and mechanical damage, i.e., to conditions that may arise during measurements performed in difficult field conditions, in which wind turbines are often installed.

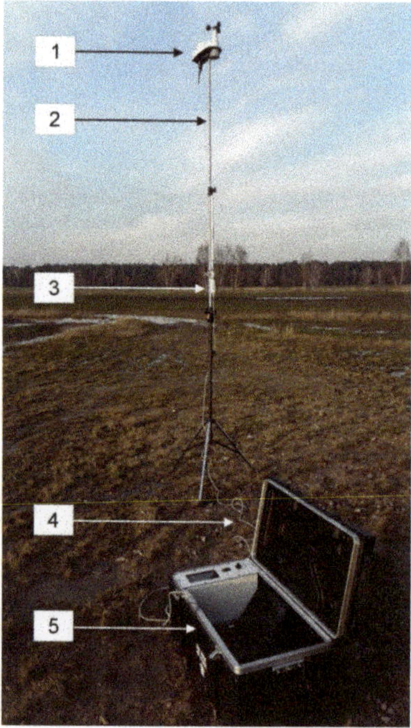

Figure 1. View of measuring station and weather stations, where: 1—Vantage Vue 6250EU weather station; 2—telescopic tripod; 3—ProEter104 omnidirectional antenna; 4—connector linking antenna to router and 5—measurement case with Wi-Fi 2.4 GHz access point.

With the purpose of ensuring a reliable and long-distance wireless connection with measurement stations, a vertical polarization antenna with 10 dBi energy gain was used, which was designed to operate in the 2.4 GHz bandwidth. The antenna used was dedicated to locations with a large number of other radio networks. Moreover, the large beam width in the vertical plane of 230 for −3 dB enables the implementation of connections in the case of even relatively large differences in height between the data collection system and individual setups. This can occur relatively frequently during measurements performed in field conditions. During the measurements, the antenna was placed on a 2.5 m high mast, which ensured a secure and stable connection with the antennas of measuring stations in the range of up to 300 m, with a real data transmission speed of 100 Mbps +, regardless of weather conditions and terrain type.

Additionally, in accordance with the recommendations of the PN-EN 61400-11 standard, the Davis Vantage Vue 6250EU (Davis Instruments, Hayward, CA, USA) wireless weather station was utilized for comparison purposes between infrasound signals and weather conditions accompanying measurements. During the registrations, its location was selected in accordance with the guidelines contained in the standard, i.e., in front of the turbine at a distance of 2D = 220 m (where D is the rotor diameter of the turbine). Wind speed and direction were measured both at a height of 10 m using a telescopic mast in accordance with the standard, and comparatively at a height of around 3 m using a telescopic tripod, similarly to the procedures applied for measurement setups. Its application offers air temperature measurements in the range from −40 °C to +65 °C with an accuracy of ±0.10 °C; relative humidity in the range from 0 to 100% with an accuracy of 1%; atmospheric pressure in the range from 540 hPa to 1100 hPa, with a bias of 0.1 hPa; wind speeds in the range from 1 m/s do 80 m/s (320 km/h) with an accuracy of 0.1 m/s (1 km/h) and its direction in the range from 0° to 360° with an accuracy of 10°; as well as the amount of liquid precipitation in the range from 1 to 1016 mm/h, with an accuracy of 0.2 mm. It is also possible to determine derivative parameters, such as: Dew point temperature and felt temperature. The system of external sensors, which the weather station was equipped with, communicated with the console wirelessly (frequency: 868 MHz) for a distance of up to 300 m in an open space, and following the use of additional signal amplifiers and antennas, the feasible range of distances cam increase up to 1 to 2 km. Throughout the measurements, the measurement station was connected to the DELL Latitude E7270 laptop via an Ethernet cable with the installed INFRA v. 1.2 software.

Each of the three measurement setups comprised of the following components: A measuring microphone; measurement card, battery power supply system, weather station, wind direction sensor, anemometer and external Wi-Fi antenna (Figures 3 and 4). For the measurement of infrasound emitted by wind turbines, pre-polarized a GRAS Sound & Vibration A/S free field GRAS 46AZ (Gras Sound & Vibration, Holte, Denmark) condenser microphone was employed. In this type of microphones, variations in acoustic pressure led to the diaphragm vibrations, which also forms the movable lining of the condenser. Hence, the capacitance value changes proportionally to the changes in the pressure of the acoustic wave. As the microphone is pre-polarized and its linings are connected to each other by a resistor, proportional voltage variations were obtained at the microphone output. Their main purpose was diagnostic measurements of acoustic signals in the low frequency band from 0.5 Hz to 20 kHz. These microphones meet the standards set out in the IEC 61094 WS3F standard. Moreover, they have an impulse response that is optimized for pressure measurement in a dispersed diffusion field. The values of the basic technical parameters of the microphones used are presented in Table 1.

We should note that the applied microphones were integrated via a standard BNC connector with the 26CI type preamplifiers with a gain of −0.35 dB (gain: −0.35 dB). They are characterized by a very low inherent noise level, which is typically 3.5 μV (Linear: 20 Hz–20 kHz), high dynamics and transmission bandwidth from 1 Hz to 200 kHz (±0.2 dB) (frequency range). In addition, they have a very high input impedance of 40 GΩ/0.4 pF and are effectively shielded by an annular shield to minimize the effects of dispersed capacity and direct interference from the coupled microphone.

Table 1. Summary of basic parameters of GRAS 46AZ microphone.

Parameter	Unit	Value
Frequency range (±1 dB)	Hz	1–10 k
Frequency range (±2 dB)	Hz	0.5–20 k
Sensitivity	dB(A)	17
Dynamic range lower limit with GRAS preamplifier	dB	138
Dynamic range upper limit with GRAS preamplifier	mV/Pa	50
Nominal sensitivity at 250 Hz	mA	2–20
Input current (CCP)	°C	od −30 do +70

For the purposes of effective wireless communication using Wi-Fi technology between three measuring setups and the base station, a Cyberbajt directional, microband antenna, LineEter 19 type (Cyberbajt, Zamość, Poland), with an energy gain of 19 dBi, was used for operation in the frequency bandwidth from 2.4 to 2.5 GHz. Its horizontal beam width in the horizontal plane is 250 for −3 dB and, respectively, 200 for −3 dB for the vertical plane.

The measuring station has a dedicated case made of aluminum and hardened polypropylene, which is waterproof, dustproof and impact resistant. The measuring case has been integrated with the cDAQ-9191 measurement module (National Instruments Corp., Austin, TX, USA), to which all measurement connectors have been connected and the four-channel, 24-bit NI 9234 measurement card (National Instruments Corp., Austin, TX, USA). The card used has built-in anti-aliasing filters for each of the four channels as standard, which can automatically adjust to the current sampling frequency. It is noteworthy that the maximum value of the sampling frequency is equal to 51.2 kS/s.

The noise parameters of the measurement card in the idle state and the noise density for the sampling frequency were respectively: 97 dBFS (50 μV_{rms}) for channel noise and 310 nV/$\sqrt{Hz}$ for noise density. The self-noise density of the entire measuring system was also analyzed. The obtained results showed insignificant noise in measuring channels (Figure 2), which was taken into account when analyzing the measured signals.

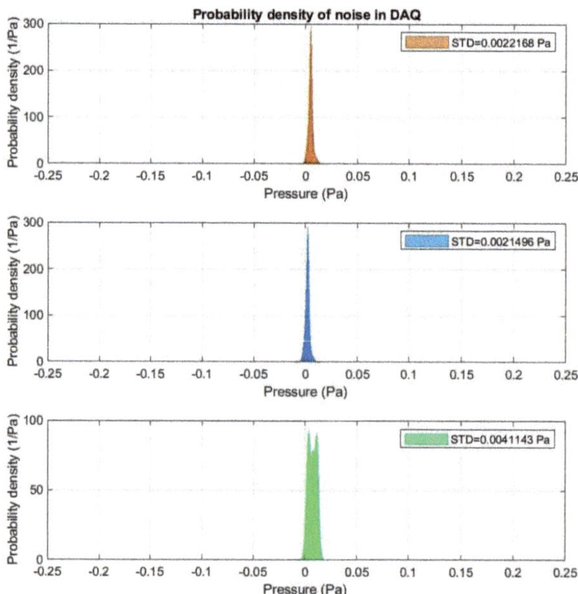

Figure 2. Probability density of noise in data acquisition system (DAQ): Orange color—measuring setup number 1; blue color—measuring setup number 2 and green color—measuring setup number 3.

The wind direction sensor, anemometer and an external Wi-Fi antenna were installed on a custom-made two-armed aluminum stand. The accuracy of the anemometer is ±0.1 m/s, and the wind speed measurement was in the range from 0 m/s to 25 m/s. In addition, the accuracy of the wind vane was in the range of ±10° in the range from 0° to 360° (Figure 3).

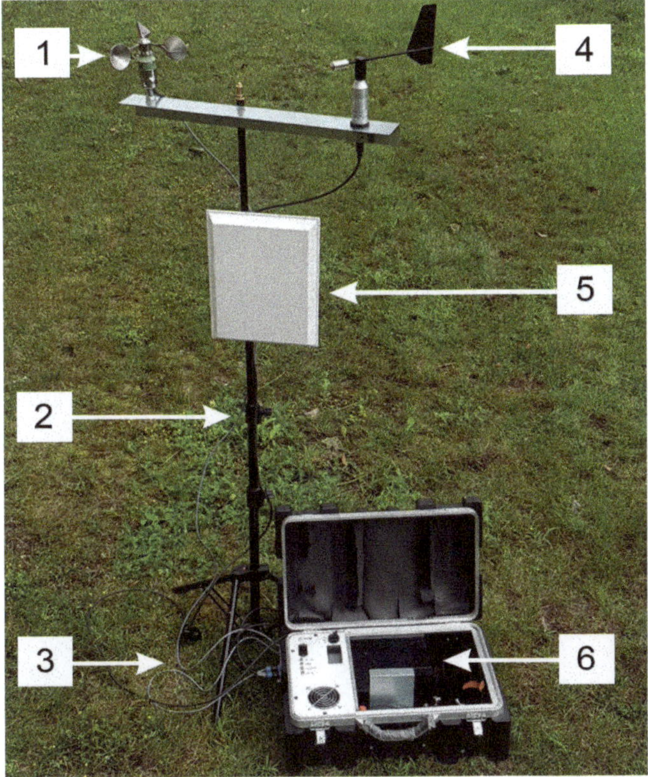

Figure 3. View of measuring station, where: 1—anemometer; 2—telescopic measurement tripod; 3—connector cables; 4—wind vane; 5—directional antenna and 6—measuring case.

In accordance with the recommendations found in the IEC/EN 61400–11 standard that applies to measurement of audible noise generated by wind turbines, two windscreens were used for recording infrasound signals. In this way, the influence of wind gusts directly on the microphone diaphragm could be effectively eliminated, and thus their influence on the obtained results of recording was excluded.

The first one was the internal cover by Brüel & Kjær, type UA-0207 (Brüel & Kjær, Naerum, Denmark), made in the shape of a hemisphere with a radius of 4.5 cm, which was placed directly on the active part of the measuring microphone. It was made of a special polyurethane foam with open pores, which is capable to suppress noise caused by wind gusts in the range from 10 to 12 dB, depending on its speed. Its use effectively reduces the excessive air pressure exerted by wind on the measuring microphone (Figure 4).

The second outer windscreen, which was also in the shape of a hemisphere with a radius of 45 cm, is part of the Brüel & Kjær UA-2133 probe set (Figure 5). The screen was made of an aluminum frame on which a nylon dome was placed. The cover was attached to a 1 m diameter round plate made of 12 mm thick waterproof plywood in order to separate the microphone from ground vibrations. In its central part, a microphone was placed in specially made plastic clamps, the measuring tip of which was directed towards the tested wind turbine. The use of the same reflecting surface during all measurements additionally minimized the effect of the soil surface on the recorded waveforms. Moreover, placing the microphone directly on the reflecting surface on the ground level additionally reduces the influence of gusts of wind on the recordings.

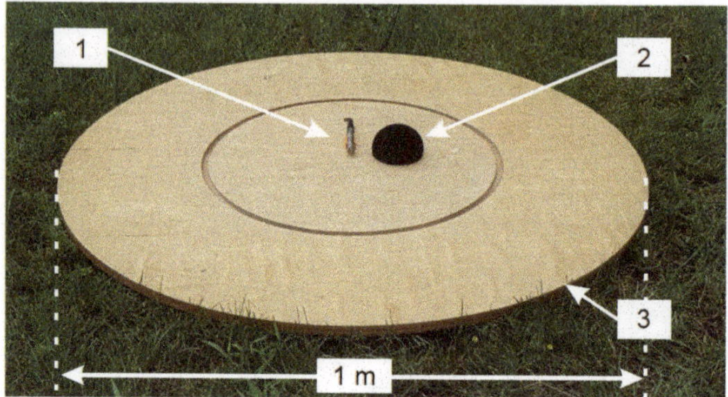

Figure 4. View of measuring setup, where: 1—GRAS 46AZ 26CI measuring microphone z with preamplifier located in a special holder; 2—external windscreen and 3—reflecting board.

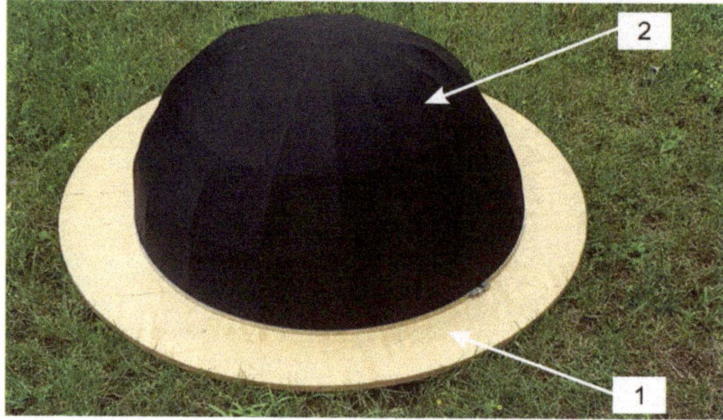

Figure 5. Image of external windscreen (2) installed on the reflecting board (1).

On the basis of the rated data regarding the equipment applied in the measurement system, the value of the type B standard uncertainty was estimated at 0.51. Moreover, the expanded uncertainty of B type was calculated and was equal to 0.92, with the assumed confidence interval of 95.5% [39].

4. Methodology of Measurements

Figure 6 shows the location of three measuring setups in the field (named: MS_1; MS_2; MS_3), which were located at a distance of around 175 m in relation to the investigated wind turbine (denoted as: WIND_TURBINE). There were plowed farming fields around the turbine. There were no obstacles between the measuring stations and the analyzed turbine, and the area was virtually flat and devoid of vegetation.

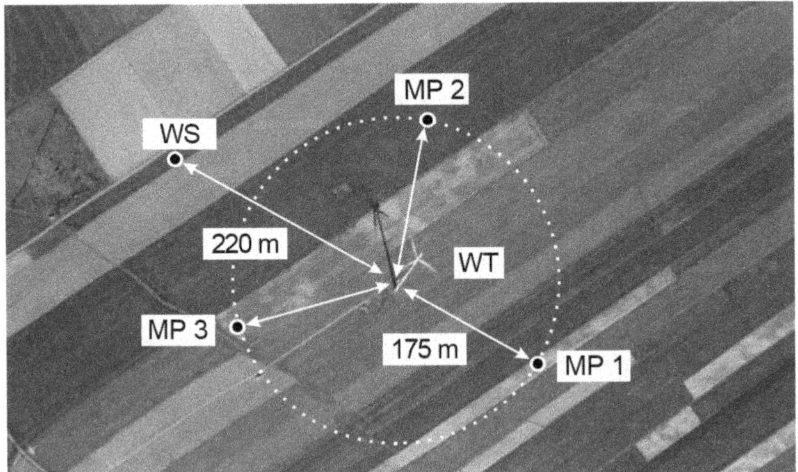

Figure 6. View of the location of the tested wind turbine and the distribution of measurement points, where: MP1-MP3—measurement points; WT—wind turbine under study and WS—weather station.

Figure 7 presents the layout of the measuring point (MP1) and the location of the weather station (WS) in relation to the tested wind turbine, which depends on the rotor diameter, D, and the height of the wind turbine tower, H, and results from the guidelines contained in the IEC/EN 61400–11 standard.

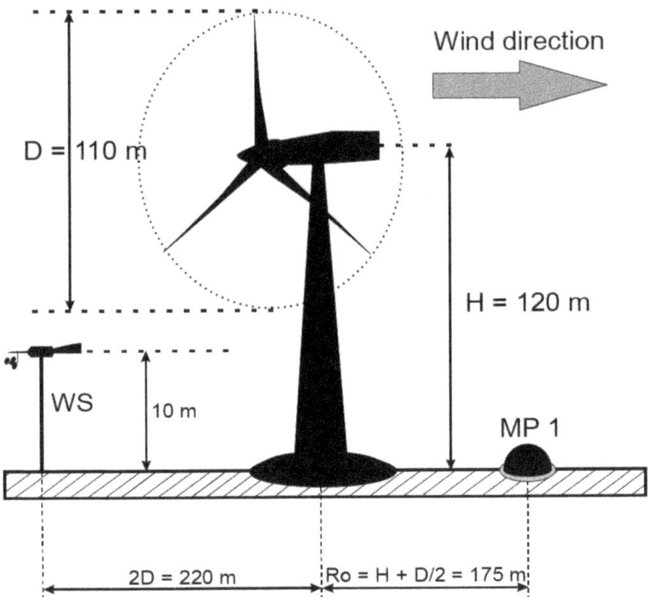

Figure 7. Location of the point of measurement (MP1) and WS in relation to investigated wind turbine on the basis of recommendations of IEC/EN 61400–11 standard, where: D—diameter of wind turbine rotor and H—height of turbine mast.

The infrasound measurements were carried out in early autumn, at the turn of September and October and took eight days. Signals were recorded in series taking total of 15 to 30 min each for every location of the measurement point so as to maintain a constant speed and steady wind direction. The measurements were performed during rainless weather and involved synchronized recording of infrasound signals from three measurement setups located in different locations in relation to the tested wind turbine. Apart from the simultaneous measurement of the emitted acoustic signals from three different measurement points, synchronized recordings of wind direction and speed were performed separately for each of the three measurement setups. The measured physical quantity involved the variations in the noise level, which were recorded simultaneously in three measurement setups. In this way, it was possible to determine the potential range of infrasound impact directly at the location of the measurement sites. The signals were recorded at a sampling rate of 51.2 kS/s. During the measurement days, the following values of the basic weather parameters were recorded: The mean value of the wind speed in the range from 3.4 to 12.3 m/s, atmospheric pressure in the range from 994.7 to 1001.2 hPa, air temperature ranged from 8 to 12 °C and its humidity was within in the range: 73–82%.

In accordance with the recommendations contained in the international ISO 389-7 standard and the recommendations contained in the IEC 61400-11 standard, prior to and after each measurement series, device calibration was performed for a given location and separately for each of the three measuring setups. For this purpose, the analysis applied a class 1 acoustic calibrator manufactured by B&K, type 4231. The calibration signal had a frequency of 1 kHz with a level of 94 dB, with its level stability being ±0.2 dB. On the other hand, the stability of the generated frequency, with distortions less than 1%, was equal to ±0.1%. The calibrator used meets the requirements of IEC 60942 class 1.

On each measurement day, background measurements were performed in the conditions of the stalled wind turbine. The values of reference signals obtained in this way were subtracted from the values of infrasound signals recorded during normal operation of the investigated wind turbine. The effect of the disturbances resulting from vehicular traffic, as well as by agricultural machinery operating in the fields and specialized vehicles used by foresters was eliminated by recording the time of their occurrence in order to remove them later from further analysis of the data recorded during measurements. Additionally, an occurrence of unusual variations in the frequency spectrum of the recorded infrasound and large fluctuations in their dynamics was recorded at a given stage of the analysis, the applied software made it possible to listen to a given section of the recorded signal, which in turn provided the possibility of identifying sources of disturbance and their effective elimination.

5. Results and Discussion

Figure 8 contains an example and representative waveforms of variations in the acoustic pressure occurring over one-minute intervals, which were recorded separated by the three measurement setups. One-minute time intervals were randomly selected from the measurement series with a duration of 15 to 30 min, during which the wind direction and speed were found to be constant during the performed registrations. At the same time, the dependencies presented in the article concern the data recorded for the wind speed of 12.3 m/s, which was the highest value for which case the infrasound signals measurements were performed. We should note that this is the value at which the tested turbine assumes its rated operating parameters. The wind direction was from the northeast.

For the waveforms presented in Figure 8, histograms of acoustic pressures were determined separately for three locations. The obtained distributions are illustrated in Figure 9.

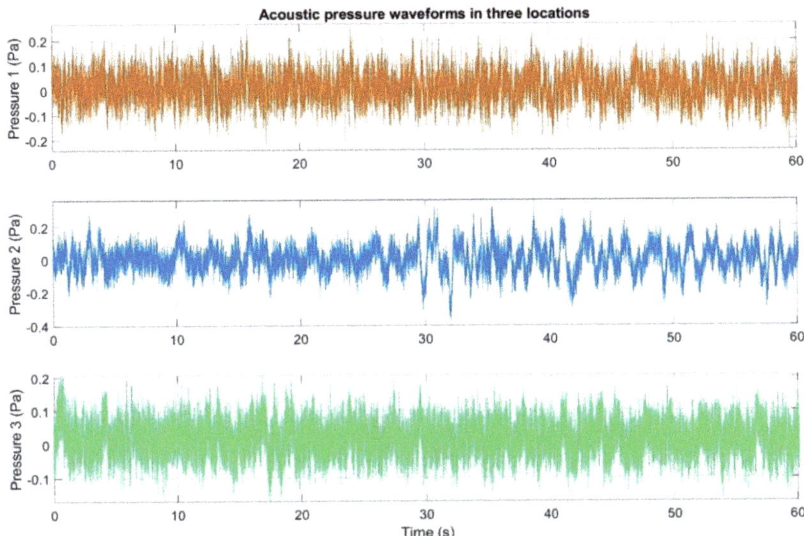

Figure 8. Acoustic pressure waveforms in three different locations measured simultaneously: Orange color—measuring setup number 1; blue color—measuring setup number 2 and green color—measuring setup number 3.

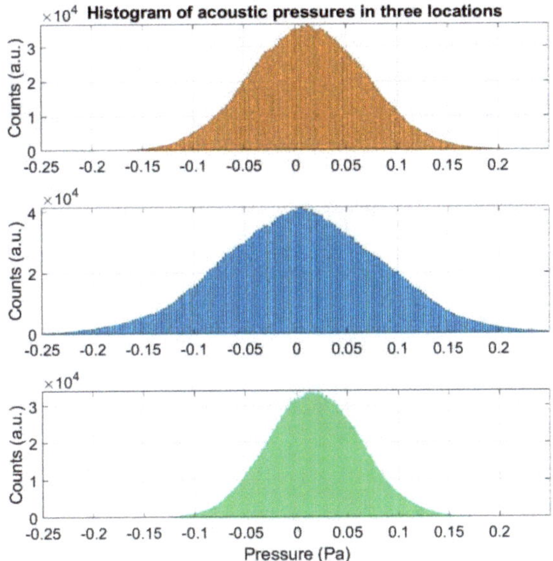

Figure 9. Histogram estimated for acoustic pressure acquired in three different locations at the same time: orange color—measuring setup number 1; blue color—measuring setup number 2 and green color—measuring setup number 3.

The first stage of the analysis involved the recording of infrasound signals by the three measuring setups, whose waveforms are presented in Figure 8, followed by the subsequent correlation analysis in the time domain. For this purpose, the waveforms of the auto-variance function (auto-correlation with the subtracted mean value) were determined separately for each of the three measuring setups

(Figures 10–12). Moreover, the courses of the cross-correlation function were calculated separately for each of the possible combinations of infrasound recorded in the setups (Figures 13–15). The use of correlation analysis in the time domain provides the possibility of identification of the characteristics of the recorded signals, and indicate the ratios of deterministic and stochastic components. Moreover, it can be used to identify the quantities accompanying the noise and interference measurements. Additionally, in order to determine the relations between the acoustic pressure changes for different time shifts, the cross-correlation functions were determined, taking into account subsequent possible combinations of two of the three measuring setups.

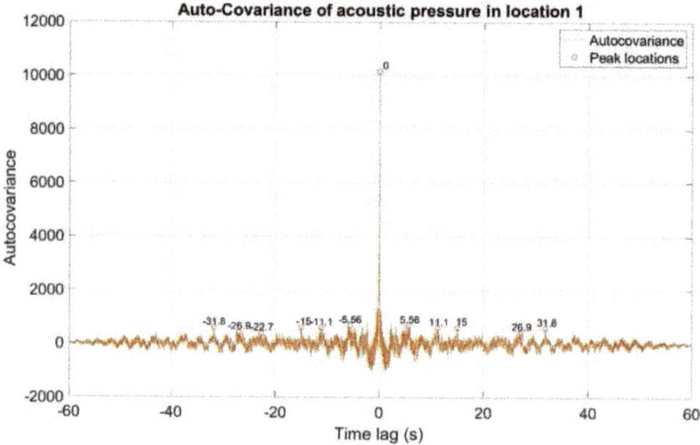

Figure 10. Course of auto-covariance function of infrasound signals generated by the investigated wind turbine, recorded in measuring setup number 1.

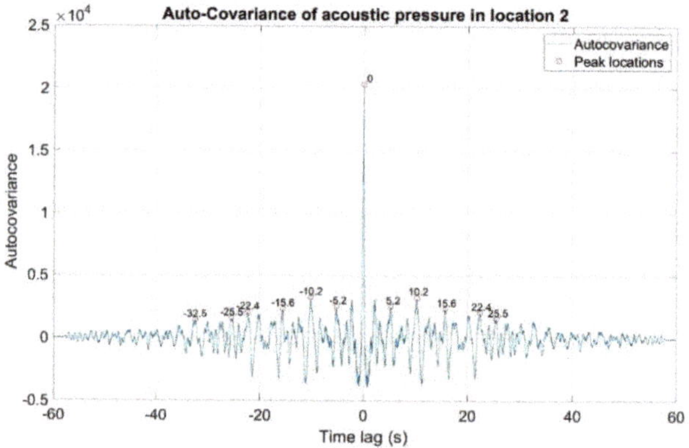

Figure 11. Course of auto-covariance function of infrasound signals generated by the investigated wind turbine, recorded in measuring setup number 2.

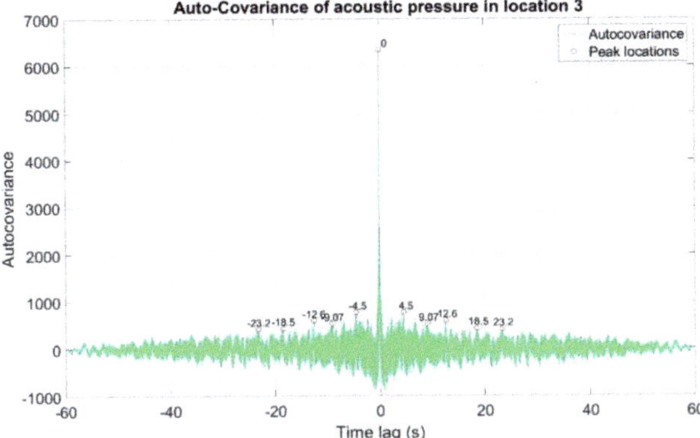

Figure 12. Course of auto-covariance function of infrasound signals generated by the investigated wind turbine, recorded in measuring setup number 3.

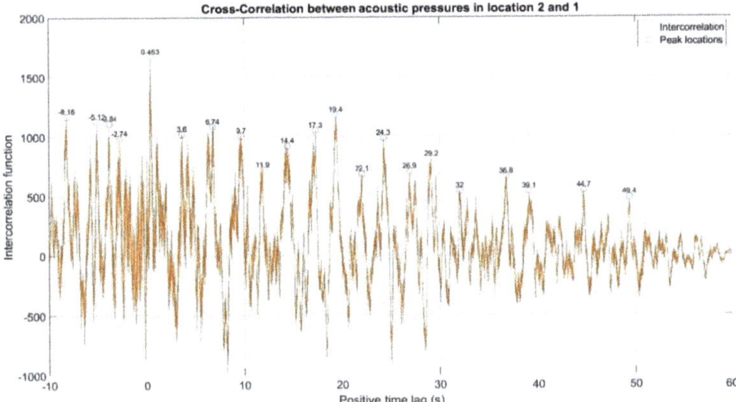

Figure 13. Inter-correlation between acoustic pressures in location 2 and 1. There are several peaks for positive lag. The highest correlation is for lag of 0.45 s.

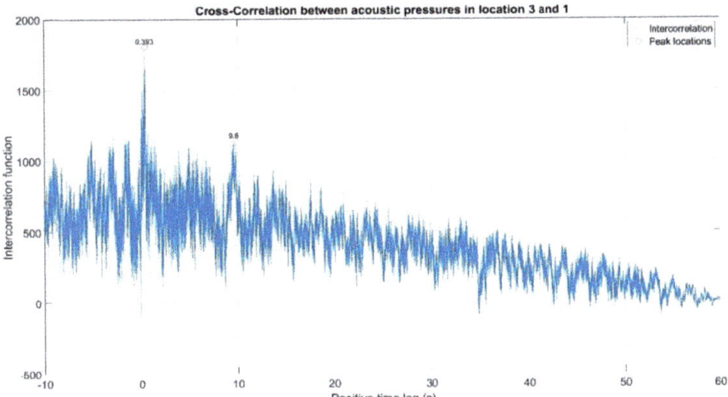

Figure 14. Inter-correlation between acoustic pressures in location 3 and 1. There are few peaks for positive lag. The highest correlation is for lag of 0.39 s.

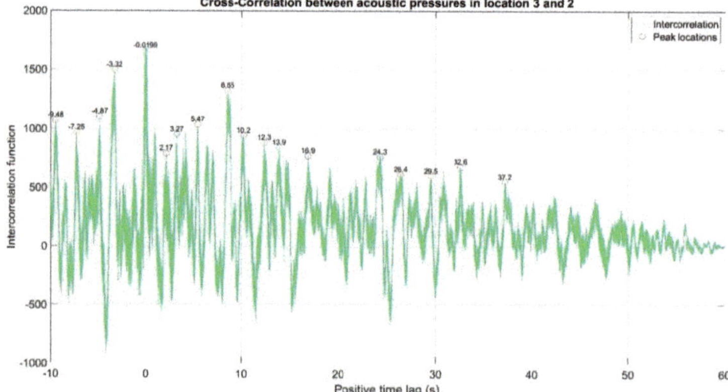

Figure 15. Inter-correlation between acoustic pressures in location 3 and 2. There are several peaks for positive lag. The highest correlation is for lag of −0.02 s.

The second stage involved the analysis of the recorded infrasound signals in the frequency domain that was carried out by determining the waveforms of the power density spectra using the Welch method and Hamming window from the length N = 512e^3. For this purpose, computational scripts that were developed in the Matlab programming environment. In detail, the methodology of calculations and the used mathematical relationships proposed by P.D. Welch for the estimation of power density spectra, e.g., in [40]. We should emphasize that this method was used in this article primarily to reduce the level of noise in the estimated power density spectra, which is the case when using the standard fast Fourier transform and to calculate the power density spectra or energy, respectively.

We can also remark that the relations developed for lower wind speeds do not differ in terms of quality, as the characteristics of the determined waveforms of frequency spectra were maintained. However, lower amplitude values of the recorded infrasound signals were recorded. Figure 16 contains averaged waveforms of power density spectra, which were determined separately for each of the three setups.

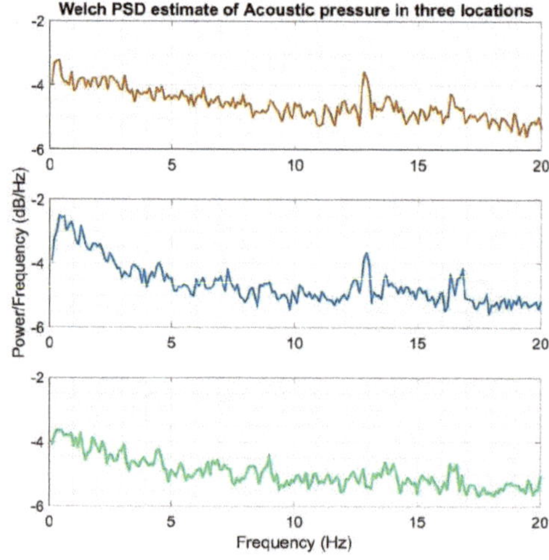

Figure 16. Welch power spectral densities estimated for acoustic pressure acquired in three different locations at the same time: orange color—measuring setup number 1; blue color—measuring setup number 2 and green color—measuring setup number 3.

Subsequently, coherence function was used in order to determine the similarities in the frequency domain for the recorded infrasound signals, whose waveforms are presented in Figures 17–19. The analysis was carried out by comparing the obtained results, pairing, successively, two measuring setups.

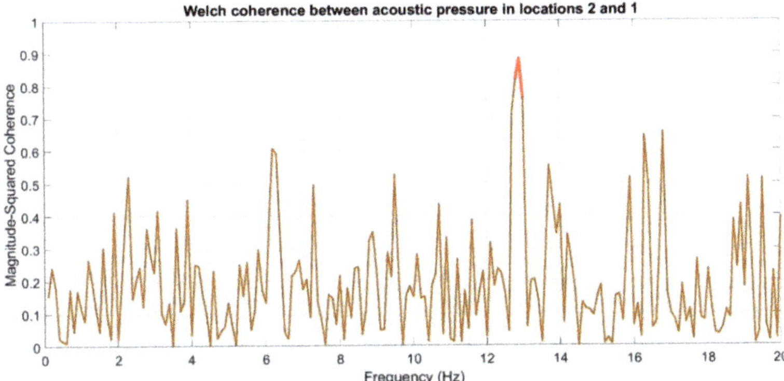

Figure 17. Coherence estimate via the Welch method for acoustic pressures in locations 2 and 1. The coherence above 0.75 is marked with red line.

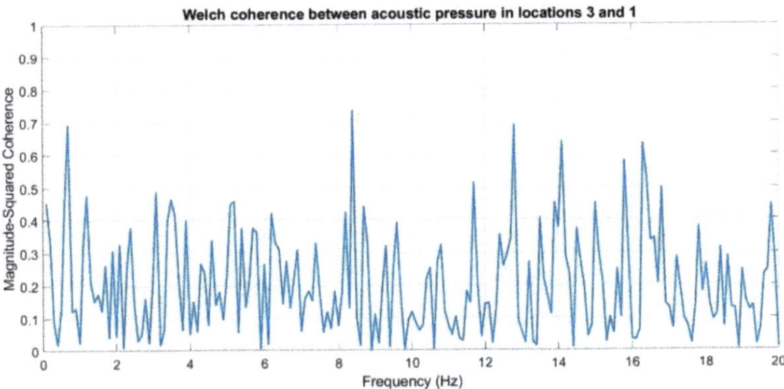

Figure 18. Coherence estimate via the Welch method for acoustic pressures in locations 3 and 1.

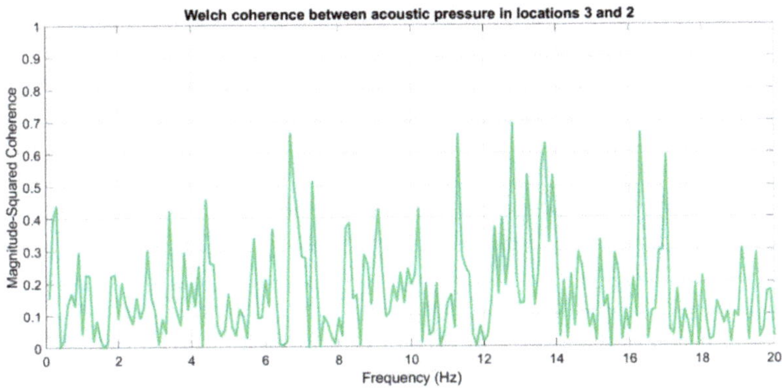

Figure 19. Coherence estimate via the Welch method for acoustic pressures in locations 3 and 2.

In the following stage, infrasound signals recorded in three measuring setups were subjected to time-frequency transformations. The waveforms of the scalograms presented in Figures 20–22 were determined using a continuous wavelet transform. Its use offers an increase the time-frequency bandwidth when the results are compared to the STFT (Short-time Fourier transform) transform, as it enables the use of narrow observation windows at high frequencies coupled with sufficiently wide for low frequencies. A Morlet wavelet was used as the base wavelet for determining the scalograms.

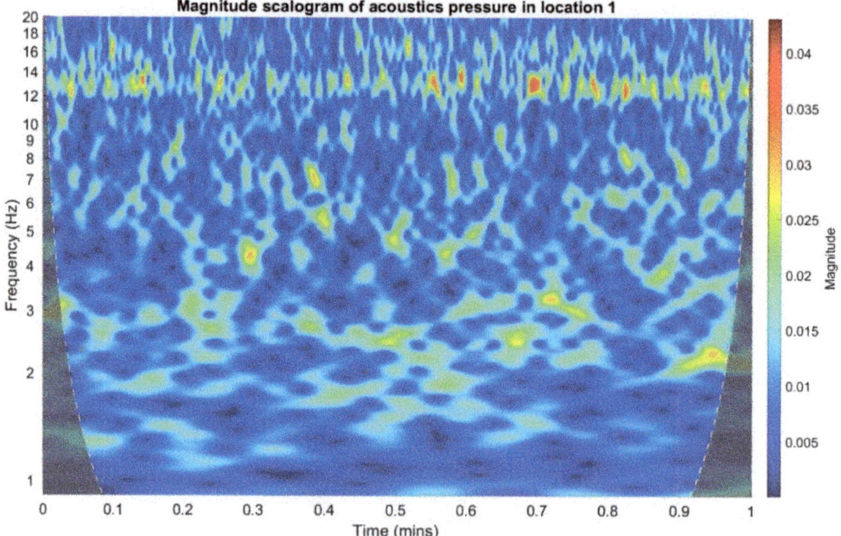

Figure 20. Magnitude scalogram of acoustic pressure in location 1.

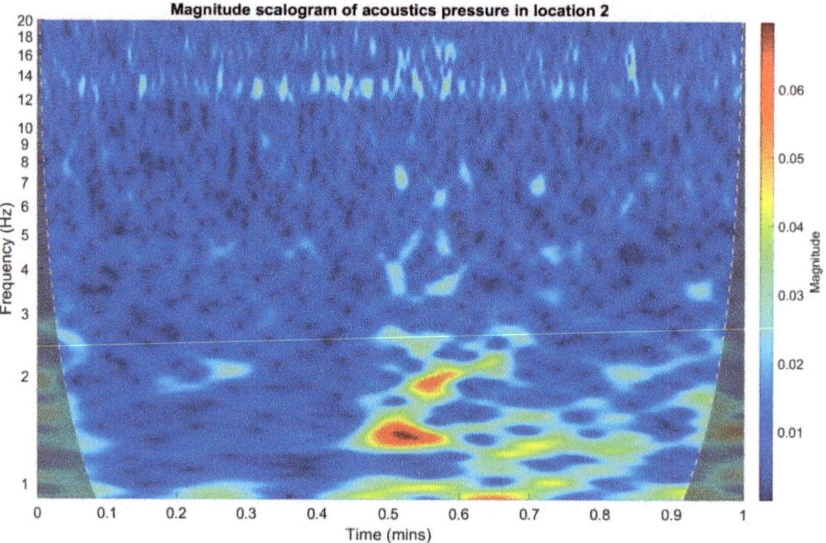

Figure 21. Magnitude scalogram of acoustic pressure in location 2.

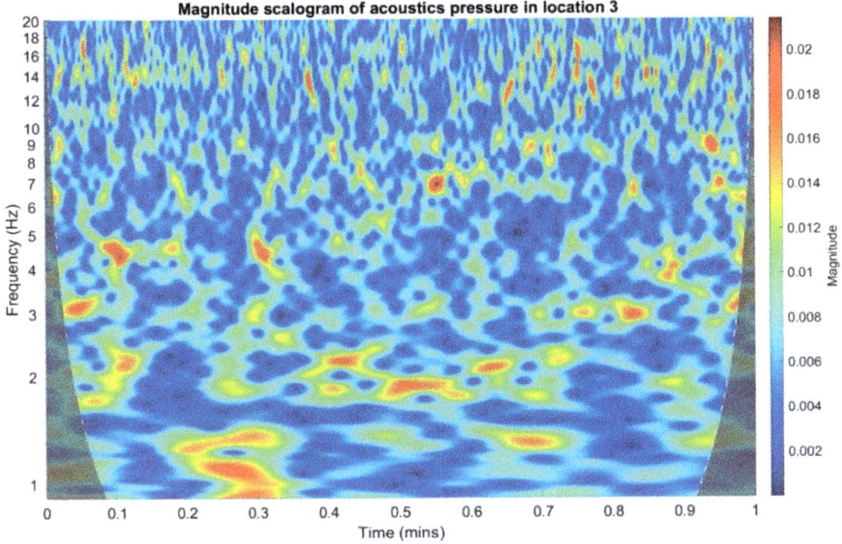

Figure 22. Magnitude scalogram of acoustic pressure in location 3.

The courses of the coherence functions demonstrate that the highest correlation of frequency occurs between the courses recorded in setups number one and two. However, for setups three and two (Figure 19) and three and one (Figure 18), there is no coherence with a value above 0.75.

In an analogous manner to the time and frequency analysis of the recorded infrasound signals, wavelet coherence functions were determined to identify the correlations of the scalograms in the time-frequency domain. Wherein, the wavelet transform coherence (WTC) is a method for analyzing the coherence and phase lag between two time series as a function of both time and frequency. The wavelet coherence is define as [41]:

$$R^2(\tau,f) = \frac{\left|S\left(\frac{1}{s}W_{xy}(\tau,\ s)\right)\right|^2}{S\left(\frac{1}{s}\left|W_x(\tau,\ s)\right|^2\right)\cdot S\left(\frac{1}{s}\left|W_y(\tau,\ s)\right|^2\right)} \tag{1}$$

$$0 \leq R^2(\tau,f) \leq 1 \tag{2}$$

where:

W_{xy} is the cross-wavelet transform:

$$\mathrm{W_{xy}}(\tau,\ \mathrm{s}) = W_x(\tau,s)W_y^*(\tau,s) \tag{3}$$

W_x is the continuous wavelet transform:

$$\mathrm{W_x}(\tau,\mathrm{s}) = \int_{-\infty}^{\infty} x(t)\psi_{\tau,s}^*(t)dt \tag{4}$$

where:

τ is the translation parameter and s is the scale parameter.

$\psi_{\tau,s}^*$ is the complex conjugate function of mother wavelet scaled and translated by τ and s.

s is the smooth operator.

The wavelet coherence is a squared correlation localized in time and frequency. Calculations were performed separately for each of the three possible combinations of measuring setups, and the results are presented in Figures 23–25. The arrows in the designated regions on the wavelets illustrate the variations in the correlation coefficient in time and frequency.

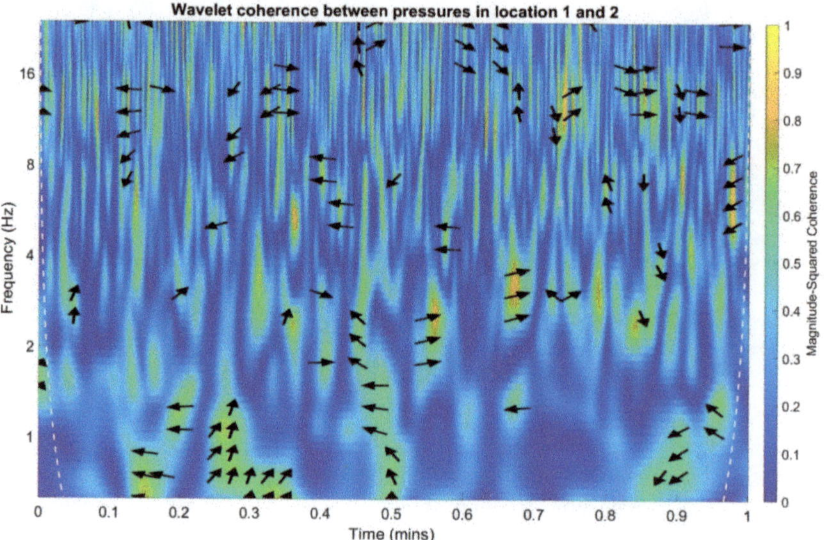

Figure 23. Wavelet coherence between acoustic pressures of infrasound measured in locations 1 and 2.

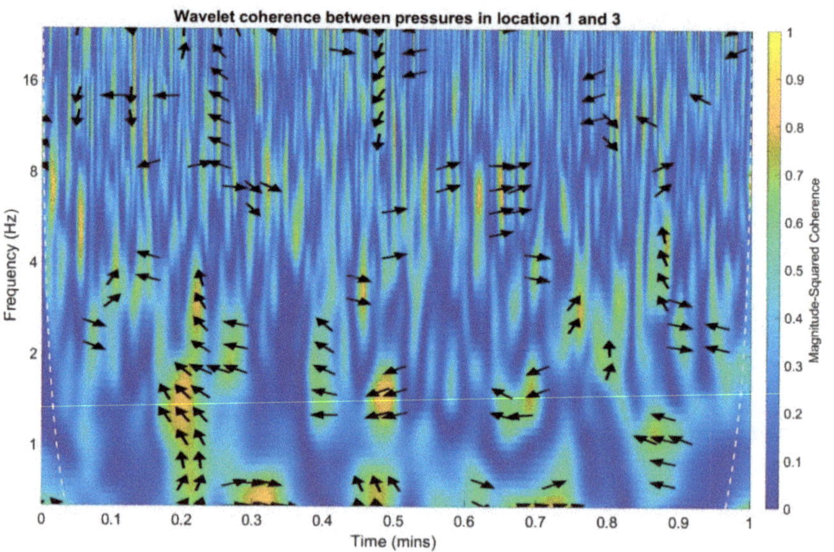

Figure 24. Wavelet coherence between acoustic pressures of infrasound measured in locations 1 and 3.

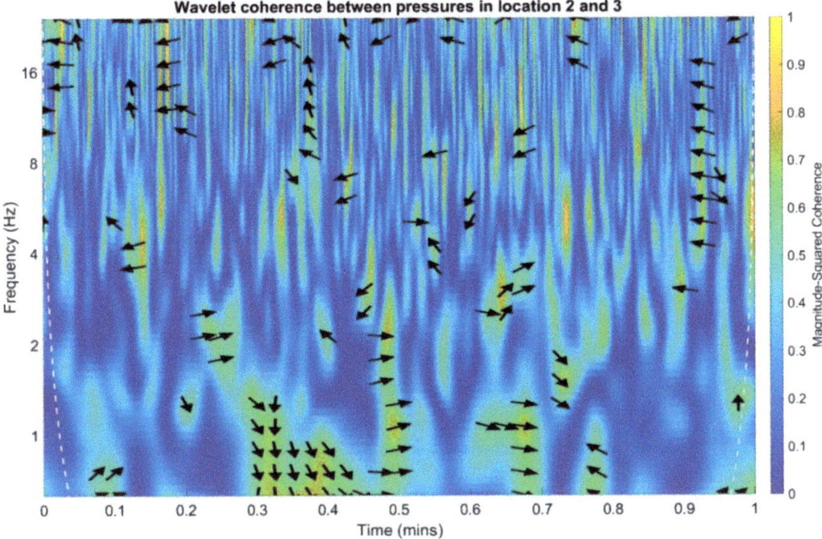

Figure 25. Wavelet coherence between acoustic pressures of infrasound measured in locations 2 and 3.

For examples arrow pointing to the right indicate that pressure in location one and pressure in the location two are positively correlated. Arrows pointing to the left indicate that pressures in the location one and two are negatively correlated. The straight up arrow implies that pressure in location one is leading in respect to location two. The straight down arrow imply that the pressure in location one is lagging in respect to pressure in location two (Figure 23).

On the basis of the conducted analysis, we can state the following conclusions:

- The waveforms containing variations in the acoustic pressure values presented in Figure 8 demonstrate the existence of relatively high levels of noise in the recorded infrasound signals, which has been confirmed by the designated histograms (Figure 9), where the shape is characteristic of Gaussian white noise.
- The calculated autocovariance waveforms (Figures 10–12) are characterized by the presence of both a broadband stochastic component with a variable amplitude regardless of the setup applied for data recording. This component is visible in the entire range of the analyzed waveforms, and is combined with the presence of a deterministic component with a relatively high value visible for a zero shift. This is true of all three autocovariance courses. In addition, the numerical values of time shifts were marked on the designated waveforms, for which individual peaks were identified, the values of which differ from the noisy, time-blurred waveform. It should be noted that these values are definitely smaller than the value determined for the zero shift (at least several times).
- For the purposes of the easier identification of the common harmonics with the highest values that correspond to the greatest time correlation of the infrasound signals recorded in individual setups, the values of the corresponding time shifts have been marked on the cross-correlation functions presented in Figures 13–15. In addition, several peaks for positive lag are visible in the determined course of the cross-correlation function between acoustic pressures in location two and one presented in Figure 13. The highest correlation was recorded for the lag of 0.45 s. In the case of inter-correlation between acoustic pressures in location three and one (Figure 14), a few peaks for positive lag were identified. The highest correlation occurs for the lag of 0.39 s. In contrast, for the course of inter-correlation function between acoustic pressures in location three and two (Figure 15). There are several peaks for positive lag. The highest correlation is for lag of −0.02 s. The presented waveforms demonstrate the occurrence of the greatest time correlation

between the waveforms recorded in setups number one and two and three and two, respectively, and relatively smaller for setups number three and one.

- The variations in wind speed in the range from 3.4 to 12.3 m/s did not affect the waveforms of the recorded power density spectra (Figure 16). These issues were the subject of other studies described in the article [42], therefore they were not analyzed in detail in this case. The recorded range of wind speed applied practically the entire range of the capacity of the examined turbine, starting from its start-up at a starting speed of 4 m/s, until the rated parameters were reached for the speed of 12.0 m/s (measured at the height of the rotor hub). As a result of increasing the wind speed, an increase in the values of the determined amplitude spectra occurred as well, which did not exceed 12 dB. These relations are identical for the data recorded in each of the three measurement setups and are independent of the rotor orientation and the angle of the turbine blades inclination.
- The determined waveforms of power density spectra take a course that is similar to an inverted, asymmetrical parabola. From the frequency of 0.1 Hz, an increase in the values of the calculated spectra is visible, up to the extreme, which is achieved at a frequency close to 1 Hz. Subsequently, a systematic, almost exponential, decrease of this value occurs until the end assumed at about 10 Hz, above which the waveform is almost flat. In the range from 10 to 20 Hz, single resonance peaks are noticeable. They are in particular visible for the frequency of 13 Hz (for setups number 1 and 2) and for approximately 16.5 Hz (for all three setups).
- The dominant frequency bands in the range from 2 to 4 Hz are visible on the scalograms obtained on the basis of the study. There is also a noticeable increase in power for frequencies in the range of around (6–8) Hz and from about 12 to 16 Hz. We can also note that the highest values in the spectra occur for infrasound frequencies. Above 20 Hz, the power density value drops sharply by more than 6 dB.
- The coherence above 0.75 is marked with red line (Figure 17). There is an evidence of correlation in the range of 13 Hz. There is no evidence of coherence higher than 0.75 (Figure 18). There is no evidence of coherence bigger than 0.75 (Figure 19).
- In Figure 23 for times in the range from 0.6 to 0.8 exist several regions of positive correlation above 0.9 in the frequency range of 2–3 Hz and 12–16 Hz. In the example, there exists a region with negative correlation in the frequency range of 6–8 Hz. In general, there are more positive correlation over the negative correlation areas. On the Figure 24 we can see the negative correlation for the time of 0.2 s and frequency range 1 to 3 Hz. Furthermore, for time 0.5 s and frequency range 1–2 Hz. Moreover, there are several smaller areas of negative correlation mainly for frequency lower than 2 Hz and higher than 8 Hz. There are few areas of positive correlation mainly for frequencies in the range from 4 to 8 Hz e.g., for 0.7 s. In general, there are more negative correlation over the positive areas. In Figure 25 we can see straight down arrows for times 0.3–0.4 s which means the pressure in location two is lagging in respect to location three. There are several time-frequency areas of positive correlation. In general, positive correlation in for lower frequencies below 2 Hz. And negative correlation is for higher frequencies from 4 to 16 Hz. Between locations there can exist both positive and negative correlation. Usually, lower frequencies are more positively correlated and for higher frequencies the correlation is more negative for higher frequencies where changing the distance of pressure sensors. There are only few areas which would evidence the lead of lag pressure signal between locations.

6. Conclusions

As a result of the application of a wireless measuring system comprising three separate setups, it was possible to record the occurrence of variations in the low-frequency signals emitted by a wind turbine, simultaneously in any three directions, at the distances up to 300 m from the place of their generation. These types of measurements are not possible using professional measuring equipment that uses existing wired communication. We can note that the tests carried out explicitly confirmed its practical usefulness for measurements carried out during normal operation of wind installations, in often very demanding field conditions.

In the summary, we can state that the measurements performed by using only a single measuring setup, in a similar way as described in the recommendations specified in the standard [PN-EN 61400-11] applicable to the methodology of acoustic signals measurements, in particular in the audible bandwidth, does not take into account the possibility of propagation of infrasound waves in different directions in relation to the analyzed source. As a consequence, the resulting information about the studied phenomenon can be incomplete or even incorrect.

However, simultaneous measurements of infrasound signals performed in three different locations in relation to the investigated wind turbine makes it possible to obtain a mean of the values of measurements followed by option of verifying the results. In addition, it is possible to assess the influence on the obtained dependencies of such parameters as: the position of the turbine rotor axis in relation to the direction of the wind blow, other operating wind installations, terrain and other environmental parameters on the recorded infrasound signals, which is not possible by application of only one measuring setup. The research carried out by these authors aims to develop a model of the propagation of infrasound waves emitted by wind turbines on the basis of the measurement programs executed during normal operation of wind turbines. The use of correlation analysis in the time, frequency and time-frequency domains according to the methodology described in the article makes it possible to find the occurring similarities and, respectively, differences in the obtained relationships illustrated on time courses, frequency spectra and wavelet scalograms.

Author Contributions: Conceptualization, T.B.; methodology, D.Z., D.W.; validation, D.Z., D.W. and M.K.; formal analysis, D.Z.; investigation, T.B., M.K. and D.W.; data curation, D.Z., M.K., D.W.; writing—original draft preparation, T.B.; writing—review and editing, M.K.; supervision, T.B.; project administration, T.B. All authors have read and agreed to the published version of the manuscript.

Funding: This research was co-financed by the National Science Centre, Poland (NCN) as a part of the OPUS Research Project No. 2015/17/B/ST8/03371.

Conflicts of Interest: The authors declare no conflict of interest. The funders had no role in the design of the study; in the collection, analyses, or interpretation of data; in the writing of the manuscript, or in the decision to publish the results.

References

1. Zagubień, A.; Wolniewicz, K. The impact of supporting tower on wind turbine noise emission. *Appl. Acoust.* **2019**, *155*, 260–270. [CrossRef]
2. Malec, T.; Boczar, T.; Wotzka, D.; Frącz, P. Comparison of low frequency signals emitted by wind turbines of two different generator types. *E3s Web Conf.* **2017**, *19*, 01001. [CrossRef]
3. Boczar, T.; Malec, T.; Wotzka, D. Studies on Infrasound Noise Emitted by Wind Turbines of Large Power. *Acta Phys. Pol. A* **2012**, *122*, 850–853. [CrossRef]
4. Tonin, R. A Review of Wind Turbine-Generated Infrasound: Source, Measurement and Effect on Health. *Acoust. Aust.* **2018**, *46*, 69–86. [CrossRef]
5. Pilger, C.; Ceranna, L. The influence of periodic wind turbine noise on infrasound array measurements. *J. Sound Vib.* **2017**, *388*, 188–200. [CrossRef]
6. Hansen, C.; Zajamšek, B.; Hansen, K. Infrasound and Low-Frequency Noise from Wind Turbines. In *Lecture Notes in Mechanical Engineering*; Springer: Berlin/Heidelberg, Germany, 2016; pp. 3–16. ISBN 9783662488669.
7. Irarrazabal, F.J.; Cook, M.R.; Gee, K.L.; Nelson, P.; Novakovich, D.J.; Sommerfeldt, S.D. Initial infrasound source characterization using the phase and amplitude gradient estimator method. In Proceedings of the Meetings on Acoustics, Louisville, KY, USA, 13–17 May 2019; Volume 36, pp. 1–12.
8. Assink, J.; Smets, P.; Marcillo, O.; Weemstra, C.; Lalande, J.M.; Waxler, R.; Evers, L. Advances in Infrasonic Remote Sensing Methods. In *Infrasound Monitoring for Atmospheric Studies: Challenges in Middle Atmosphere Dynamics and Societal Benefits*; Le Pichon, A., Blanc, E., Hauchecorne, A., Eds.; Springer International Publishing: Cham, Switzerland, 2019; pp. 605–632. ISBN 978-3-319-75140-5.
9. Hanning, C.D.; Evans, A. Wind turbine noise. *BMJ* **2012**, *344*, e1527. [CrossRef]

10. Doolan, C. A Review of Wind Turbine Noise Perception, Annoyance and Low Frequency Emission. *Wind Eng.* **2013**, *37*, 97–104. [CrossRef]
11. Shen, S.V.; Cain, B.E.; Hui, I. Public receptivity in China towards wind energy generators: A survey experimental approach. *Energy Policy* **2019**, *129*, 619–627. [CrossRef]
12. Rand, R.W.; Ambrose, S.E.; Krogh, C.M.E. Occupational Health and Industrial Wind Turbines. *Bull. Sci. Technol. Soc.* **2011**, *31*, 359–362. [CrossRef]
13. Ambrose, S.E.; Rand, R.W.; Krogh, C.M.E. Wind Turbine Acoustic Investigation. *Bull. Sci. Technol. Soc.* **2012**, *32*, 128–141. [CrossRef]
14. Thorne, B. The relevance of the precautionary principle to wind farm noise planning. In Proceedings of the INTERNOISE 2014—43rd International Congress on Noise Control Engineering Improving the World through Noise Control, Melbourne, Australia, 16–19 November 2014; pp. 1–10.
15. Krogh, C.M.E. Industrial Wind Turbine Development and Loss of Social Justice? *Bull. Sci. Technol. Soc.* **2011**, *31*, 321–333. [CrossRef]
16. Inagaki, T.; Li, Y.; Nishi, Y. Analysis of aerodynamic sound noise generated by a large-scaled wind turbine and its physiological evaluation. *Int. J. Env. Sci. Technol.* **2015**, *12*, 1933–1944. [CrossRef]
17. Graham, J.B.; Stephenson, J.R.; Smith, I.J. Public perceptions of wind energy developments: Case studies from New Zealand. *Energy Policy* **2009**, *37*, 3348–3357. [CrossRef]
18. Bell, J.A. Annoyance from wind turbines: Role of the middle ear muscles. *Acoust. Aust.* **2014**, *42*, 57.
19. Jeffery, R.D.; Krogh, C.M.E.; Horner, B. Industrial wind turbines and adverse health effects. *Can. J. Rural Med.* **2014**, *19*, 21–26.
20. Nissenbaum, M.A.; Aramini, J.J.; Hanning, C.D. Effects of industrial wind turbine noise on sleep and health. *Noise Health* **2012**, *14*, 237–243. [CrossRef] [PubMed]
21. McMurtry, R.Y. Toward a Case Definition of Adverse Health Effects in the Environs of Industrial Wind Turbines: Facilitating a Clinical Diagnosis. *Bull. Sci. Technol. Soc.* **2011**, *31*, 316–320. [CrossRef]
22. Landeta-Manzano, B.; Arana-Landín, G.; Calvo, P.M.; Heras-Saizarbitoria, I. Wind energy and local communities: A manufacturer's efforts to gain acceptance. *Energy Policy* **2018**, *121*, 314–324. [CrossRef]
23. Phadke, R. Resisting and Reconciling Big Wind: Middle Landscape Politics in the New American West. *Antipode* **2011**, *43*, 754–776. [CrossRef]
24. Salt, A.N.; Kaltenbach, J.A. Infrasound From Wind Turbines Could Affect Humans. *Bull. Sci. Technol. Soc.* **2011**, *31*, 296–302. [CrossRef]
25. McCunney, R.J.; Mundt, K.A.; Colby, W.D.; Dobie, R.; Kaliski, K.; Blais, M. Wind turbines and health: A critical review of the scientific literature. *J. Occup. Env. Med.* **2014**, *56*, e108–e130. [CrossRef] [PubMed]
26. O'Neal, R.D.; Hellweg, R.D.; Lampeter, R.M. Low frequency noise and infrasound from wind turbines. *Noise Control Eng. J.* **2011**, *59*, 135. [CrossRef]
27. Pleban, D.; Radosz, J.; Smagowska, B. Noise and Infrasonic Noise at Workplaces in a Wind Farm. *Arch. Acoust.* **2017**, *42*, 491–498. [CrossRef]
28. Knopper, L.D.; Ollson, C.A. Health effects and wind turbines: A review of the literature. *Env. Health* **2011**, *10*, 78. [CrossRef] [PubMed]
29. Kurpas, D.; Mroczek, B.; Karakiewicz, B.; Kassolik, K.; Andrzejewski, W. Health impact of wind farms. *Ann. Agric. Environ. Med.* **2013**, *20*, 595–604.
30. Berger, R.G.; Ashtiani, P.; Ollson, C.A.; Whitfield Aslund, M.; McCallum, L.C.; Leventhall, G.; Knopper, L.D. Health-Based Audible Noise Guidelines Account for Infrasound and Low-Frequency Noise Produced by Wind Turbines. *Front. Public Heal.* **2015**, *3*, 1–14. [CrossRef]
31. Leventhall, G. Infrasound from wind turbines-Fact, fiction or deception. *Can. Acoust. Acoust. Can.* **2006**, *34*, 29–36.
32. Marshall, J.P. Psycho-social disruption, information disorder, and the politics of wind farming. *Energy Res. Soc. Sci.* **2018**, *45*, 120–133. [CrossRef]
33. Sardaro, R.; Faccilongo, N.; Roselli, L. Wind farms, farmland occupation and compensation: Evidences from landowners' preferences through a stated choice survey in Italy. *Energy Policy* **2019**, *133*. [CrossRef]
34. Freiberg, A.; Schefter, C.; Girbig, M.; Murta, V.C.; Seidler, A. Health effects of wind turbines on humans in residential settings: Results of a scoping review. *Environ. Res.* **2019**, *169*, 446–463. [CrossRef]
35. Pohl, J.; Gabriel, J.; Hübner, G. Understanding stress effects of wind turbine noise–The integrated approach. *Energy Policy* **2018**, *112*, 119–128. [CrossRef]

36. Taylor, J.; Klenk, N. The politics of evidence: Conflicting social commitments and environmental priorities in the debate over wind energy and public health. *Energy Res. Soc. Sci.* **2019**, *47*, 102–112. [CrossRef]
37. Pierpont, N. *Wind Turbine Syndrome: A Report on a Natural Experiment*; K-Selected Books: Lowell, MA, USA, 2009; ISBN 978-0984182701.
38. Farboud, A.; Crunkhorn, R.; Trinidade, A. 'Wind turbine syndrome': Fact or fiction? *J. Laryngol. Otol.* **2013**, *127*, 222–226. [CrossRef] [PubMed]
39. Boczar, T.; Zmarzły, D.; Kozioł, M.; Wotzka, D. The use of advanced signal processing methods for the analysis of infrasound generated by high-power wind turbines. *Przegląd Elektrotechniczny* **2020**, *96*, 68–75.
40. Gupta, H.R.; Mehra, R. Power Spectrum Estimation using Welch Method for various Window Techniques. *Int. J. Sci. Res. Eng. Technol.* **2013**, *2*, 389–392.
41. Torrence, C.; Webster, P.J. Interdecadal Changes in the ENSO–Monsoon System. *J. Clim.* **1999**, *12*, 2679–2690. [CrossRef]
42. Boczar, T.; Zmarzły, D.; Kozioł, M.; Wotzka, D. Analiza wpływu czynników zewnętrznych na wyniki pomiarów infradźwięków emitowanych przez turbiny wiatrowe (Analysis of the influence of external factors on the measurement results of infrasounds emitted by wind turbines). *Przegląd Elektrotechniczny* **2020**, *1*, 115–122. [CrossRef]

Publisher's Note: MDPI stays neutral with regard to jurisdictional claims in published maps and institutional affiliations.

Article

Evaluation of the Possibility of Identifying a Complex Polygonal Tram Track Layout Using Multiple Satellite Measurements

Andrzej Wilk [1], Cezary Specht [2], Wladyslaw Koc [1], Krzysztof Karwowski [1], Jacek Skibicki [1], Jacek Szmagliński [3], Piotr Chrostowski [3,*], Pawel Dabrowski [2], Mariusz Specht [4], Marek Zienkiewicz [5], Slawomir Judek [1], Marcin Skóra [6] and Sławomir Grulkowski [3]

1. Department of Electrified Transportation, Gdańsk University of Technology, Gabriela Narutowicza 11-12, 80-233 Gdańsk, Poland; andrzej.wilk@pg.edu.pl (A.W.); wladyslaw.koc@pg.edu.pl (W.K.); krzysztof.karwowski@pg.edu.pl (K.K.); jacek.skibicki@pg.edu.pl (J.S.); slawomir.judek@pg.edu.pl (S.J.)
2. Department of Geodesy and Oceanography, Gdynia Maritime University, Morska 81-87, 81-225 Gdynia, Poland; c.specht@wn.am.gdynia.pl (C.S.); p.dabrowski@wn.am.gdynia.pl (P.D.)
3. Department of Rail Transportation and Bridges, Gdańsk University of Technology, Gabriela Narutowicza 11-12, 80-233 Gdańsk, Poland; jacek.szmaglinski@pg.edu.pl (J.S.); slawomir.grulkowski@pg.edu.pl (S.G.)
4. Department of Transport and Logistics, Gdynia Maritime University, Morska 81-87, 81-225 Gdynia, Poland; m.specht@wn.am.gdynia.pl
5. Department of Geodesy, Gdańsk University of Technology, Gabriela Narutowicza 11-12, 80-233 Gdańsk, Poland; marek.zienkiewicz@pg.edu.pl
6. Department of Navigation and Marine Hydrography, Polish Naval Academy, Inżyniera Jana Śmidowicza 69, 81-103 Gdynia, Poland; m.skora@amw.gdynia.pl

* Correspondence: piotr.chrostowski@pg.edu.pl; Tel.: +48-58-348-6090

Received: 7 July 2020; Accepted: 5 August 2020; Published: 7 August 2020

Abstract: We present the main assumptions about the algorithmization of the analysis of measurement data recorded in mobile satellite measurements. The research team from the Gdańsk University of Technology and the Maritime University in Gdynia, as part of a research project conducted in cooperation with PKP PLK (Polish Railway Infrastructure Manager), developed algorithms supporting the identification and assessment of track axis layout. This article presents selected issues concerning the identification of a tramway line's axis system. For this purpose, the supporting algorithm was developed and measurement data recorded using Global Navigation Satellite System (GNSS) techniques was evaluated and analyzed. The discussed algorithm identifies main track directions from multi-device data and repeated position recordings. In order to observe the influence of crucial factors, the investigated route was carefully selected. The chosen tramway track was characterized by its location in various field conditions and a diversified and complex geometric layout. The analysis of the obtained results was focused on the assessment of the signal's dispersion and repeatability using residuals in relation to the estimated track's direction. The presented methodology is intended to support railway infrastructure management processes, mainly in planning and maintenance through an efficient inventory of the infrastructure in service.

Keywords: geometric layout of railway track; track axis identification; tramway loop; route's polygon; GNSS mobile measurements

---

1. Introduction

In recent years, the managers of rail infrastructure have been increasingly recreating the course of railroads, as part of the overall inventory, in the global coordinate system. Initially, the geometric system in the global respect—the concept of Absolute Track Geometry (ATG)—was applied only to selected lines, which were characterized by specific operating conditions, such as railway lines of high operational priority (high speed rail lines) or sections coexisting with other structures of transport infrastructure, such as bridges, tunnels or squares. Other lines were measured using conventional measurement methods [1,2]. This was due to the relatively large labor input associated with the measurement and reproduction of the coordinates of the track courses in the global systems of spatial references. At present, in the processes of the modernization of railway lines, railway geodetic control networks are mounted more and more often, which is related to the state of geodetic networks. They allow the effective reconstruction of the track axis in the global system by tacheometric measurement (backward cutting to several reference points).

The real revolution in the field of track axis reconstruction was brought about by mobile surveying methods, which allow a continuous measurement at a higher speed than in the case of static or quasi static tacheometry. Due to techniques such as Global Navigation Satellite System (GNSS), Inertial Navigation System (INS) and Mobile Laser Scanning (MLS), the effective inventory of all lines at the railway network scale is now possible [3–8]. Moreover, these techniques enable a mobile measurement of coordinates with a high density of data recording at a relatively high speed of the measurement process. In extreme cases, it can even be the measurement with the speed of regular trains [9]. Contemporary methods of data processing are also worth mentioning. With frequencies of data recording of tens of hertz, it is necessary to use efficient algorithms for processing the measurement signal. The signal achieved is the so-called point cloud, which without further interpretation does not contribute valuable information. Hence, the signals should be interpreted in such a way that the data set can be converted into the model representation of geometric systems of the track, mainly in the horizontal plane (straight sections, transition curves and circular arcs), and also in the vertical alignment (sections of a constant slope and sections in vertical arcs) [10–12]. Because of the large number of collected coordinates, the effective analysis must include methods of automatic selection of particular sections of geometric systems [13–16].

Until recently, a questionable issue was the accuracy of GNSS measurements. The methods using tacheometers and railway control networks guarantee an accuracy of up to several millimeters. On the other hand, until recently, such accuracy in GNSS measurements was difficult to achieve. At present, it has been reported that an accuracy of about a 1 cm level is achievable also in this survey technique. Generally, the positions along the track are obtained based on the conversion of measurement signals from the different devices, based on the data fusion algorithms [17,18]. Furthermore, the accuracy often analyzed relates to certain specific measurement conditions, but railroads' conditions are varied.

For more than 10 years, the authors have been developing the technique of satellite measurements in order to evaluate implementation possibilities of this method [19,20]. At present, as part of the project InnoSatTrack shared with the Polish Railways, current possibilities of using GNSS, INS and MLS systems are subject to assessment for the purposes of track axis inventory (position and shape) [21,22] as well as for train positioning possibilities [23]. In the works under way, the biggest accent is put on GNSS technology, because the positioning of devices with high accuracy is the key to reconstruct the track axis in the global coordinates system [24]. In turn, measurements using additional devices, such as marine satellite compasses, inertial devices or laser scanners perform auxiliary functions, in particular for the purpose of determining corrections on the basis of additional information about the movement of measurement platform relative to the track [25,26].

In this paper we present the assessment of the possibility of determining the main directions of a polygon system, based on the example of a complex geometric system of the track. The main directions of the route are defined as tangents to horizontal curves in their boundaries with straight sections. It could be assumed that in specific cases the main direction covers a straight track section or a tangent between reverse curves. According to the above definition, the polygon of the route is understood as a system of main directions defined in the Cartesian coordinate system.

Geometric layouts in railway lines are usually characterized by mild changes of curvature (long straight lines and transition curves, large radii). For this reason, the authors tested both the measurement methods and data analysis methodology in relation to layouts of more diverse geometry. The examples of such systems are railway lines of agglomeration importance and mainly tramway lines, which in conditions of urban development require the use of short geometric sections [27,28]. In such situations, the exact reconstruction of the route's polygon is a prerequisite for the proper reconstruction of the system of non-linear sections by using analytical methods of design [29–32].

An additional advantage of the analysis of the system on the tramway network is a varied degree of the occurrence of field screens. In such a situation the assessment of the accuracy of the reconstruction of the polygon is more reliable and indicates the potential possibilities for planning measurement campaigns on the lines of rail transport.

2. Materials and Methods

2.1. Characteristics of the Measured Tramway Route

For the analysis of the main directions, a tramway line including a street terminus in the district of Gdańsk Brzeźno (coastal town in northern Poland) was selected. Choosing the route, the following assumptions were taken into account:

- The route should be characterized by diversified visibility conditions of GNSS satellites, in order to verify and assess the impact of the number of visible satellites (their distribution on the horizon) on the possibility of the reconstruction of track geometry;
- The measuring section should have a diversified condition of track (structural and geometrical) so as to assess the impact of the state on the reconstruction quality;
- The measuring section should allow multiple measurements, in order to assess the reproducibility of the results obtained related to the parameters of the identified geometric system of the axis of track.

The measurement campaign was carried out as a part of the project InnoSatTrack from 28.11.2018 until 29.11.2018 on a section of the tramway network in Gdańsk [22,24]. A large street terminus in Brzeźno district was selected (shown in Figure 1). The length of the test section was approximately 3 km, allowing multiple measurements of the same section overnight. Additionally, the system of the route in the form of a street terminus assured that there was no need to perform any maneuvers to prolong the duration of measurement sessions. Finally, the selected street terminus was characterized by diverse pavement conditions and the presence of field obstacles along particular sections.

Detailed characteristics of sections of the selected terminus:

Section *a*: Section of a length of about 850 m, starting at the terminus Brzeźno Plaża. Double-track tramway route is located on a strip dividing four-lane alleys. The section is characterized by a small number of field screens (single trees near the track at the terminus) and a very good technical condition of the track's structure. The section was chosen as the reference section in relation to others.

Section *b*: Section of a length of about 400 m. One-track tramway route is located asymmetrically on the north side of a four-lane avenue. The fragment is characterized by a small number of field screens (single trees near the track), and a very good technical condition of the track's structure. This section was also chosen as a reference section.

Section *c*: Section of a length of about 1000 m. One-track tramway route is located asymmetrically on the west side of a two-lane street. This section is characterized by a larger number of field screens than in previous sections (groups of trees and buildings located along the track) and very poor technical condition of the pavement (high values of twist and irregularities of the track in both the horizontal and vertical planes). This section was selected as the section to study the effect of track conditions on the accuracy of measurement.

Section *d*: This section is the tramway route carried out independently of the road system with a length of approximately 150 m. This track is characterized by a complex geometry and the presence of the most extensive field screens (trees with crowns located over the streetcar track), and at the same time with a very good quality of the track. It was chosen to assess the impact of field screens on the possibility to reconstruct a complex geometric system of the track (field obstacles are shown in Figure 2). This section includes a diverging track to the Brzeźno terminus.

Section *e*: Section of a length of about 330 m. One-track tramway route located asymmetrically on the south side of an access one-lane road. It is characterized by the appearance of very extensive field screens (tall buildings adjacent to the track) and the good condition of the track. It was chosen to assess the impact of field screens on the possibility of the reconstruction of a typical geometric system of the track (see Figure 3)

Section *f*: This section is independent of the road system, and its length is about 270 m. It is characterized by an average number of field screens and good track condition. This section was chosen as a reference for section *c*.

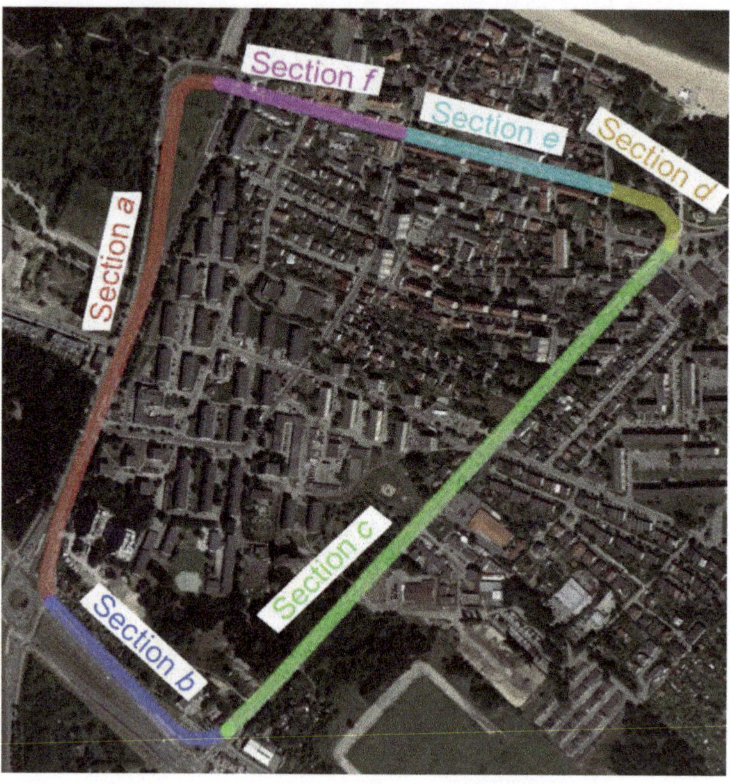

Figure 1. Measuring testing ground based on Google Maps.

Section *a* (South–North view)

Section *a* (North–South view)

Section *b* (South East–North West view)

Section *b* (North West–South East view)

Figure 2. View of the streetcar track in the area of sections *a* and *b*.

Section *c* (South West–North East view)

Section *c*, mid part (North East–South West view)

Section *c* (South West–North East view)

Section *d*

Section *e*

Section *f*

Figure 3. View of the tramway track in the area of sections *c*, *d*, *e* and *f*.

Figure 2 shows the track in sections *a* and *b*, while Figure 3 shows examples of locations on sections *c*, *d*, *e* and *f*. Sections *a* and *b* are characterized by rather good conditions due to the visibility to satellites, while sections *c*–*f* are located in close proximity to high-rise buildings or trees. Here, disturbances in satellite signals from the sections of the northern edge of the terminus are expected.

2.2. *Characteristics of Measurement Unit*

To carry out the measurement, 10 GNSS receivers installed on a measurement platform were used. For the investigation, Leica GS18 and Trimble R10 receivers were selected. The corrections for real time kinematic measurement were obtained from the closest reference stations of the SmartNet and VRSNet networks for the Leica and Trimble receivers, respectively.

As measurement cars, the bogies of the GWF 300 car series were adopted. They are characterized by favorable conditions for matching their ride axes into the track axis (no extensive amortization, no clearances) and a relatively low height of the measurement deck. On each of both bogies, five receivers were installed in a rectangular formation. In the centers of both bogies, directly above the bogie's pivot, central antennas were placed. A diagram of the layout of the receivers is shown in Figure 4.

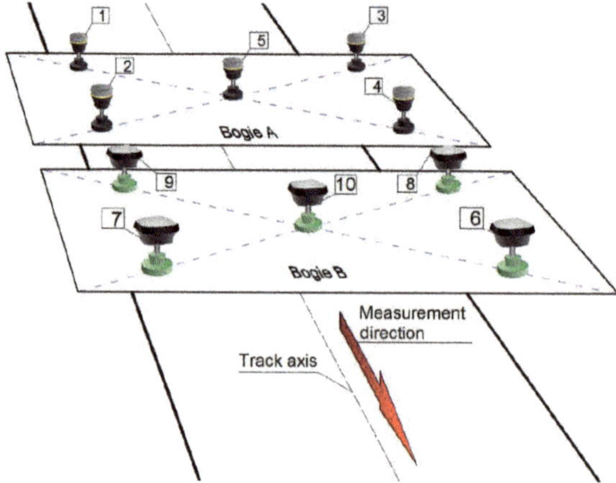

Figure 4. Diagram of the layout of GNSS receivers on the bogies.

As the tramway bogies do not have their own drive device, there is a problem concerning the proper selection of a motor vehicle. The vehicle should meet the following requirements:

- The vehicle should provide both constant and low velocity of about 30 km/h. Only relatively modern tramways, equipped with power electronics drive control fulfill this requirement;
- The vehicle should provide a 230 V AC power supply to measuring devices and recording computers from the onboard network (preferably equipped with a standard 230 V socket).

The vehicle that met the above-mentioned requirements proved to be a Bombardier NGT-6/2. Finally, three measurement bogies were attached to the engine car. On the first bogie (the closest to tramway) the devices recording additional parameters of a run (such as acceleration, inclination and direction of the run) were installed. On the other two, a set of 10 GNSS receivers were placed. To avoid a negative influence from naturally occurring obstacles, i.e., tramway cabs, an additional bogie was located between the tramway and the bogies with GNSS antennas. A diagram of the measuring train is shown in Figure 5.

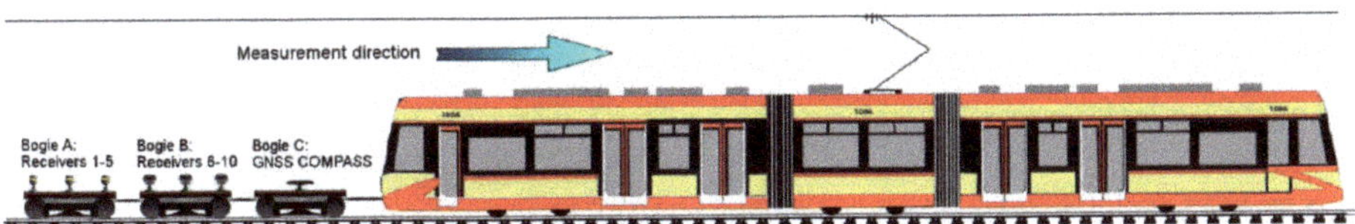

Figure 5. Diagram of the configuration of the measuring train.

In the present measurement methodology, the fact is taken into account, that the position of the axis of the track relative to the rail vehicle is not unambiguously determined. One must keep in mind that each rail vehicle has a small degree of freedom in the lateral direction due to the bogie structural clearances and clearances between the rails and wheels. These clearances ensure the correct fitting of the bogie's axles into the track. Moreover, besides of the constructional clearances of the bogies, displacements due to the spring amortization of the rolling stock occur. Taking into account also conicity in a rail and wheel interface, some kind of typical oscillation (so-called hunting oscillation) should be expected. However, this kind of movement is hard to assess without additional reference measurements (for example vision methods) and is not discussed in detail in this article.

It is therefore assumed, that the most reliable results with respect to the track axis are obtained by situating the antenna in the axis of symmetry with respect to the construction of the wheel set. This point for the construction of the bogie is in the pivot axis. This operation is difficult due to the limited access to the structural elements of pivot from the outside. Thus, there is the first factor related to the uncertainty of measurement, i.e., determination of the precise location of the base receiver on the vehicle's platform. Due to the above problem, the positioning of the GNSS receivers' centers was performed in conditions that minimize positioning errors. Installation of equipment was carried out in the hall of the tram depot. The process of measuring devices installation is shown in Figure 6. Special support structures enabled the precise installation of each tribrach according to the scheme assumed. Figure 7 presents the bogies instrumented with controllers and receivers during measurements.

During the measurement campaign, six passages on the specified street terminus were performed between 0:09 and 3:37. At this time, 10 receivers operated in continuous recording mode at a frequency of 20 Hz. During the study, a total of 1,186,518 positions were collected. All of the records met the assumed accuracy of designation (uncertainty radius not greater than 0.2 m).

2.3. Methodology for Determining the Main Directions of the Routes

The axis of the railway or tramway track physically does not have any representation. Considering a typical cross-section of a track, its axis is located at a point distant by a length of half the width (track gauge) between the rail courses at a height of the line connecting running surfaces of rail heads. In general, the axis of the track does not have to coincide with the central axis (so-called centerline) of a railway track. This is the case in curves with small radii, in which a widening of the track is often made. Figure 8 shows the definition of the axis of a railway track and its centerline.

Reconstruction of the track position in straight sections does not require additional adjustments related to the transverse or longitudinal inclinations. It is assumed that the track in the straight section is designed and built as a track without a lateral inclination (cant). Therefore, any found deformations resulting from the technical degradation of the track will be removed from the signal if necessary. Thus, the main directions will be assessed based on the analysis of the recorded positions on the straight sections of the track in particular passing measurements.

Figure 6. Setting out the position of receivers in relation to the bogie's structure and the tramway track in the hall of the Wrzeszcz tram depot.

Figure 7. The measuring set used in the mobile satellite measurements duringmeasurements in Gdańsk.

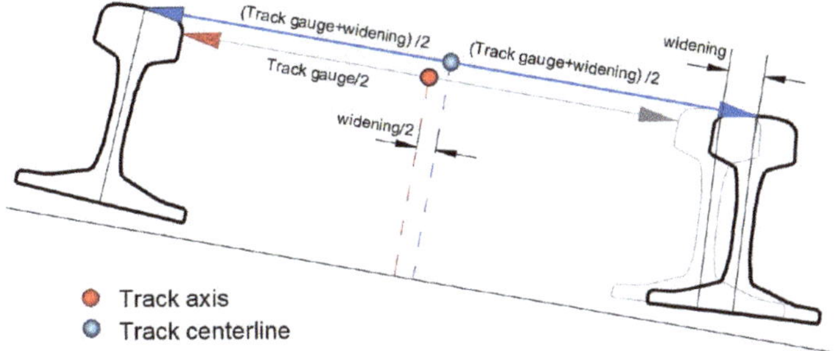

Figure 8. A schema of the location of the track axis and the track centerline.

Prior to the analysis, the following assumptions were made:

- Multiple repeated measurement of a precisely indicated track section will enable to evaluate the error of the identification of the main directions. The directions will be defined on the grid of horizontal coordinates;
- Generally, the main directions will be designated by antennas positioned in the pivot axis of the measurement bogies;
- The geometric layout of the selected test section will be identified as a system of straight lines, regardless of the nature of deformation occurring in the track;
- Obvious deformations observed in the measurement signal, such as the so-called broken straights, will be removed, as they distort the assessment of the main directions;
- It was assumed that the assessment of the accuracy of the identification in the horizontal coordinate system will be made on the basis of the analysis of the reproduction of straight sections signals.

2.3.1. Data

The selected straights located in the designated sections of the tramway terminus were analyzed. Figure 9 shows the location of individual straights in the form of coordinates on the grid of the PL-2000 system in force in Poland (Gauss–Krüger projection for Geodetic Reference System '80). These straights were separated from GNSS measurement data based on the visual assessment of the results. Table 1 shows the signatures of straight lines in the corresponding sections of the tested track. Data in the form of coordinates and additional information, such as the accuracy recorded by the particular receiver as well as the number of satellites used to calculate the position at a given moment of time, were used in the analysis. Table 2 shows an example of the scope of measurement data.

Table 1. Selected straights in the scope of the *a–f* sections of the geometric layout analyzed.

Straight Line	Location in the Route	Assessment of Track Condition/Quality	Assessment of Visibility
Str 1	Sect. a	Fair	Good
Str 2	Sect. a	Excellent	Fair
Str 3	Sect. a	Excellent	Good
Str 4	Sect. a	Excellent	Excellent
Str 5	Sect. a	Excellent	Good
Str 6	Sect. b	Excellent	Excellent
Str 7	Sect. c	Bad	Poor
Str 8	Sect. c	Bad	Poor
Str 9	Sect. c	Bad	Fair
Str 10	Sect. c	Bad	Poor
Str 11	Sect. d	Good	Bad
Str 12	Sect. e	Good	Poor

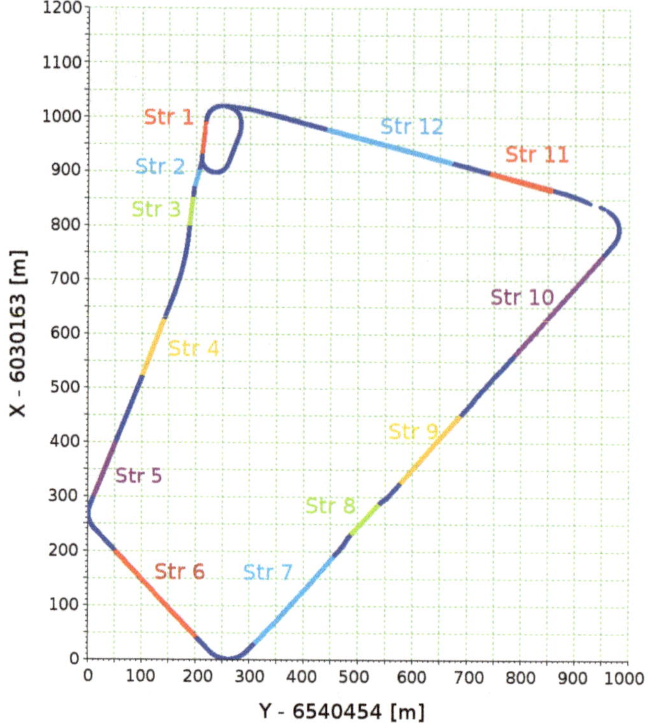

Figure 9. Measurement of the coordinates of the route in the PL-2000 system with the designation of individual straight sections.

Table 2. Input data for the analysis of the main directions for a selected straight.

Record No.	GPS Time (s)	Horizontal Coordinate Y (m)	Vertical Coordinate X (m)	Two Dimensional Position Error (m)	Number of Satellites
358670	2,029,347,009.00	6,540,504.8138	6,030,362.9699	0.0400	8
358671	2,029,347,009.05	6,540,504.9345	6,030,362.8442	0.0400	8
358672	2,029,347,009.10	6,540,505.0561	6,030,362.7132	0.0400	8
358673	2,029,347,009.15	6,540,505.1887	6,030,362.5797	0.0400	8

2.3.2. Main Directions

The main direction of the existing railway track is defined as a straight line that describes the rectilinear section of the track best or as a tangent to the reverse curves system at the point of their contact. The main directions form a system called a route's polygon.

As defined, the main direction is described by the parameters of a linear model equation of the straight line. In this analysis, these parameters were estimated by the least squares method. Suitable sets of points were selected for the analysis with YX coordinates in the PL-2000 system, which are the measured positions of GNSS receiver.

Signals from 10 receivers recorded during six measurement runs were analyzed. It can therefore be assumed, that each of the 60 signals is an independent description of the location of the analyzed section of track. Obviously, the most reliable signals are the signals recorded by antennas situated over the axis. In these places, transverse shifts of the tram bogies in relation to the track should be minimal. The antennas located at the corners of measuring platforms should also draw straight lines with the same direction coefficient. However, these positions can be subject to increased uncertainty due to the shifts and vibrations of the platform during test runs.

The signals recorded by individual antennas in each of the runs were analyzed in the local systems of coordinates, of which the abscissa axis is in accordance with the least squares line. This methodology allows for a clear interpretation of the differences between the signals recorded. A relatively large number of points used in the regression calculation makes it so that the straight lines are determined at a very low matching error, especially in situations of long straight lines.

The methodology of transformation that uses homogeneous coordinates was accepted. The transformation of the YX coordinates of the PL-2000 system to homogeneous coordinates can be written as:

$$(Y', X') \rightarrow \begin{bmatrix} Y & X & 1 \end{bmatrix} \tag{1}$$

Thus, transformations of coordinates can be described with a matrix equation in the form:

$$\begin{bmatrix} Y'_1 & X'_1 & 1 \\ Y'_2 & X'_2 & 1 \\ \cdots & \cdots & \cdots \\ Y'_n & X'_n & 1 \end{bmatrix} = \begin{bmatrix} Y_1 & X_1 & 1 \\ Y_2 & X_2 & 1 \\ \cdots & \cdots & \cdots \\ Y_n & X_n & 1 \end{bmatrix} \cdot \boldsymbol{M} \tag{2}$$

where $\boldsymbol{M}$ is a matrix of transformation in the form:

$$\boldsymbol{M} = \begin{bmatrix} m_{11} & m_{12} & 1 \\ m_{21} & m_{22} & 1 \\ m_{31} & m_{32} & 1 \end{bmatrix} \tag{3}$$

The elements of matrix $\boldsymbol{M}$ (m_{ij}, $i = 1,2,3$, $j = 1,2$) depend on a character of transformation and in specific cases of a single transformation this matrix takes the form of a translation matrix $\boldsymbol{T}(t_h, t_v)$, with rotation $\boldsymbol{R}(\varphi)$ or conversion of symmetry $\boldsymbol{F}$ performed with respect to the Ox or Oy axes of coordinates:

$$\boldsymbol{T}(t_h, t_v) = \begin{bmatrix} 1 & 0 & 0 \\ 0 & 1 & 0 \\ t_h & t_v & 1 \end{bmatrix}; \boldsymbol{R}(\varphi) = \begin{bmatrix} \cos\varphi & \sin(\varphi) & 0 \\ -\sin(\varphi) & \cos\varphi & 0 \\ 0 & 0 & 1 \end{bmatrix}; \boldsymbol{F} = \begin{bmatrix} \pm 1 & 0 & 0 \\ 0 & \pm 1 & 0 \\ 0 & 0 & 1 \end{bmatrix} \tag{4}$$

where: t_h, t_v are the distance from the origin to a translated point along the horizontal and vertical directions, and φ is the rotation angle of the vector of the transformed point with respect to the origin of the coordinate system.

2.4. Automatic Identification of the Polygon-Algorithm for Identifying Main Directions

The geometric layout of the railway in the horizontal plane consists of straight-line sections and the sections arranged in curves, consisting of circular arcs and transition curves with variable curvature along the length of the elements. The analysis presented in this paper concerns the polygon, which is formed by the main directions in accordance with the straight sections of the track or which are tangent to reverse curves. It is also assumed that two successive directions in places of their intersections are formed by the vertices of horizontal curves. The angle through which the direction changes from direction i to $i + 1$ is the intersection angle.

For the description of polygon system, the equations of the straights of the YX coordinates (in accordance with the PL-2000 system) as $X = a_{md} \cdot Y + b_{md}$ was used. Parameters of the equation (a_{md}, b_{md}) refer to the main directions of the analyzed route.

The basic steps of the algorithm supporting the identification of main directions' system are described below.

(1) The first step is the discretization of the data to a finite number of sections. Fragmentation of the measurement signal took place on the basis of assigning each coordinate to intervals of a fixed length.

At this stage the geometric system is not known, so the road is calculated by summing up distances from point to point. This is the first estimate of the distance along the route, which—when carried out in an appropriate way (filtering out concentrations of points recorded at standstills etc.)—will enable further analyses to identify the polygon. Figure 10 shows a schematic diagram of the route discretization, while Equation (5) shows a basic method of counting the distance along the route. The length of each section depends on the density of records and it is generally set in the phase of tuning the algorithm to the needs arising from the characteristics of the measurement signal.

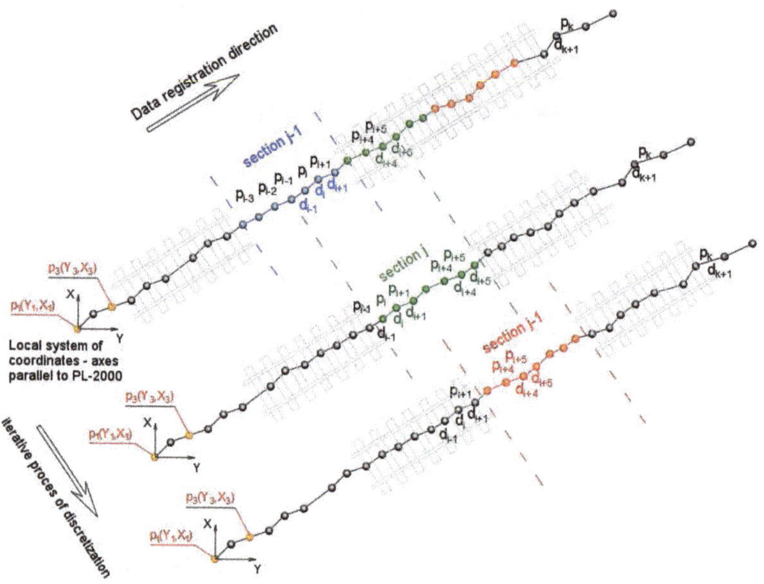

Figure 10. Schematic diagram of the route discretization into sections; p_i—measured points, d_i—distance between points.

$$D_i = D_{i-1} + d_i = D_{i-1} + \sqrt{(Y_{i+1} - Y_i)^2 + (X_{i+1} - X_i)^2} \tag{5}$$

where: D_i—distance from the starting point to the current one; Y_i, X_i—flat coordinates of measuring points p_i in the local system with directions in accordance with the PL-2000.

In each of the sections, there are a number of points with coordinates YX. When testing the algorithm, the minimum number of points in a particular interval was fixed; it should be remembered, that in continuous measurements there are places with no records of a measurement signal. Then, for each of the intervals the analysis of regression is made between the Y and X coordinates, due to which the estimate of the directions characterizing individual intervals of discretized route takes place, as shown in Figure 11. Due to the auxiliary character of estimated directions, the calculations were performed according to the least squares algorithm method for linear correlation. The method in this case involves the determination of the parameters of simple regression by minimizing the deviations of the measurement values of vertical coordinate X of the reference system PL-2000 (or the local system in accordance with the directions compatible with PL-2000) with respect to the linear model. The method accepted can be described with the equation:

$$f(a_j, b_j) = \sum_{i=1}^{m} \left(X_i - \left(a_j \cdot Y_i + b_j\right)\right)^2 \Rightarrow min \tag{6}$$

where: a_j, b_j—parameters of linear model in section j, and Y_i, X_i—coordinates of measurement points p_i in the local system with directions in accordance with the PL-2000.

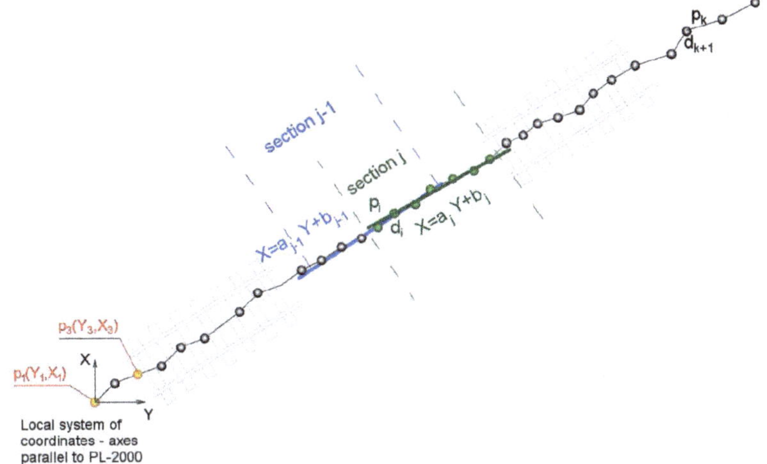

Figure 11. Schematic diagram of determining directions characterizing individual sections.

Each section has its own estimator of direction, which can be the coefficient of the least squares straight slope or the angle to the axis of abscissae Y. It could also be the azimuth of the designated direction.

(2) With estimators of the direction in the individual sections of the route, variation of this direction is analyzed in the passages from one interval to another, i.e., in the series of consecutive sections. It is assumed that the angle representing the direction in each section maps the variation of the shape of the axis of the route analyzed. On the straight sections, a slight oscillation of the estimator value is expected, and in the areas of curves beside the oscillation a clear increasing trend of this value should appear. In order to examine these trends, the algorithm groups the adjacent sections with one another (of which each has a calculated estimator of direction in the previous step) in a minimum number and carries out the next analysis on the sets thus designated. The number of sections in each group depends on the geometric system (length of homogeneous sections of the route) and is determined by the analysis of Pearson's correlation coefficient. This time, the analysis of regression concerns the values of the angles of individual sections. If we assume the angle as azimuth (Az) then the analysis of the grouped regressions of section can be written as follows:

$$f(a_g, b_g) = \sum_j \left(Az_j - \left(a_g \cdot \mathrm{D}_{mid_j} + b_g\right)\right)^2 \Rightarrow min \tag{7}$$

where:

a_g, b_g—parameters of a linear model of the group of adjacent sections;
Az_j—angles characterizing the individual sections in the group;
D_{mid_j}—middle values of the distance covered in each section (equivalent to mileage).

Schematic diagram of the analysis of the regression of the sequence of angles indicated in the individual sections of discretized route is shown in Figure 12.

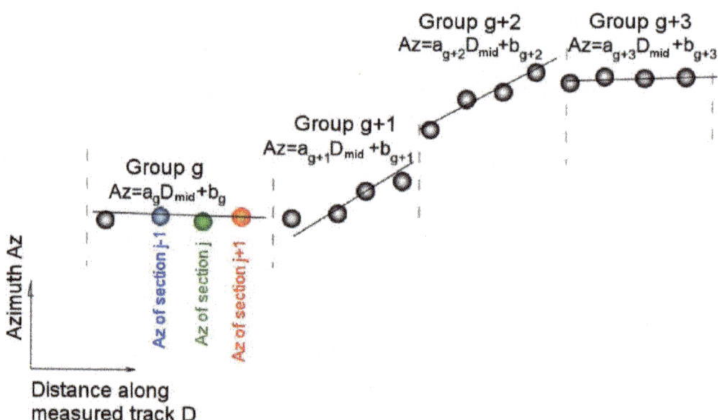

Figure 12. Schematic diagram of the analysis of the variability of azimuths along the discretized route.

According to the assumptions, if the absolute value of the direction coefficient a_g does not exceed the threshold value assumed in a particular group of sections, the algorithm assigns to a particular group (at this distance) the attribute of a straight section. The total length of the straight line is calculated as the sum of the lengths of the next (consecutive) groups of sections characterized by the stability of angle, and thus the direction coefficient of regression did not exceed a certain value close to zero. This approach makes it possible to find the main direction as well in combination with reverse curves. Then, the location with the value of regression close to zero will appear in the analysis.

(3) Once the records belonging to the straight sections are identified, the total, i.e., the linear regression defined on the whole set of points, is determined. Due to the fact that the straight sections can vary in length and for longer sections precision parameters of determining the position can undergo dynamic changes, the authors used a weighted regression analysis, in which the weights are defined by measurement uncertainties recorded for flat coordinates by GNSS receivers. The method implemented in this way maps the main directions on the sets of points with the differentiated distribution of position error more accurately than the simple linear regression (unweighted).

(4) The coefficients of the equations of straights constitute the basis for building the polygon, in which the coordinates of vertices are calculated as points of the intersection of adjacent (by indexation) straight lines. The coordinates of the points of the start and end points of the polygon are calculated based on measured extreme coordinates. A schematic diagram of the reconstructed system of vertices is shown in Figure 13.

(5) The last step of algorithm is the analysis of the slopes of the adjacent straight lines. In this step, it is possible to merge the adjacent straight lines with one another. This is the moment in which the two adjacent straight lines with different direction slopes should be accepted (or not) as the area of the change of main directions. The adjacent two straight lines can therefore be designated as two tangents of a horizontal curve or connected together in one tangent (main direction) of the polygon.

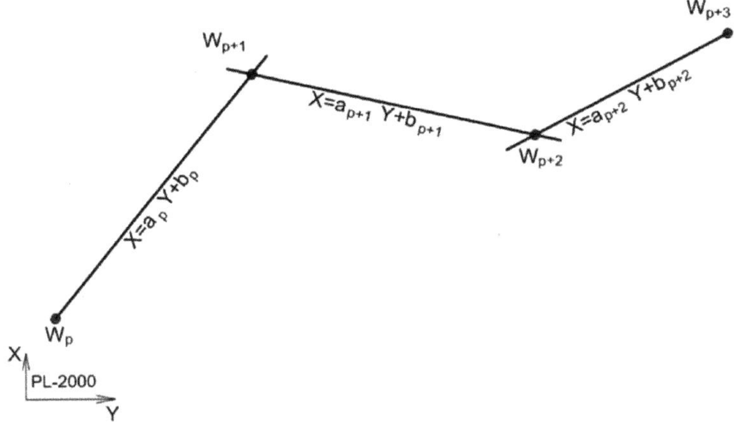

Figure 13. Schematic diagram of the reconstructed vertices system.

3. Results

For detailed analyses, the data obtained from receivers number 5 and 10 were selected. These receivers were mounted on the pivots of the tramway bogies, which means that the trajectory recorded by them in the horizontal plane maps the axis of the track. The exceptions are sections located in the curves with a cant, but the test section was selected in such a way that the impact of the cant is as small as possible. These receivers recorded a total number of 253,435 coordinates during the passage through the test section.

Then, the analysis of the resulting measurement signal was made, according to the algorithm shown in Section 2.4. For input data in the form of the measurement signal recorded by two receivers (with a frequency of 20 Hz) during the six passages of the route, the algorithm detected the presence of 536 straight sections, giving an average of about 45 straight sections making up the whole geometrical system. Figures 14–16 show the calculation results of the algorithm. They represent scalable windows of a program for data analysis, implemented in a Scilab environment. The automatically developed computer program defined the beginnings and ends of the straight sections, and in half the reconstructed section a label with the next number of section was pasted. Since the algorithm operated simultaneously on all 12 measuring signals, these fragments, which a number of straight lines overlap, can be treated—as the way of visual assessment of the result in the graph—as repeatedly reconstructed. If the numbers of straight lines are offset with respect to one another, it means that the algorithm was not able to reconstruct the beginnings and ends of straight or divided sections into shorter ones at a high level of repeatability, or indicated that the straight lines as false positive, recognizing, for example, the distortion of a signal or inequality of a track in a straight section as a curvilinear section.

It can be noted that on sections *a* and *b*, the algorithm identified the vertices most accurately. Figure 15 shows the vertices determined on the basis of identified straight lines in sections *b* and *c*. Section *c* compared to section *b* featured less precise characteristics of the straight lines reconstructed. The result was a clear dispersion of the designations of vertex in the whole series of measurements. In addition, Figure 14 shows that on the other sections of the test track (mostly *c*, *d*, *e* and *f*) a series of very short straight sections were measured, hindering the overall analysis (expressed as a loss of clarity of description at the scale assumed). In addition, in section *c* there were fragments of the track with very short reverse curves (shown in Figure 16), which constituted an additional problem for the automatic identification of the algorithm.

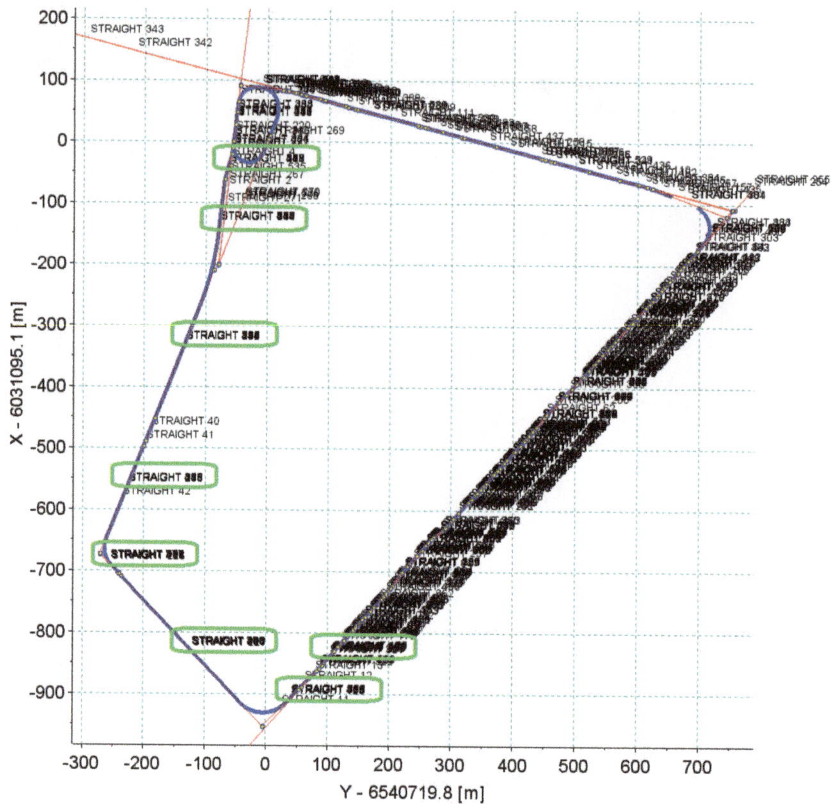

Figure 14. Automatic division of the measuring route into straight sections.

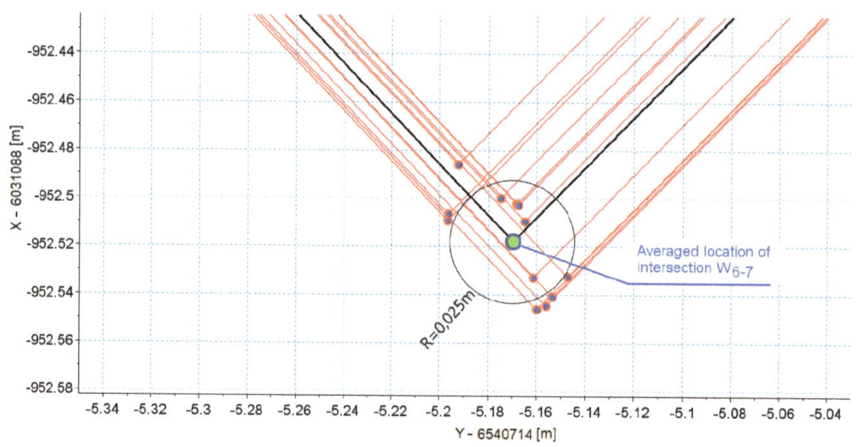

Figure 15. Result of the multiple reconstruction of the selected vertex (intersection) of the polygon.

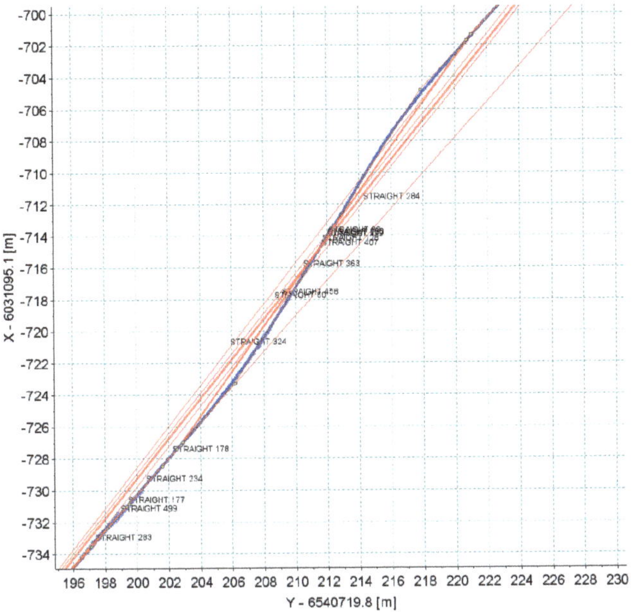

Figure 16. Close-up of the system with reverse curves.

4. Discussion

4.1. Analysis of Main Directions

Bearing in mind that in the case of tramway lines, deformations of the track axis in the horizontal plane can be of similar length to the short straight sections—full automation turned out to be very difficult. In this situation, for the further analysis of the main directions only those sections were selected for which there was little doubt that they constitute the main directions of the route.

Among the sections identified by the algorithm as straight sections, *a*, *b* and *c* were selected for further analysis of eight straight of sections. Figure 17 shows the final selected straights and straights selected initially (on the basis of a visual evaluation of the measurement data). To distinguish the straight lines from the earlier selected sections (Figure 9), for the first ones *Str′* markings were introduced.

Comparing both sets of straight lines, it can be concluded that there was a large convergence of both methods. There were significant differences in the division of *Str 6* to *Str′ 5* and *Str′ 6* and division of *Str 7* to *Str′ 7* and *Str′ 8*. It is worth noting that the reconstructions of straight sections of relatively short length and the longest ones were different. In addition, existing GNSS measurement uncertainties affect the repeatability of the reconstruction of very short sections. However, the uncertainty of designating the angle in the case of short sections will cause small deviations of the model course of the straights with respect to the actual course. However, in the case of long sections, even a small error in the designation of angle can have a significant effect on the results obtained at the ends of the reproduced geometric layout. That is why the indicator taking into account the length of a section was introduced. This indicator, designated as $\sigma_{Az(rel)}$, is the standard deviation of the reconstructed azimuth multiplied by the length of section. The obtained parameters of the sections examined are shown in Tables 3 and 4.

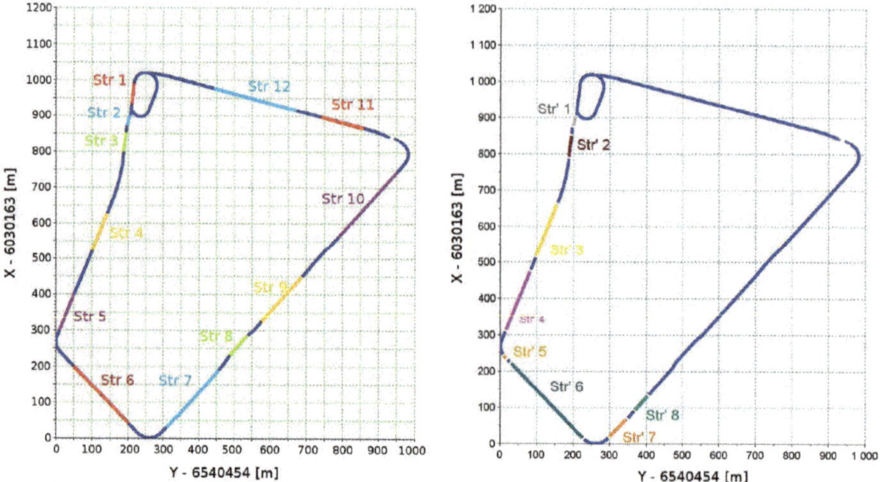

Figure 17. Comparison of the set of selected straights through a visual assessment (**left**) to a set of straights identified automatically (**right**).

For each measurement, a set of coordinates was determined. Then, the mean and standard deviation of the corresponding values were calculated. Based on the coordinates of vertices the distances $|\Delta W|$ between these vertices were calculated (of the individual measurements $i = 1, \ldots, n$) as well as the averaged vertex X_{av}, Y_{av}. The measure of uncertainty determined the vertex based on the straights reconstructed were therefore the distance $|\Delta W|_{av}$ and its standard deviation $\sigma_{|\Delta W|}$. All parameters describing the vertices, together with intersection angles α, are shown in Table 4.

The values in Tables 1 and 2 clearly show that Section 5 and the associated vertex W5–6 were reproduced with the highest uncertainty. Section 5 is a straight insert between the crossover needle (the measurement train moved in the reverse crossover direction) and the curve behind the crossover. This section is very short, and the intersection angle at this point is very small (less than 1 degree). The measurement on such a short section is burdened with considerable uncertainty to designate its direction, and also coordinates of vertices W4–5 and W5–6. Additionally, the vertices analyzed featured by a very small intersection angle, comparable to the value with the uncertainty to determine the angle. The coincidence of a short length of Section 5 and the small intersection angle in vertex W5–6 generated a very high level of uncertainty (mean value of error and standard deviation at 1.3 m). However, this was not the result of an incorrectly accepted measurement technique, but of geometric relationships in the measured geometric system. The coordinates of vertices with the favorable system of tangents were determined with very high reproducibility.

Table 3. Comparison of the reconstructed parameters of straight sections.

Section Number	Number of Measurements n	Average Length L_{av} [m]	Average Azimuth Az_{av} [deg]	Standard Deviation σ_{Az} [deg]	Indicator $\sigma_{Az(rel)}$ [deg*m]
1	12	19.1	196.139	0.026	0.500
2	12	71.5	186.536	0.002	0.116
3	11	194.9	201.427	0.001	0.176
4	12	187.1	202.196	0.002	0.332
5	12	0.7	135.792	0.346	0.233
6	12	274.5	136.415	0.002	0.670
7	12	50.9	44.051	0.012	0.603
8	7	37.3	41.706	0.008	0.313

Table 4. Comparison of the reconstructed parameters of the polygon's vertices.

Marking of Vertex	Angle of Return α_{av} [o]	Number of Measurements n	Vertical Coordinate X_{av} [m]	Standard Deviation σ_X [m]	Horizontal Coordinate Y_{av} [m]	Standard Deviation σ_Y [m]	Distance $\|\Delta W\|_{av}$ [m]	Standard Deviation $\sigma_{\|\Delta W\|}$ [m]
W1-2	9.603	12	−69.656	0.011	−62.422	0.064	0.049	0.041
W2-3	14.891	11	−86.535	0.007	−209.746	0.025	0.023	0.011
W3-4	0.769	11	−181.861	0.152	−452.648	0.391	0.337	0.226
W4-5	66.404	12	−271.796	0.098	−673.074	0.247	0.150	0.215
W5-6	0.623	12	−240.944	1.284	−704.802	1.339	1.261	1.306
W6-7	92.364	12	−5.170	0.017	−952.518	0.020	0.024	0.008
W7-8	2.345	7	86.052	0.126	−858.219	0.125	0.137	0.099

4.2. Accuracy of Measurement Based on the Repeatability of Signals

All originally selected straight sections were analyzed. The existence of numerous unfavorable deformations of the track and the field conditions for GNSS measurement should appear in the analysis of signal reproducibility. Figure 18 shows the results of all measurement series—for each straight section—recorded by one of the antennas positioned on the pivot (receiver 5). In most cases, a very good repeatability of the trajectory of this device was observed. It is clear that in cases of bends on the straight lines all runs recorded the same qualitative character of deformation. This observation makes it possible to trim signals so that they do not distort the precision assessment. Of course, cutting out only the deformed parts from the straight line disturbs the actual direction of the straight line; however, to assess the accuracy of the measurement itself such an operation is necessary. Figure 19 shows the results recorded by all receivers during the measurements of section *Str 9*. This section features the poor track conditions. Numerous irregularities and deformations caused the occurrence of vibrations on the measurement vehicle. Graphs of recorded signals indicated a large number of bends (deformations) along the section of this track. This is one of the sections where the algorithm of the detection of main directions selected a series of short straights. The signals recorded by all receivers revealed the same deformation of the track.

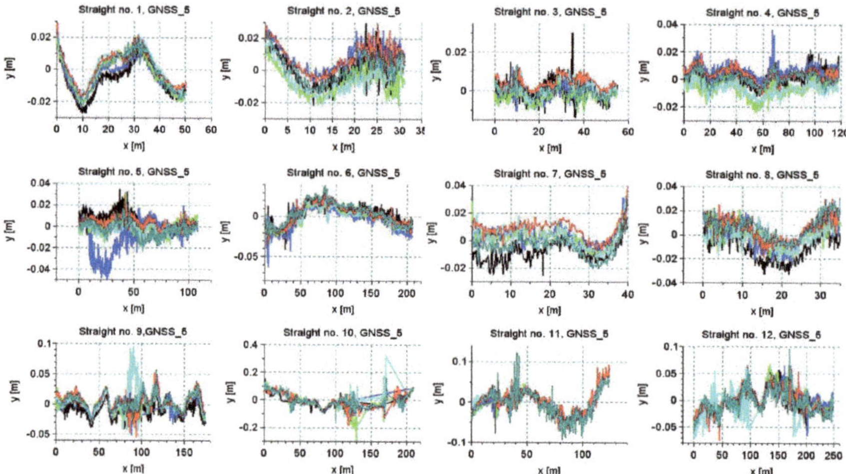

Figure 18. The results of a series of six measuring sections *Str 1–Str 12* recorded by receiver no. 5.

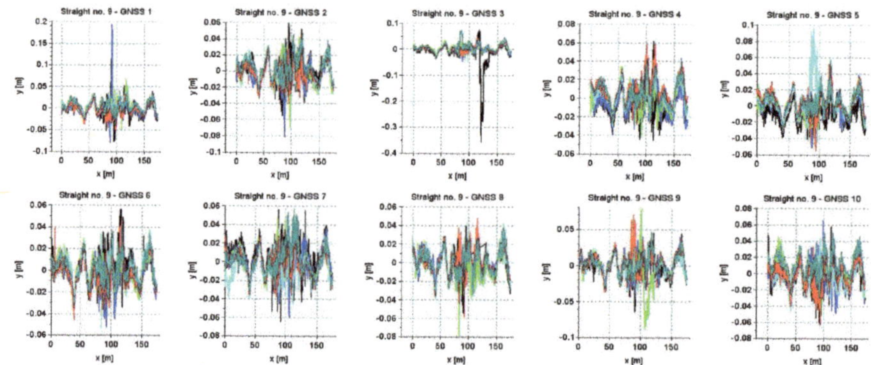

Figure 19. The results of a series of six measuring section of *Str 9* characterized by poor track conditions, recorded by all receivers of the measurement system.

Figures 20 and 21 show the results of the analysis of the reproducibility of the data recorded by receivers 5 and 10 for sections *Str 12* and *Str 6*. As seen in Table 1, these sections were characterized by a very unfavorable environment in the form of buildings and trees. This resulted in a discontinuity in data recording during part of the particular run. The analysis was done for a short segment that was recorded during all six crossings. As regards the accuracy in the reconstruction of direction, the standard deviation was at a level of 1–2 cm, depending on the recorded segment of a particular section, but reliability of measurement under these conditions was relatively low—every crossing showed disturbances of signal loss by different receivers.

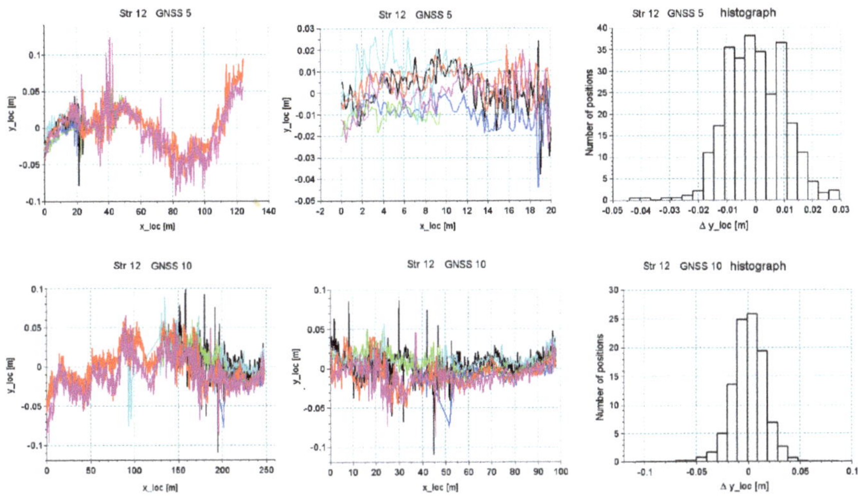

Figure 20. Analysis of the accuracy of the reproduction of the main direction in section *Str 12* featuring extremely unfavorable field screens.

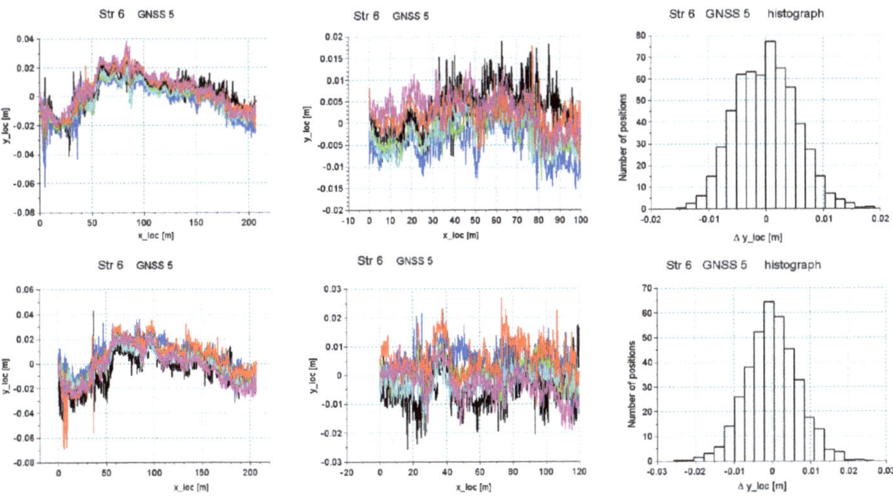

Figure 21. Analysis of the accuracy of the reproduction of the main direction in section *Str 6* featuring very good visibility of satellites.

The example of section *Str 6* (Figure 21) shows that in the accepted methodology of measurement, a very high reproducibility of independent measurements can be achieved, which directly proves a high precision of the reconstruction of the main directions of the tested route. However, the overall assessment ultimately depends on the field conditions in which the measurement is made. Systems based on several receivers operating independently increase the reliability. Therefore, even in difficult measurement conditions, data collection is still possible. However, the scope can be limited to certain sections of the route.

Figure 22 shows a graph of least squares lines in the local coordinate system for section *Str 6*. Also in this case, reference straights were brought to the local system, which facilitated the assessment of mutual relationships with individual straights. The black line represents the estimator of the expected value, that is, the real position of the track (reduced to level 0 of the local system). The lines in other colors are six estimated straight lines from the six individual runs. The graphs show that the maximum values of the straight lines most distant from one another did not exceed 1 cm.

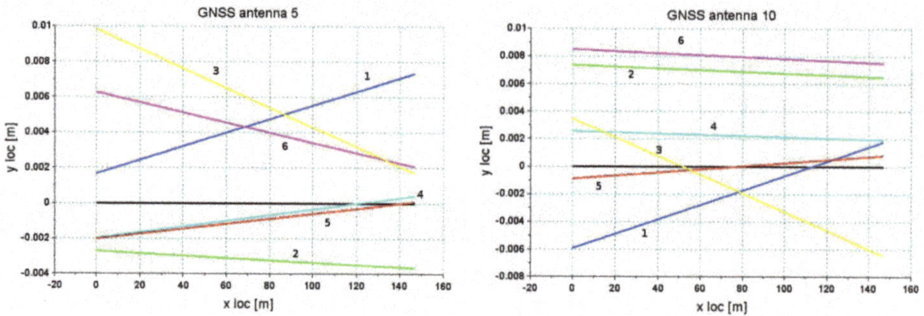

Figure 22. The relative position of straight line *Str 6* estimated from individual runs around the least squares straight line—antenna no. 5 (**left**) and no. 10 (**right**).

5. Conclusions

The selected test track met the assumptions as regards different conditions of satellite measurements. The analysis of the results clearly showed the impact of urban infrastructure on the possibilities of the reproduction of main directions along tramway lines. The algorithm presented using the regression analysis of measured points (representing the track axis) with the appropriate approach allowed efficient determination of the position of straight sections in a global system of references. However, the specification of geometric systems of tramway tracks (short geometric elements) led to the analysis of the main directions requiring a sufficiently high density of data. In the case of a large number of points, which are burdened with a high level of uncertainty, there is a risk of incorrect estimation of the identified parameters of the route directions, which in turn leads to the incorrect reconstruction of the polygon. The biggest problems with the polygon reconstruction are the intersection angles with very short tangents. The occurrence of small values of intersection angles (about 3 degrees or less) generates a relatively large uncertainty of the reproduction of the intersection of tangents (vertices). Finally, the coincidence of small intersection angles with short tangents makes that even a small deviation of the direction of the tangent causes very significant differences in the coordinates of the polygon's vertices.

As shown, the tramway infrastructure is characterized by an extremely unfavorable shape from the point of view of the reproduction of geometric systems in the global system. It is dominated by very short-length straight sections and curves with a complex system. In the measurements made, the shortest uniquely located straight section had a length of 0.7 m. It was on the extension of the main track in the crossing (straight section inserted within the turnout construction). Since railway systems have much longer straight sections, a successful measurement made on the tramway tracks is a positive prognostic for further research related essentially to the inventory of railway lines.

The analysis also showed that the geometric condition of track (rails' deformations and geometric imperfections) has a significant impact on the identification of the main directions of the route. One of the straight lines in the street terminus analyzed featured a high degree of geometric deformation, with the result that the algorithm of the identification of the main directions did not distinguish real directions (design) from the short straight lines resulting from the deformations present. The automatic detection of straight sections in such conditions turned out to be very difficult, requiring more advanced computational algorithms. Sections characterized by a good geometric condition were identified properly and the polygon designated on this basis was characterized by a relatively small error of the reproduction of vertices, at the level of a few centimeters.

Author Contributions: Conceptualization, A.W., C.S., W.K., K.K., J.S. (Jacek Szmagliński), and P.C.; data curation, C.S., J.S. (Jacek Szmagliński), P.C., and P.D.; formal analysis, J.S. (Jacek Szmagliński). and P.C.; investigation, A.W., C.S., W.K., K.K., J.S. (Jacek Skibicki), J.S. (Jacek Szmagliński), P.C., P.D., M.S. (Mariusz Specht), M.Z., and S.J., M.S. (Marcin Skóra) and S.G.; methodology, A.W., C.S., W.K., K.K., J.S. (Jacek Szmagliński), P.C., P.D., and M.Z.; resources, A.W., C.S., and K.K.; software, J.S. (Jacek Szmagliński) and P.C.; supervision, W.K.; validation, J.S. (Jacek Skibicki); visualization J.S. (Jacek Szmagliński) and P.C.; writing—original draft, W.K., J.S. (Jacek Szmagliński), and P.C.; writing—review and editing, W.K. and P.C. All authors have read and agreed to the published version of the manuscript.

Funding: Work were carried out as part of the research project "Development of innovative methods for determining the precise trajectory of rail vehicle" InnoSatTrack (POIR.04.01.01-00-0017/17), financed by NCBiR and PKP PLK SA.

Acknowledgments: The authors wish to thank the Company Gdańskie Autobusy i Tramwaje for making this measurement campaign possible.

Conflicts of Interest: The authors declare no conflict of interest.

References

1. NR (Network Rail). NR/L3/TRK/0030 Iss. 1, Reinstatement of Absolute Track Geometry (WCRL Routes), Network Rail 2008. Available online: https://standards.globalspec.com/std/10216330/nr-l3-trk-0030 (accessed on 5 August 2020).
2. CEN (European Committee for Standardization). BS EN 13803:2017 Railway Applications–Track–Track Alignment Design Parameters–Track Gauges 1435 mm and Wider. CEN. 2017. Available online: https://shop.bsigroup.com/ProductDetail/?pid=000000000030279796 (accessed on 5 August 2020).
3. Szwilski, T.B.; Begley, R.; Dailey, P.; Sheng, Z. Determining rail track movement trajectories and alignment using HADGPS. In Proceedings of the AREMA 2003 Annual Conference, Chicago, IL, USA, 5–8 October 2003; pp. 1–10.
4. Szwilski, A.B.; Dailey, P.; Sheng, Z.; Begley, R.D. Employing HADGPS to survey track and monitor movement at curves. In Proceedings of the 8th International Conference "Railway Engineering 2005", London, UK, 29–30 June 2005.
5. Glaus, R. The Swiss Trolley- A Modular System for Track Surveying. *Geodätisch-Geophys. Arb. in der Schweiz.* 2006, p. 184. Available online: https://www.sgc.ethz.ch/publications.html (accessed on 5 August 2020).
6. Akpinar, B.; Gülal, E. Multisensor railway track geometry surveying system. *IEEE Trans. Instrum. Meas.* **2012**, *61*, 190–197. [CrossRef]
7. Stein, D.; Spindler, M.; Kuper, J. Rail detection using lidar sensors. *Int. J. Sustain. Dev. Plan.* **2016**, *11*, 65–78. [CrossRef]
8. Chen, Q.; Niu, X.; Zuo, L.; Zhang, T.; Xiao, F.; Liu, Y.; Liu, J. A Railway Track Geometry Measuring Trolley System Based on Aided INS. *Sensors* **2018**, *18*, 538. [CrossRef] [PubMed]
9. RILA Track. Available online: https://www.fugro.com/our-services/asset-integrity/raildata/rila-track-rila-360 (accessed on 14 May 2020).
10. Gikas, V.; Daskalakis, S. Determining rail track axis geometry using satellite and terrestrial geodetic data. *Surv. Rev.* **2008**, *40*, 392–405. [CrossRef]
11. Lenda, G. Determining the geometrical parameters of exploited rail track using approximating spline functions. *Arch. Civ. Eng.* **2014**, *60*, 295–305. [CrossRef]

12. Li, W.; Pu, H.; Schonfeld, P.; Song, Z.; Zhang, H.; Wang, L.; Wang, J.; Peng, X.; Peng, L. A method for automatically recreating the horizontal alignment geometry of existing railways. *Comput.-Aided. Civ. Inf.* **2018**, *34*, 71–94. [CrossRef]
13. Elberink, S.O.; Khoshelham, K. Automatic Extraction of Railroad Centerlines from Mobile Laser Scanning Data. *Remote Sens.* **2015**, *7*, 5565–5583. [CrossRef]
14. Singh, K.; Swarup, A.; Agarwal, A.; Singh, D. Vision based rail track extraction and monitoring through drone imagery. *ICT Express* **2019**, *5*, 250–255. [CrossRef]
15. Xiong, Z.; Li, Q.; Mao, Q.; Zou, Q.A. 3D Laser Profiling System for Rail Surface Defect Detection. *Sensors* **2017**, *17*, 1791. [CrossRef]
16. Yang, B.; Fang, L. Automated extraction of 3-d railway tracks from mobile laser scanning point clouds. *IEEE J. Sel. Top. Appl. Earth Obs. Remote Sens.* **2014**, *7*, 4750–4761. [CrossRef]
17. Akpinar, B.; Gülal, E. Railway track geometry determination using adaptive Kalman filtering model. *Measurement* **2013**, *46*, 639–645. [CrossRef]
18. Zhou, Y.; Chen, Q.; Niu, X. Kinematic Measurement of the Railway Track Centerline Position by GNSS/INS/Odometer Integration. *IEEE Access* **2019**, *7*, 157241–157253. [CrossRef]
19. Koc, W.; Specht, C. Selected problems of determining the course of railway routes by use of GPS network solution. *Arch. Transp.* **2011**, *23*, 303–320. [CrossRef]
20. Specht, C.; Koc, W.; Chrostowski, P.; Szmaglinski, J. Accuracy Assessment of Mobile Satellite Measurements Relation to the Geometrical Layout of Rail Tracks. *Metrol. Meas. Syst.* **2019**, *26*, 309–321. [CrossRef]
21. Wilk, A.; Specht, C.; Koc, W.; Karwowski, K.; Chrostowski, P.; Szmagliński, J.; Dąbrowski, P.; Specht, M.; Judek, S.; Skibicki, J.; et al. Research project BRIK: Development of an innovative method for determining the precise trajectory of a railway vehicle. *Transp. Overv.* **2019**, *7*, 32–47. [CrossRef]
22. Dabrowski, P.; Specht, C.; Koc, W.; Wilk, A.; Czaplewski, K.; Karwowski, K.; Specht, M.; Chrostowski, P.; Szmaglinski, J.; Grulkowski, S. Installation of GNSS receivers on a mobile platform—methodology and measurement aspects. *Sci. J. Marit. Univ. Szczec.* **2019**, *60*, 18–26. [CrossRef]
23. Specht, M.; Szmagliński, J.; Specht, C.; Koc, W.; Wilk, A.; Czaplewski, K.; Karwowski, K.; Dąbrowski, P.S.; Chrostowski, P.; Grulkowski, S. Analysis of Positioning Methods Using Global Navigation Satellite Systems (GNSS) in Polish State Railways (PKP). *Sci. J. Marit. Univ. Szczec.* **2020**, *62*, 26–35. [CrossRef]
24. Specht, M.; Specht, C.; Wilk, A.; Koc, W.; Smolarek, L.; Czaplewski, K.; Karwowski, K.; Dąbrowski, P.S.; Skibicki, J.; Chrostowski, P.; et al. Testing the Positioning Accuracy of GNSS Solutions during the Tramway Track Mobile Satellite Measurements in Diverse Urban Signal Reception Conditions. *Energies* **2020**, *13*, 3646. [CrossRef]
25. Koc, W.; Specht, C.; Szmaglinski, J.; Chrostowski, P. A method for determination and compensation of a cant influence in a track centerline identification using GNSS methods and inertial measurement. *Appl. Sci.* **2019**, *9*, 4347. [CrossRef]
26. Dąbrowski, P.S.; Specht, C.; Felski, A.; Koc, W.; Wilk, A.; Czaplewski, K.; Karwowski, K.; Jaskólski, K.; Specht, M.; Chrostowski, P.; et al. The Accuracy of a Marine Satellite Compass under Terrestrial Urban Conditions. *J. Mar. Sci.* **2020**, *8*, 18. [CrossRef]
27. Specht, C.; Chrostowski, P.; Dąbrowski, P.; Szmagliński, J.; Specht, M.; Dera, M.; Koc, W.; Skóra, M. Mobile satellite measurements on the Pomeranian Metropolitan Railway. *Transp. Overv.* **2016**, *5*, 24–35. [CrossRef]
28. Specht, C.; Koc, W.; Chrostowski, P.; Szmaglinski, J. The Analysis of Tram Tracks Geometric Layout Based on Mobile Satellite Measurements. *Urban Rail Transit* **2017**, *3*, 214–226. [CrossRef]
29. Koc, W. Design of rail-track geometric systems by satellite measurement. *J. Transp. Eng.* **2012**, *138*, 114–122. [CrossRef]
30. Koc, W. Analytical method of modelling the geometric system of communication route. *Math. Probl. Eng.* **2014**, *2014*, 1–13. [CrossRef]
31. Koc, W.; Chrostowski, P. Computer-aided design of railroad horizontal arc areas in adapting to satellite measurements. *J. Transp. Eng.* **2014**, *140*, 1–8. [CrossRef]
32. Chrostowski, P.; Koc, W.; Palikowska, K. Prospects in elongation of railway transition curves. *P I Civil Eng-Transp.* **2017**, 1–12. [CrossRef]

Article

Window-Modulated Compounding Nakagami Parameter Ratio Approach for Assessing Muscle Perfusion with Contrast-Enhanced Ultrasound Imaging

Huang-Chen Lin [1] and Shyh-Hau Wang [1,2,*]

1 Department of Computer Science and Information Engineering, Institute of Medical Informatics, National Cheng Kung University, No. 1, University Road, East District, Tainan City 70101, Taiwan; lukaslin886@gmail.com

2 Intelligent Manufacturing Research Center, National Cheng Kung University, No. 1, University Road, East District, Tainan City 70101, Taiwan

* Correspondence: shyhhau@mail.ncku.edu.tw; Tel.: +886-06-2757575 (ext. 62519)

Received: 19 May 2020; Accepted: 23 June 2020; Published: 24 June 2020

Abstract: The assessment of microvascular perfusion is essential for the diagnosis of a specific muscle disease. In comparison with the current available medical modalities, the contrast-enhanced ultrasound imaging is the simplest and fastest means for probing the tissue perfusion. Specifically, the perfusion parameters estimated from the ultrasound time-intensity curve (TIC) and statistics-based time–Nakagami parameter curve (TNC) approaches were found able to quantify the perfusion. However, due to insufficient tolerance on tissue clutters and subresolvable effects, these approaches remain short of reproducibility and robustness. Consequently, the window-modulated compounding (WMC) Nakagami parameter ratio imaging was proposed to alleviate these effects, by taking the ratio of WMC Nakagami parameters corresponding to the incidence of two different acoustic pressures from an employed transducer. The time–Nakagami parameter ratio curve (TNRC) approach was also developed to estimate perfusion parameters. Measurements for the assessment of muscle perfusion were performed from the flow phantom and animal subjects administrated with a bolus of ultrasound contrast agents. The TNRC approach demonstrated better sensitivity and tolerance of tissue clutters than those of TIC and TNC. The fusion image with the WMC Nakagami parameter ratio and B-mode images indicated that both the tissue structures and perfusion properties of ultrasound contrast agents may be better discerned.

Keywords: contrast-enhanced ultrasound; muscle perfusion; Nakagami parameter; compounding Nakagami imaging

1. Introduction

As the largest of the soft tissues, muscle is mainly functioning for the protection and composition of almost all essential organs in the human body. In response to a variety of external attacks, most of the muscle diseases are still difficult to be fully differentiated, since muscles belong to heterogeneous tissues [1]. Consequently, only a few muscle symptoms may be diagnosed with a specific myopathy that is predominantly associated with proximal muscle weakness [2,3]. To date, several medical imaging modalities have been developed that are capable of noninvasively detecting the compositions and variations of muscles, including the tissue edema, fatty, and atrophic changes [3,4]. Nevertheless, it remains difficult for the majority of medical imaging modalities to precisely diagnose a specific muscular disease, since the morphological changes of tissues are usually non-disease-specific. Therefore, it certainly is desirable to further explore and develop alternate diagnostics for better assessing the states of muscular disease or treatment effect covering the microvascular changes in a local muscle tissue.

Among current available medical imaging modalities, nuclear medicine imaging [5], positron emission tomography [6,7], and contrast-enhanced magnetic resonance imaging [8,9] have been utilized to detect and assess the microcirculation of muscle tissues. For example, the positron emission tomography incorporated with scintigraphy methods has been applied to detect the local tissue perfusion by tracing the injected radioactive agents in the bloodstream. However, positron emission tomography has not been routinely implemented in clinical diagnosis for assessing muscle perfusion, owing to several such considerations as the availability of scanners and radiopharmaceuticals, exposure of ionizing radiation, and insufficient spatial resolution of images. The contrast-enhanced magnetic resonance imaging is without the hazard of ionizing radiation and that has been demonstrated to be able to detect tissue perfusion [10]. It, however, is still encountering insufficient imaging sampling rate to acquire changes of the tissue perfusion thoroughly. In contrast to those just-mentioned modalities, ultrasound imaging is a relatively safe modality that is frequently employed to diagnose the tissue structures [3,11], movement [12], and blood flow [10,13]. The frequency of most of the diagnostic ultrasounds is less than 10 MHz, and that tends to result in a short-of-sufficient spatial resolution to measure the blood flow or perfusion in the capillary beds. Certainly, it is straightforward to increase the resolution of the ultrasound image by the increase of employed ultrasound frequency. For instance, a 50 MHz high-frequency ultrasound is with micrometer resolution capable of assessing the burn degree of wounded muscle tissue [14]. Nevertheless, the increase of ultrasound frequency tends to unavoidably increase the acoustic attenuation and then decrease the depth and contrast of image greatly [15,16].

To further alleviate these hurdles, the air-filled microbubbles were developed as the ultrasound contrast agents (UCAs); they are able to generate much more scattering energy than those of rigid spheres of the same size and to stably flow throughout the capillary vessels [17,18]. The administrated UCAs in the bloodstream allow the acquisition of contrast-enhanced ultrasound (CEUS) imaging over a certain duration, to estimate the tissue perfusion, as well as to estimate perfusion parameters by utilizing the time-intensity curve (TIC) [13,19] technique, similarly to that of contrast-enhanced magnetic resonance imaging. The CEUS imaging and TIC have been shown to be able to detect and localize an altered microcirculation for the diagnosis and treatment monitoring of such muscular diseases as myocardial infarction and ischemia [20], renal diseases [21], and advanced diabetes mellitus [22]. The impaired skeletal muscle perfusion of the lower legs was also found to be a hallmark of various diseases, microvascular involvement in diabetes mellitus [23,24], and compartment syndrome [10,25]. Furthermore, the perfusion parameters estimated from the corresponding TIC, typically including the arrival time (AT), time-to-peak (TTP), peak value (MAX), blood flow velocity (BFV), and area under the curve (AUC), have been applied to assess the muscle microcirculation [13,19]. These perfusion parameters have been applied to assess patients with peripheral arterial occlusive disease (PAOD) [22]. After engaging in exercise, the PAOD patients tended to have a significantly longer TTP than those of the control subjects. The absolute flow reserve and TIC perfusion parameters were found capable of better assessing the state and severity of impaired and diseased muscles. Moreover, patients with advanced stages of peripheral arterial disease resulted in a longer TTP, lower MAXs, and slower BFVs [23,24]. In general, TTP, among other perfusion parameters estimated from CEUS imaging, was reported to be a more reproducible and suitable parameter for clinically assessing the muscle perfusion [26,27]. However, the accuracy of ultrasound-intensity-based TIC approach with the CEUS imaging for estimating the microcirculation may still be affected by the variations of microbubble scattering corresponding to ultrasound scanner settings (such as gain, time–gain compensation, and dynamic range) and acoustic properties of tissues (such as attenuation and clutter signals associated with non-perfused area of tissues, and vessel walls) [28,29].

In comparison with conventional ultrasound-intensity-based TIC approach, the statistical analysis of ultrasound backscattering signals, using Nakagami statistics, has been validated to be capable of further reducing the effects of tissue clutter and attenuation [30]. Specifically, the Nakagami parameter is a shape parameter that is able to quantitatively differentiate the probability density function (PDF) of ultrasonic backscattered envelopes into Rayleigh, pre-Rayleigh, or post-Rayleigh distributions [30,31]. Subsequently, instead of ultrasound intensity, the Nakagami parameter was implemented into the time–Nakagami parameter curve (TNC) for estimating the flow velocities, ranging from 2 to 18 mm/s, from the flow conduit of a tissue-mimicking phantom with a bolus of UCAs administration [32,33]. The TNC approach has achieved consistent results to those of the TIC approach, and it has further demonstrated a better tolerance of tissue clutter signals in the region of interest (ROI), without the need to utilize an additional wall filter. However, the TNC approach may still be encountered with the subresolvable effect [32] corresponding to the condition as that of the tissue clutter signals are sufficiently larger than the perfused UCAs signals. The subresolvable effect may lead to the underestimated Nakagami parameters and also affect the reproducibility of perfusion parameters derived from the TNC approach [32]. Moreover, the TNC approach with CEUS imaging might also be affected by the ultrasound spectral and energy characteristics associated with the variations of transmitted ultrasound (such as pressure amplitude, frequency, phase, pulse length, and pulse repetition rate), as well as UCAs properties (including shell viscoelasticity and gas solubility) and behaviors (including oscillation, coalesce, jet, or fragment) [17,18]. Therefore, prior the estimation of perfusion parameters with TNC approach to be regularly applied in clinical applications, the sensitivity and uniformity of this approach associated with the influence of UCAs behaviors, tissue clutters, and measurement conditions are essential to be further investigated.

In the present study, extensive efforts for better assessing the muscle perfusion were made by proposing the window-modulated compounding (WMC) Nakagami parameter ratio imaging with CEUS imaging, to alleviate the effects of tissue clutter, UCAs behaviors, and ultrasound system factors. The WMC Nakagami ratio imaging was realized by taking the ratio of Nakagami parameters from signals of the same region of interest insonified by two different acoustic pressures alternatively. The measurements from flow phantom and animal subjects were arranged and carried out to verify the proposed approach. Subsequently, the time–Nakagami parameter ratio curve (TNRC) implemented from a sequence of backscattering signals corresponding to a constant flow and bolus injection of UCAs in the flow phantom and animal bloodstream, respectively, was acquired and utilized for perfusion parameters' estimation. The WMC Nakagami parameter ratio images were subsequently fused with corresponding B-mode images, to better visualize and assess the perfusion areas of UCAs and the surrounding tissues.

2. Materials and Methods

2.1. Ultrasound Imaging System

The ultrasound imaging system for the present study was arranged and is schematically shown in Figure 1. The system mainly comprises a 7.5 MHz single-element ultrasound transducer (Panametrics V321, Waltham, MA, USA), pulser/receiver (Model 5072PR, Panametrics, Waltham, MA, USA), stage positioner (Model SGSP26-200, Sigma Koki, Tokyo, Japan) and controller (Model CSG-602R, Sigma Koki, Tokyo, Japan), and an 8-bit analog-to-digital converter (PXI-5152, National Instruments, Austin, TX, USA). The data-acquisition system was synchronized with the motor stage and controlled by the program developed with LabVIEW software (National Instruments, Austin, TX, USA). The pulse-echo response and characteristics of the transducer are given in Figure 2 and Table 1, respectively. The ultrasound transducer was driven by the bipolar signals generated and amplified, respectively, by an arbitrary function generator (AFG3252, Tektronix, Beaverton, OR, USA) and a broadband power amplifier (325LA, E&I, Rochester, NY, USA). Two acoustic pressures of 2.86 and 4.37 MPa, transmitted by the transducer and calibrated by a needle hydrophone (HNP-0200, ONDA Corporation, Sunnyvale, CA, USA), were prepared for flow phantom and animal experiments. The ultrasound signals were acquired and digitized at a 100 MHz sampling frequency. Subsequently, ultrasound B-mode images with a 42 dB dynamic range were formed, following a sequence of such processes as the filtering, Hilbert transform, logarithmic compression, and grayscale mapping. As a result, the imaging system achieves the axial and lateral image resolutions of 460 and 580 μm, respectively, measured from a 10 μm tungsten wire.

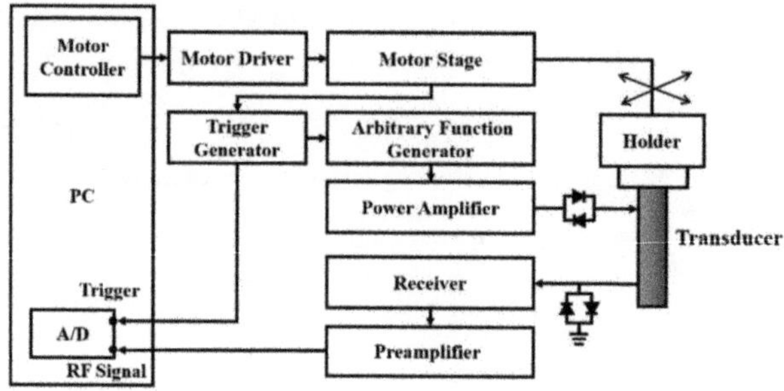

Figure 1. Schematic diagram of the ultrasound imaging system.

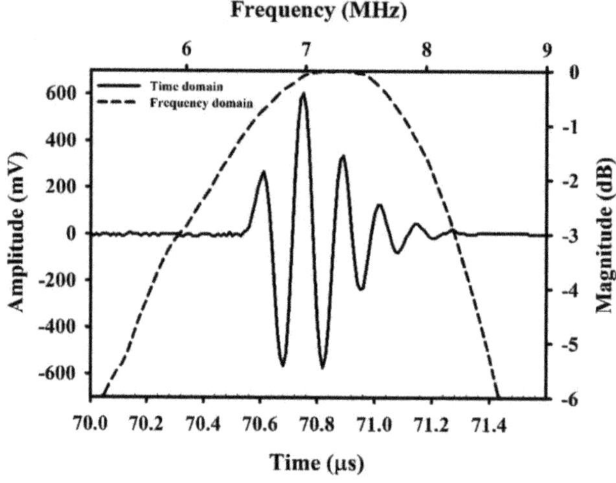

Figure 2. Pulse-echo response of the applied 7.5 MHz transducer.

Table 1. Characteristics of the applied transducer.

Center frequency	7.3 MHz
−6 dB bandwidth	3.2 MHz
f-number	2.8
Depth of focus	53.3 mm
Aperture size	19 mm

2.2. Flow Phantom Experiments

The arrangement of flow phantom experiments includes an ultrasound imaging system, flow phantom, and a syringe pump (KDS100, KD Scientific, New Hope, PA, USA), as given in Figure 3. The flow phantom is composed of a 1.5 × 1.5 mm^2 flow conduit embedded in a tissue-mimicking material that was made of a mixture of gelatin, graphite powder, and deionized distilled water (DD water). A total of four concentrations of UCAs (USphere™, Trust Bio-Sonics, Hsin-Chu, Taiwan) were prepared by mixing a 50 μL of 2.5 × 10^{10} UCAs/mL with 25, 50, 100, and 200 mL Ringer's solution.

The mean diameter of UCAs, 0.6 μm approximately, was measured by a dynamic light-scattering system (Delsa™ Nano C, Beckman Coulter, Brea, CA, USA). The UCAs suspensions of various concentrations were continuously infused into the flow phantom, using a syringe pump, at an infusion rate of 140 mL/h or mean flow velocity of 4.75 mm/s. For each measurement, a sequence of 500 ultrasound images was acquired at a rate of 4 frames/s for 125 s. Each frame of 10×15 mm^2 image is composed of 150 A-lines signals with a 100 μm scanning interval. The perfusion of UCAs was estimated from signals within a 2.5×4 mm^2 ROI. A total of ten repeated experiments for each concentration of UCAs was measured.

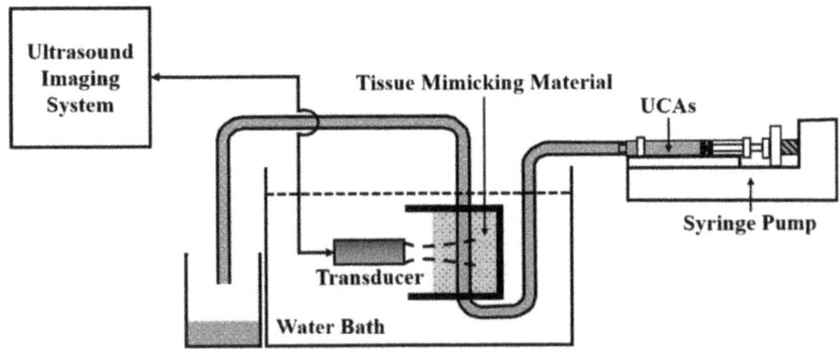

Figure 3. The arrangement for flow phantom experiments.

2.3. Animal Experiments

In vivo animal experiments were subsequently performed from 20 male Sprague Dawley rats (10 weeks old and 433.3 ± 43.1 (mean ± standard deviation) g weight). The rats were taken care of in a typical rodent vivarium in which a fixed temperature and a 12 h light/dark cycle environment were equipped. For each animal experiment, the rat was anesthetized, and then the hair on the right gastrocnemius muscle was subsequently removed. The arrangement of animal experiments is detailed in Figure 4, in which the bottom of the rectangular Plexiglas container was placed with a *p*-Xylene membrane. This allows the transmission and reception of ultrasound waves throughout the DD water, coupling gel, and the muscle tissue. The distance, 50.9 mm, between the transducer and the gastrocnemius muscle was adjusted, and that was subjective to the transducer's focal length. During the experiment, a bolus of UCAs suspension was injected into the vein of the rat's tail. For each measurement, a series of 600 images covering the process of UCAs perfusion was acquired. The size of each frame of image was 58×25 mm^2, and that was composed of 250 A-lines signals at a 100 μm scanning interval. The time perfusion curve was estimated from the ROI of 3.5×6.9 mm^2 within each frame of collected signals. The whole process of animal experiments was approved by the Ethical Committee for Animal Study (approval number: 108180), National Cheng Kung University, Tainan, Taiwan.

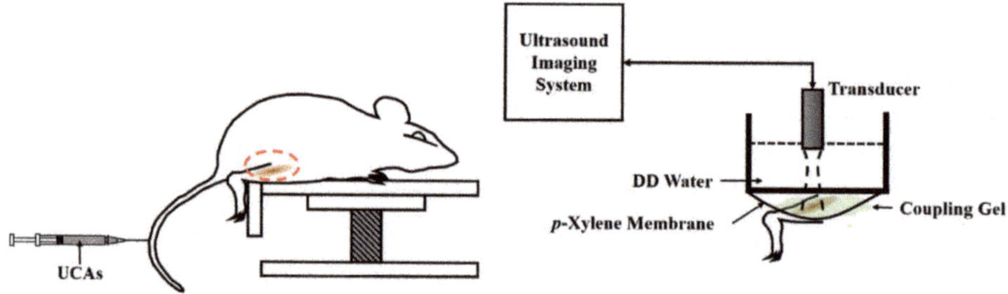

Figure 4. The arrangement for in vivo animal experiments.

2.4. Perfusion Parameters Estimation

The perfusion parameters were estimated from the Nakagami statistical distribution of UCAs backscattered signals. The Nakagami distribution, $f(R)$, of the backscattered envelope (R) is formulated [30,34] as follows:

$$f(R) = \frac{2m^m R^{2m-1}}{\Gamma(m)\Omega^m} \cdot e^{\left(-\frac{m}{\Omega}R^2\right)} \cdot U(R) \tag{1}$$

where $\Gamma(\cdot)$ and $U(\cdot)$ represent, respectively, the gamma function and unit step function. Ω and m denote, respectively, the scaling parameter and Nakagami parameter, and which may be formulated as follows:

$$\Omega = E\left(R^2\right) \tag{2}$$

and

$$m = \frac{\left|E\left(R^2\right)^2\right|}{E\left|R^2 - E\left(R^2\right)\right|^2} \tag{3}$$

in which $E(\cdot)$ is the statistical mean. The scaling parameter and Nakagami parameter correspond to the average power of the backscattered envelope and shape of statistical distribution, respectively. Furthermore, the WMC Nakagami imaging has been developed and validated to be able to better visualize the scattering properties in local tissues and to preserve the backscattered strengths within a certain size of the gated window [35]. The WMC Nakagami imaging, denoted as $M_{com}(x, y)$, may be implemented by summing and averaging all Nakagami images, $M(x, y)$, over different sliding windows ranging from 1 to N times of the pulse length, and which may be formulated as follows:

$$M_{com}(x, y) = \frac{1}{N}\sum_{i=1}^{N} M_i(x, y) \tag{4}$$

In the present study, WMC Nakagami parameter ratio imaging was further developed and aimed to reduce the effect of tissue clutter on WMC Nakagami imaging for improving the reproducibility of UCAs perfusion measurements. The WMC Nakagami parameter ratio imaging, denoted as $M_{cr}(x, y)$, is formulated as follows:

$$M_{cr}(x, y) = \frac{1}{N}\sum_{i=1}^{N} \frac{M_{high_i}(x, y)}{M_{low_i}(x, y)} - 1 \tag{5}$$

where $M_{high_i}(x, y)$ and $M_{low_i}(x, y)$ represent the acquired Nakagami images with respect to the incidence of a certain higher and lower acoustic pressures, respectively. Subsequently, TNRC may be estimated from a sequence of $M_{cr}(x, y)$ corresponding to UCAs perfusions. Those of M_{com} and M_{cr} within the ROI of WMC sliding windows were further averaged and denoted, respectively, as $\overline{m_{com}}$ and $\overline{mr_{com}}$.

The pseudo-color for displaying the Nakagami imaging was designated to reflect the magnitude of both $\overline{m_{com}}$ and $\overline{mr_{com}}$ ranging from 0 to 0.5 and 0 to 0.04, respectively.

Perfusion parameters, as mentioned in the Introduction section, including AT, TTP, MAX, BFV, and AUC [10,13], for further assessing the muscle perfusion, were estimated from the perfusion curves of backscattered power-based TICs, as well as those of Nakagami-parameter-based TNCs and TNRCs. Detail descriptions about the perfusion parameters are given in Table 2. Furthermore, BFV been verified by multivessel model [36] is proportional to the MAX, with the relationship given as follows:

$$\text{BFV} = d \times k \times \frac{2}{3}\text{MAX} \tag{6}$$

where d and k denote, respectively, the ultrasound beam width and the slope of perfusion curve over the perfusion time between AT and TTP. The variation of perfusion parameters was assessed by the relative standard deviation (RSD) [37] given as follows:

$$\text{RSD} = \frac{\sigma}{\mu} \times 100\% \tag{7}$$

where μ and σ represent, respectively, the mean and standard deviation of the corresponding perfusion parameters. The one-way ANOVA test was analyzed for assessing the significance, with the p-value smaller than 0.05, of results.

Table 2. Perfusion parameters derived from the perfusion curve.

Perfusion Parameter	Descriptions
Arrival time (AT)	Time at the arrival of UCAs
Peak value (MAX)	Peak value of the perfusion curve corresponding to the flow-in UCAs
Time-to-peak (TTP)	Interval between AT and the time of MAX
Blood flow velocity (BFV)	Mean blood flow velocity within the duration of UCAs perfusion
Area under perfusion curve (AUC)	Area under the curve of UCAs perfusion

3. Results

3.1. Flow Phantom Experiments

Figure 5a,b shows typical B-mode images corresponding to the incidence of 2.86 and 4.37 MPa acoustic pressures, respectively, into the flow phantom in which different UCAs concentrations of 0, 6.7, 13.3, 26.7, and 53.3 × 10^6 UCAs/mL were administered. Apparently, the largest concentration of UCAs administration contributes to the highest echogenicity in the conduit area. The yellow rectangular area indicates the radio-frequency signals in the ROI of 2.5 × 4 mm^2 for calculating the backscattered power, WMC Nakagami parameter, and WMC Nakagami parameter ratio parameter. The mean signal-to-noise ratios (SNR) corresponding to both incident acoustic pressures are higher than 22 dB and which satisfies the minimum SNR of 20 dB to estimate the Nakagami parameter for biological tissue characterization [38]. Moreover, the corresponding radio-frequency signals and envelopes acquired in the center stream of the flow conduit are given in Figure 6, in which the black arrow indicates the wall of the flow conduit. In response to the increase of incident acoustic pressures and UCAs concentrations, the strengths and variation of UCAs backscattered signals tended to increase accordingly. The corresponding PDFs and Nakagami parameters capable of describing the physical concentration, distribution, and properties of scatterers were given in Figure 7, in which the Nakagami parameters of backscattered envelopes generally increase with the increase of UCAs concentration and acoustic pressure. Furthermore, the average backscattered power ($\overline{BP}$), $\overline{m_{com}}$, and $\overline{mr_{com}}$ as a function of UCAs concentration are given in Figure 8, where these parameters apparently also tend to increase with the increase of UCAs concentrations. Specifically, the backscattered power of UCAs is highly affected by the incidence of acoustic pressure, as can be seen in Figure 8a; the $\overline{m_{com}}$, on the other hand, is less sensitive to that of the employed acoustic pressure. These results suggest that the Nakagami parameter is less sensitive to the backscattered power in response to different incident acoustic pressures, and thus that the blood perfusion in biological tissues may be better estimated by $\overline{mr_{com}}$, to reduce the influence of tissue clutters in biological tissues.

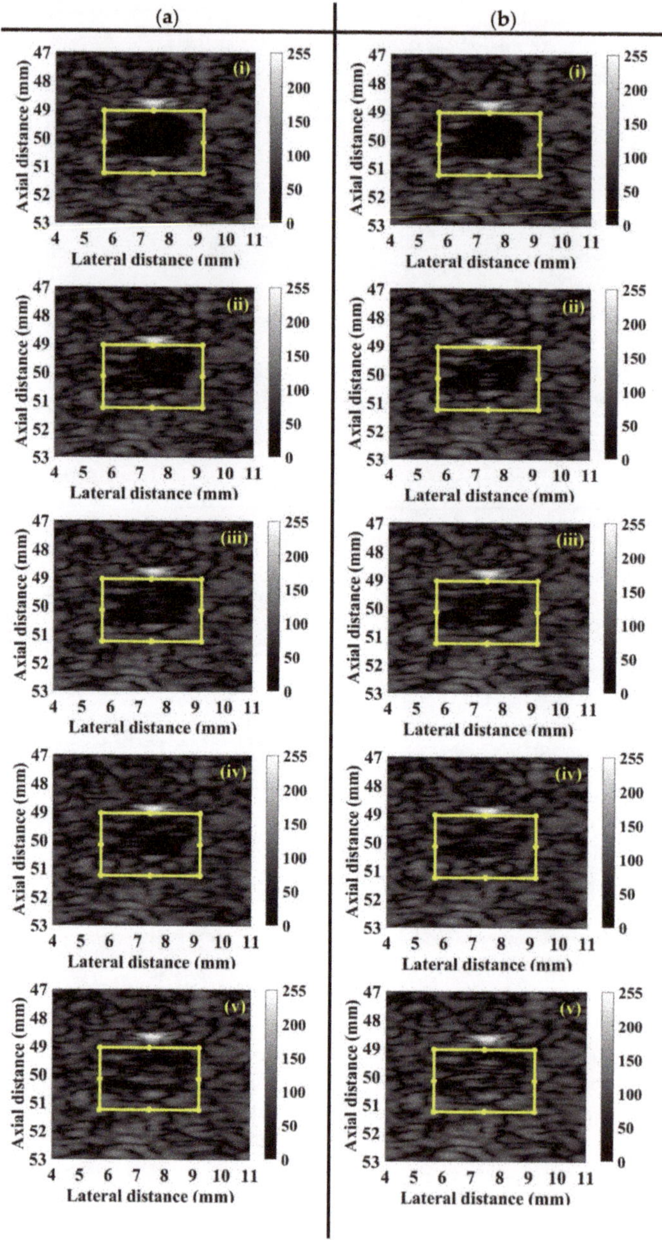

Figure 5. Typical B-mode images corresponding to the incidence of (**a**) 2.86 and (**b**) 4.37 MPa acoustic pressures into the flow phantom in which UCAs concentrations of (i) 0, (ii) 6.7, (iii) 13.3, (iv) 26.7, and (v) 53.3 × 106 UCAs/mL were administered.

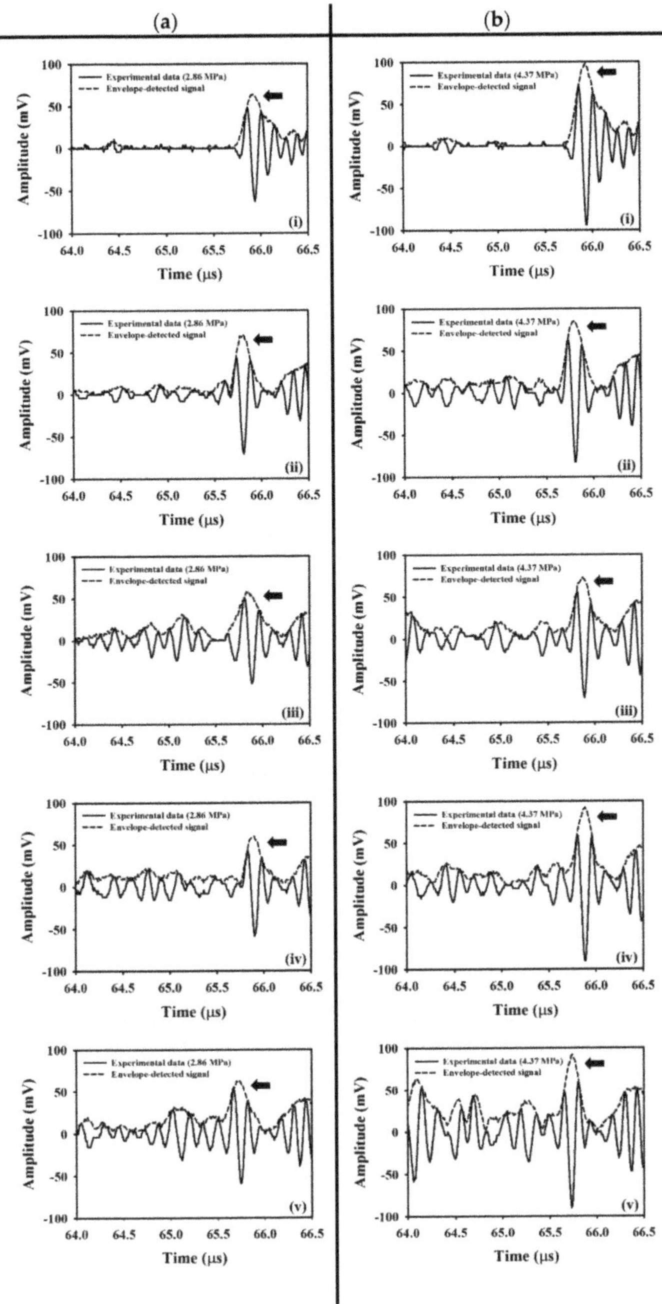

Figure 6. Typical radio-frequency signals and envelopes corresponding to the incidence of (**a**) 2.86 and (**b**) 4.37 MPa acoustic pressures into the flow phantom, where UCAs of various concentrations of (i) 0, (ii) 6.7, (iii) 13.3, (iv) 26.7, and (v) 53.3 × 10^6 UCAs/mL were administered. The black arrow indicates the back wall of the flow conduit.

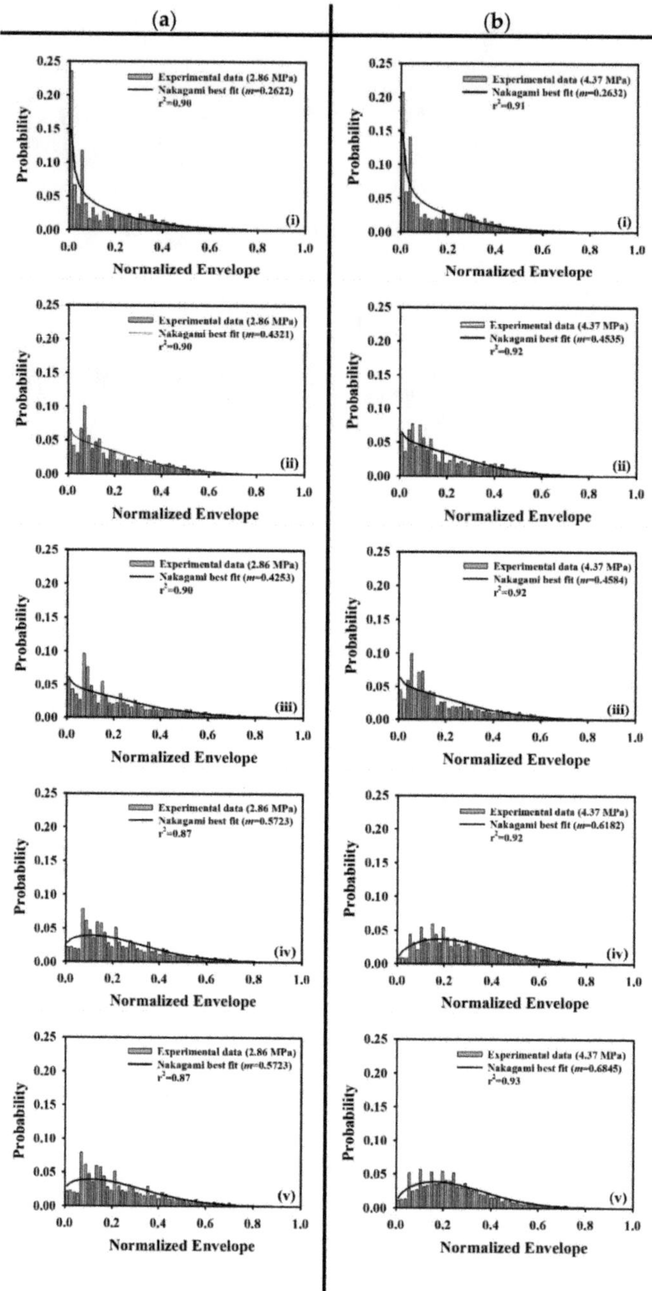

Figure 7. Typical probability density functions (PDFs) and fitted Nakagami distribution corresponding to the incidence of (**a**) 2.86 and (**b**) 4.37 MPa acoustic pressures into the flow phantom in which UCAs of various concentrations of (i) 0, (ii) 6.7, (iii) 13.3, (iv) 26.7, and (v) 53.3 × 10^6 UCAs/mL were administered.

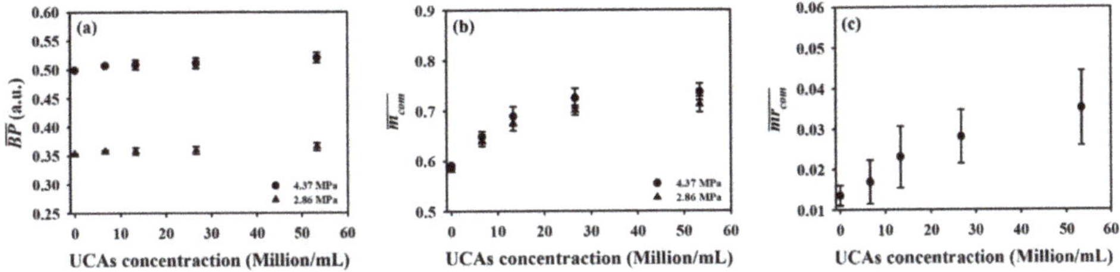

Figure 8. Average (**a**) backscattered power, (**b**) window-modulated compounding (WMC) Nakagami parameter, and (**c**) WMC Nakagami parameter ratio parameter as a function of UCAs concentration.

3.2. In Vivo Animal Experiments

For each animal experiment, a bolus of 10 μL UCAs suspensions with the concentration of 1.25×10^8 UCAs/mL was manually injected into the tail vein of the rat, using a 26G syringe. A series of typical B-mode images, depicting the muscle tissue of pre-contrast, 55 s, 75 s, and 110 s, following UCAs administration associated with the incidence of (a) 2.86 and (b) 4.37 MPa acoustic pressures, was given in Figure 9. The ultrasonic signals of ROI achieved, respectively, the SNRs of 31.12 ± 0.06 dB and 32.39 ± 0.06 dB, with respect to the insonification of 2.86 and 4.37 MPa acoustic pressures. The B-mode images in the area of hindlimb GM apparently increased in response to the wash-in of UCAs at approximately 55 s, and then gradually decreased till that of the wash-out at approximately 110 s. The corresponding WMC Nakagami images, given in Figure 10, were implemented and that the Nakagami parameters were estimated from the sliding square windows, with the window lengths varying from one to eight times of the transducer pulse length. The pre-contrast WMC Nakagami images, as shown in Figure 10a(i),b(i), of UCAs perfusion regions depicted with dark-blue color and indicated by the white arrows, were resultant from the incidence of 2.86 and 4.37 MPa acoustic pressures, and which are with the mean $\overline{m_{com}}$ smaller than 0.15. The flow-in UCAs tended to increase the resultant mean $\overline{m_{com}}$ of the WMC Nakagami images in Figure 10a,b of (ii) 55 s, (iii) 75 s, and (iv)110 s to be larger than 0.2.

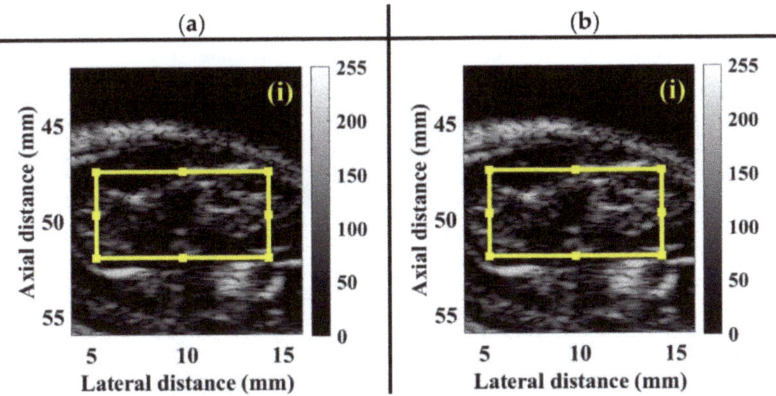

Figure 9. *Cont.*

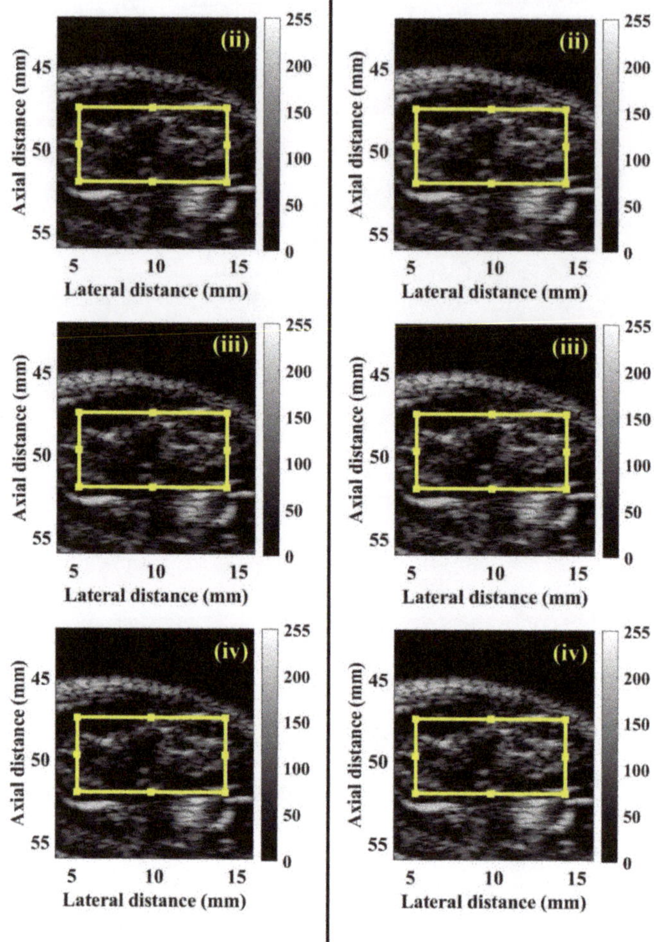

Figure 9. Typical B-mode images of muscle tissue at (i) pre-contrast, (ii) 55 s, (iii) 75 s, and (iv) 110 s, following UCAs administration corresponding to the incidence of (**a**) 2.86 and (**b**) 4.37 MPa acoustic pressures. The region of interest (ROI) is indicated by the yellow rectangular area.

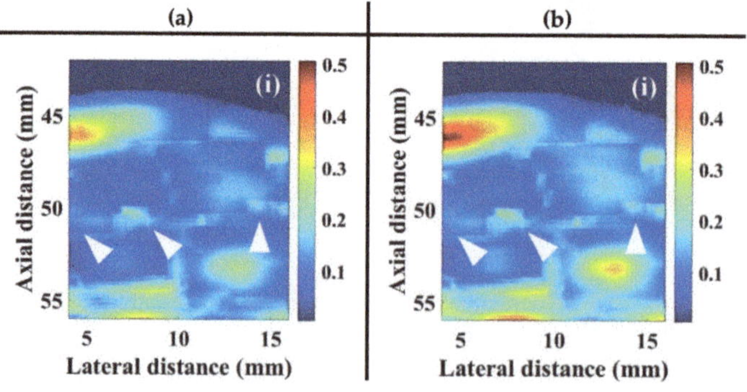

Figure 10. *Cont.*

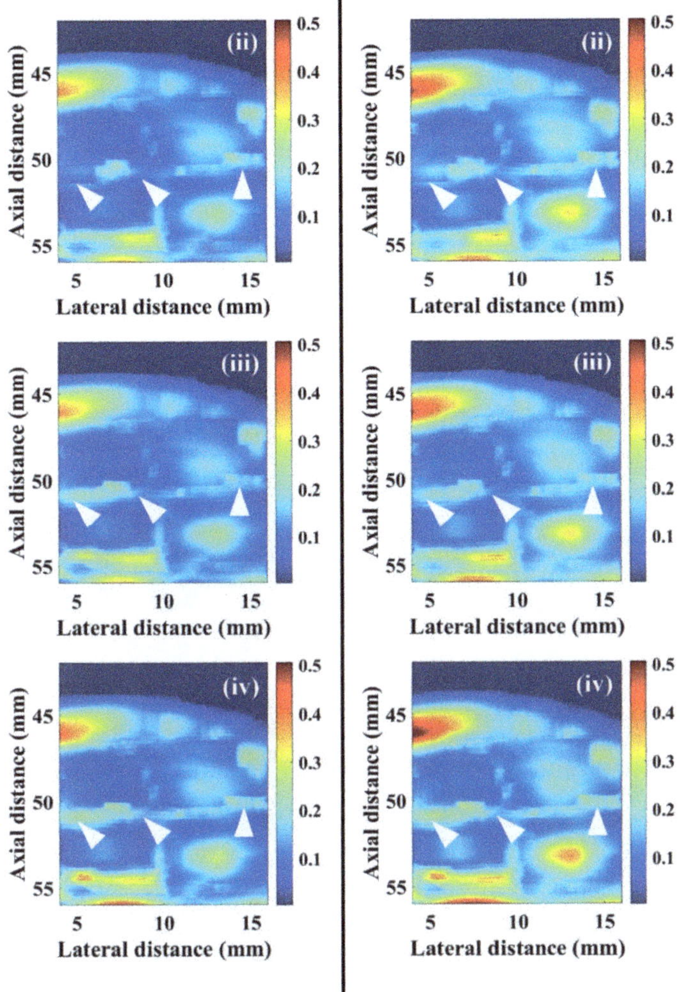

Figure 10. WMC Nakagami images of muscle tissue at (i) pre-contrast, (ii) 55 s, (iii) 75 s, and (iv) 110 s, following UCAs administration corresponding to the incidence of (**a**) 2.86 and (**b**) 4.37 MPa acoustic pressures. The perfusion areas of UCAs are indicated by white arrows.

In the present study, the WMC Nakagami parameter ratio images were implemented by taking the ratio of WMC Nakagami parameters corresponding to two different acoustic pressures, to reduce the effects of variations of scattering and subresolvable effect of UCAs, for better assessing muscle perfusion. Figure 11a is a series of typical WMC Nakagami parameter ratio images of muscle tissues, following the UCAs administration at (i) pre-contrast, (ii) 55 s, (iii) 75 s, and (iv) 110 s. The regions near the bottom of the lateral and medial GM muscle (as indicated white arrows) in the pre-contrast WMC Nakagami parameter ratio image may be discerned, as shown by the deep-red shadings in Figure 11a(i), with the $\overline{mr_{com}}$ smaller than 0.01. The perfusion areas (as indicated white arrows) following the first-pass of UCAs can also be readily discerned, as the bright-yellow color in Figure 11a(ii), with the mean $\overline{mr_{com}}$ larger than 0.025. The perfusion areas (as indicated white arrows) in Figure 11a(iii) become darker after the peak of the first pass of UCAs and that correspond to the mean $\overline{mr_{com}}$ larger than 0.015. The perfusion areas eventually fade to the deep-red shadings corresponding the flow-out of UCAs with the mean $\overline{mr_{com}}$ returning to values of smaller than 0.01 similarly to that of the pre-contrast. Furthermore, to better visualize and assess tissue perfusion, the image fused the WMC Nakagami parameter ratio images with the corresponding B-mode images was proposed and implemented. The fusion image provides not only the visualization of the anatomical structures by B-mode images, but also the additional Nakagami-parameter-based perfusion information with less subresolvable effect affected by tissue clutters. Figure 11b(i–iv) shows the corresponding WMC Nakagami parameter ratio/B-mode fusion images, in which the perfusion areas of UCAs are indicated as white arrows. Note that the fusion images provide abundant information about the anatomical tissue structure and areas of UCAs perfusion, and that the perfusion areas in the muscle tissues are with an appearance less blurring than those of WMC Nakagami and WMC Nakagami parameter ratio images.

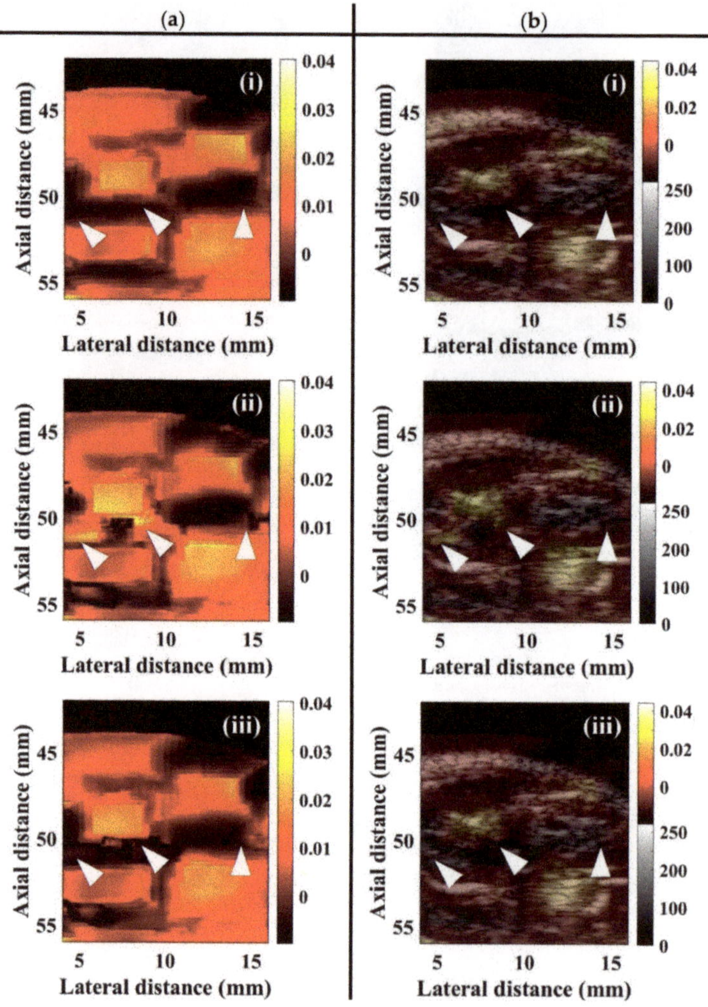

Figure 11. *Cont.*

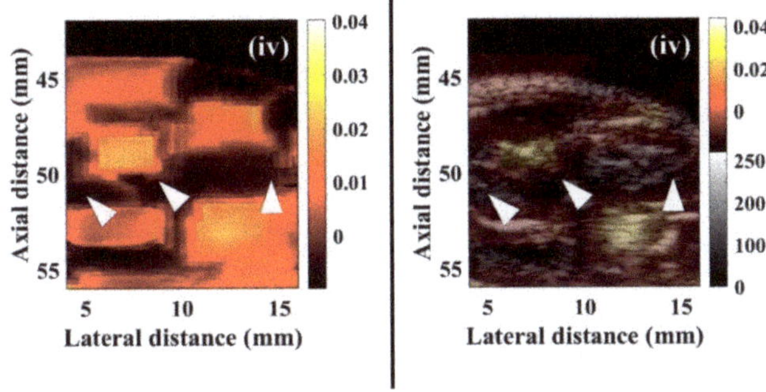

Figure 11. Typical (**a**) WMC Nakagami parameter ratio images and (**b**) fusion of WMC Nakagami parameter ratio and B-mode images at (i) pre-contrast, (ii) 55 s, (iii) 75 s, and (iv) 110 s, following UCAs administration. The areas of UCAs perfusion are indicated by white arrows.

To further estimate the tissue perfusion by UCAs, the $\overline{BP}$, $\overline{m_{com}}$, and $\overline{mr_{com}}$ as a function of time, as given in Figure 12, were measured and then applied to estimate the perfusion parameters with respect to TIC, TNC, and TNRC, respectively. The TIC in Figure 12a clearly shows that backscattered strengths of signals resultant from the insonification of the higher acoustic pressure of 4.37 MPa are much larger than that of 2.86 MPa, and that the shapes and intervals of TIC remain consistent. The TIC tended to decrease gradually with the decrease of perfused UCAs concentration, and with that, the maximum $\overline{BP}$ corresponding to the first arrival of UCAs may be clearly discerned. It yet is difficult to discern the signals backscattered from the recirculated UCAs, as they were mixing with backscattering signals of tissue clutters, as well as affected by the incidence of acoustic pressures, transducer beam characteristics, imaging system factors, and tissue properties. On the other hand, the TNC in Figure 12b tends to be able to preserve the first recirculation of UCAs, as indicated by the arrows; however, as a result of the subresolvable effect [32], the second recirculation of UCAs was barely observable. Results of Figure 12c demonstrated that the TNRC approach in comparison with those of TIC and TNC approaches is able to better reduce the subresolvable effect and to preserve detailed information about the UCAs perfusion in the tissue. Furthermore, the corresponding perfusion parameters, including AT, TTP, and BFV in Figure 13, estimated from TNRC, are generally in good agreement with those of from TNC and TNRC. The average AT (26.73 s) estimated from TNC was generally much shorter than those of calculated by TIC (32.54 s, $p < 0.001$) and TNRC (30.81 s, $p < 0.001$), as shown in Figure 13a. This could be partially due to a large variation of backscattering signals around the onset flow-in UCAs corresponding to the largest RSD of AT estimated from TNC, as given in Table 3. In particular, the AT and BFV parameters derived from the TNRC approach achieved a smaller RSD than those from the TIC and TNC approaches.

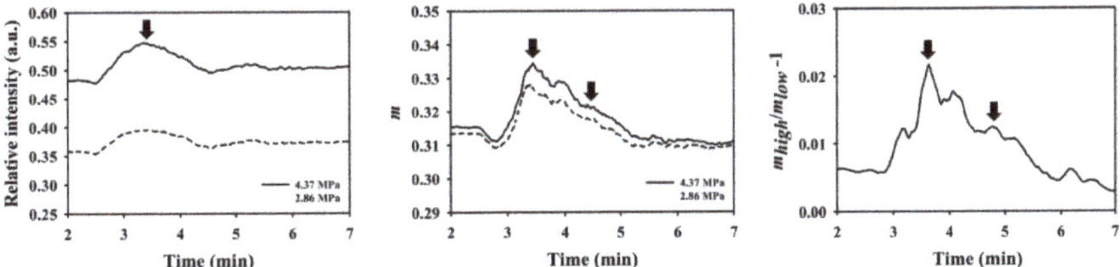

Figure 12. Typical (**a**) backscattered power, (**b**) WMC Nakagami parameter, and (**c**) WMC Nakagami parameter ratio as a function of time measured from the hindlimb GM of a rat. The black arrows indicated the peak concentrations corresponding to the first arrival and the first recirculation of administrated UCAs.

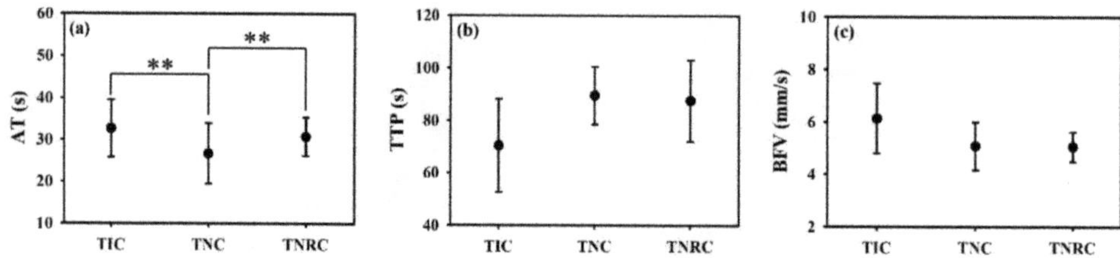

Figure 13. The comparison of (**a**) arrival time, (**b**) time-to-peak, and (**c**) blood flow velocity estimated from TIC, TNC, and TNRC (N = 20, **: $p < 0.01$, data were presented as mean ± standard deviation).

Table 3. Relative standard deviation (RSD) of perfusion parameters estimated from different time curves of contrast-enhanced ultrasound (CEUS) imaging.

Parameter \ Method	TIC	TNC	TNRC
AT	19.27%	53.78%	10.07%
TTP	25.20%	12.29%	17.12%
MAX	33.33%	11.11%	26.08%
BFV	19.59%	14.93%	9.24%
AUC	24.04%	8.62%	17.48%

4. Discussion

It is of importance to efficiently reduce the effects of tissue clutters to allow CEUS imaging to be able to better visualize and assess the perfusion of a local tissue for in situ diagnosing the muscular diseases. The variation of backscattering signals of UCAs and such ultrasound system factors as gain, time-gain compensation, and dynamic range are also primary factors to affect the accuracy of perfusion measurement by conventional TIC approach. These factors have been found to be able to be eliminated by the statistical analysis of signals, using Nakagami statistics [30,31]. Furthermore, by replacing the ultrasound intensity with the Nakagami statistical parameter, the TNC approach was found capable of better tolerating tissue clutters and UCAs attenuations, without the need to utilize an additional wall filter for estimating the flow from phantom experiments [32,33]. The accuracy of the TNC approach on the flow estimation was improved because this approach is less affected by UCAs concentrations, incident acoustic pressures, attenuation, and intrinsic noise of the system [32]. Nevertheless, the reproducibility and robustness of the TNC approach for assessing tissue perfusion could still be affected by a variety of small vessels, hydrostatic pressures, and heterogeneous tissue clutters [39,40]. These studies utilized a 25 MHz ultrasound in an attempt to increase the spatial resolution of CEUS images. The incident frequency, however, is far from the resonance frequency of the UCAs [31]. The mismatch between the employed ultrasound frequency and the resonance frequency of UCAs tended to limit a thorough study on the pressure-dependent behavior of UCAs' scattering properties [41,42]. The abovementioned issues were further alleviated by the utilization of WMC Nakagami parameter ratio approach with CEUS imaging in the present study. Specifically, the WMC Nakagami parameter ratio image is capable of preserving the smoothness of parametric image and reducing the tissue clutters effect by utilizing both the window-modulated compounding imaging and the ratio of Nakagami parameters resultant from the acquired signals associated with the incidence of two different acoustic pressures. Consequently, it allows the Nakagami-parameter-based CEUS imaging to be a suitable candidate approach for better assessing the tissue perfusion than that of conventional ultrasound-intensity-based CEUS imaging. Moreover, the spatial resolution of the Nakagami image typically is degraded and insufficient to assess and characterize the properties of a local tissue in detail than that of B-mode image due to the use of a minimum window size of three times of the transducer pulse length for Nakagami imaging [14,43]. This issue may be further resolved by the proposed fusion imaging, using ultrasound B-mode and WMC Nakagami parameter ratio images. The fusion imaging provides the information of not only the tissue structures in the perfusion area and surroundings, but also their scattering and perfusion properties.

In general, the estimations of perfusion parameters in flow phantom experiments by the TNRC approach are consistent with those of the TIC and TNC approaches, as well as those reported in previous studies [32,33]. The subsequent animal experiments resulted in a slight increase of echogenicity of B-mode images between the pre-contrast and post-contrast of UCAs, and that however is short of sensitivity to detail the variations of echogenicity between the wash-in and wash-out of UCAs. The estimation of perfusion parameters by using TIC approach was easily contaminated by the clutter signals from the surrounding tissues and vessel walls. The variation of perfusion parameters in animal experiments also tended to be higher, owing to the fact that slight movement of the animal was occasionally found during data acquisition in experiments. The WMC Nakagami images are generally less affected by the tissue clutter to discern the area of tissue images with or without the perfusion of UCAs than that of the B-mode images. However, although the SNRs of ultrasound signals of both flow phantom and animal experiments were higher than 20 dB, the resultant WMC Nakagami images appeared to have blurred features and reduced spatial resolution. This is partially due to the subresolvable effect that may reduce the dynamic range of Nakagami imaging to be between 0.24 and 0.36, as reported in previous studies [32]. The subresolvable effect on WMC Nakagami images may be readily discerned in Figure 10, in which it led the images at 75 and 110 s post-contrast of UCAs, as the indicated areas with bright-blue color near the bottom of the lateral and medial GM muscles, to be with the blurred appearance and reduced spatial resolution. The predominant cause of this result is due to the variation of backscattered statistics associated with the fascia of GM muscle.

The WMC Nakagami parameter ratio imaging and TNRC approach in the present study have further demonstrated themselves to be able to better describe and assess the perfusion of UCAs in the muscle, by reducing the subresolvable effect of tissue clutters with the normalization of WMC Nakagami parameters corresponding to the incidence of two acoustic pressures alternatively. The perfusion parameters estimated from the newly developed WMC Nakagami parameter ratio and TNRC approach were generally comparable to those of the TIC and TNC approaches. Specifically, the AT (26.73 s) estimated from the TNC approach is significantly shorter than that of the TIC (32.54 s) and TNRC (30.81 s) approaches, and that may readily correspond to the dramatic decrease of WMC Nakagami parameters to the onset flow-in of UCAs in Figure 12b. The RSD of AT and BFV estimated from the TNRC approach is smaller than that of the TIC and TNC approaches. This indicates that the TNRC approach has better tolerance on speckle noise and tissue clutters for reducing the subresolvable effect. The WMC Nakagami parameter ratio imaging is not yet suitable to be adopted for assessing the non-perfusion areas with CEUS imaging, since the surrounding tissues tend to be more rigid and exhibit less pressure-dependent behavior than those of UCAs [17,18]. Apparently, this issue may be alleviated by the introduced WMC Nakagami parameter ratio/B-mode fusion image to preserve both the anatomical structure and perfusion information. Consequently, it is worthwhile to further design and develop a dedicated compound ultrasound system to combine the CEUS images and perfusion parameters into a single metric for comprehensively diagnosing tissue perfusion. The metric containing certain pathological meanings may also be served as inputs for automatically classifying the pathological tissue by machine- and deep-learning classifiers in a future study.

5. Conclusions

This study employed phantom measurements and in vivo experiments on the rat leg muscles, to assess the UCAs perfusion, utilizing the newly developed WMC Nakagami parameter ratio imaging and TNRC approach. The animal experiments achieved some important findings and practical considerations, given as follows. Firstly, the subresolvable effect leads the WMC Nakagami imaging to have a blurred appearance and reduced spatial resolution, and that can be readily discerned near the bottom of the lateral and medial GM muscle. It thus can further reduce the reproducibility and robustness of the TNC approach for assessing tissue perfusion. Secondly, the AT of 26.73 s derived from the TNC approach is much shorter than those of TIC (32.54 s) and TNRC (30.81 s), and that corresponds to a dramatic decrease of WMC Nakagami parameters to the onset flow-in of UCAs. Thirdly, the RSDs of both AT and BFV derived by the newly developed TNRC approach are smaller than those of TIC and TNC approaches, indicating that the WMC Nakagami parameter ratio imaging may effectively suppress the subresolvable effect. On the other hand, the WMC Nakagami parameter ratio imaging is not suitable to assess the non-perfusion areas from tissues, owing to their limited pressure-dependent behavior. This issue was alleviated by fusing the newly developed WMC Nakagami parameter ratio and B-mode image to preserve both the anatomical structure and perfusion information. Furthermore, various parameters and CEUS images corresponding to a certain pathological meaning are worthwhile to be introduced into machine- and deep-learning classifiers for automatically classifying and diagnosing the perfusion of tissues.

Author Contributions: H.-C.L. designed and conducted the experiments, developed the algorithm for signal and image processing, and wrote the manuscript; and S.-H.W. supervised the whole work in this study and proof-read the manuscript. All authors have read and agreed to the published version of the manuscript.

Funding: This work was supported by the Ministry of Science and Technology (grant numbers: 107-2221-E-006-153; 108-2221-E-006-116) and the "Intelligent Manufacturing Research Center" (iMRC) from The Featured Areas Research Center Program within the framework of the Higher Education Sprout Project by the Ministry of Education (MOE) in Taiwan.

Acknowledgments: The authors would like to thank the support of the laboratory, university, government, and the Intelligent Manufacturing Research Center (iMRC).

Conflicts of Interest: The authors declare no conflict of interest.

References

1. Marini, M.; Veicsteinas, A. The exercised skeletal muscle: A review. *Eur. J. Transl. Myol.* **2010**, *20*, 105–120. [CrossRef]
2. Aspelin, P.; Ekberg, O.; Thorsson, O.; Wilhelmsson, M.; Westlin, N. Ultrasound examination of soft tissue injury of the lower limb in athletes. *Am. J. Sports Med.* **1992**, *20*, 601–603. [CrossRef] [PubMed]
3. Draghi, F.; Zacchino, M.; Canepari, M.; Nucci, P.; Alessandrino, F. Muscle injuries: Ultrasound evaluation in the acute phase. *J. Med. Ultrasound* **2013**, *16*, 209–214. [CrossRef]
4. Lovitt, S.; Marden, F.A.; Gundogdu, B.; Ostrowski, M.L. MRI in myopathy. *Neurol. Clin.* **2004**, *22*, 509–538. [CrossRef]
5. Matin, P.; Lang, G.; Carretta, R.; Simon, G. Scintigraphic evaluation of muscle damage following extreme exercise: Concise communication. *J. Nucl.* **1983**, *24*, 308–311.
6. Ament, W.; Lubbers, J.; Rakhorst, G.; Vaalburg, W.; Verkerke, G.J.; Paans, A.M.; Willemsen, A.T. Skeletal muscle perfusion measured by positron emission tomography during exercise. *Pflügers Arch.* **1998**, *436*, 653–658. [CrossRef]
7. Nuutila, P.; Kalliokoski, K. Use of positron emission tomography in the assessment of skeletal muscle and tendon metabolism and perfusion. *Scand. J. Med. Sci. Sports* **2000**, *10*, 346–350. [CrossRef]
8. Garcia, J. MRI in inflammatory myopathies. *Skeletal Radiol.* **2000**, *29*, 425–438. [CrossRef]
9. Lutz, A.M.; Weishaupt, D.; Amann-Vesti, B.R.; Pfammatter, T.; Goepfert, K.; Marincek, B.; Nanz, D. Assessment of skeletal muscle perfusion by contrast medium first-pass magnetic resonance imaging: Technical feasibility and preliminary experience in healthy volunteers. *J. Magn. Reson. Imaging* **2004**, *20*, 111–121. [CrossRef]
10. Weber, M.A.; Krix, M.; Delorme, S. Quantitative evaluation of muscle perfusion with CEUS and with MR. *Eur. Radiol.* **2007**, *17*, 2663–2674. [CrossRef] [PubMed]
11. Peetrons, P. Ultrasound of muscles. *Eur. Radiol.* **2002**, *12*, 35–43. [CrossRef] [PubMed]
12. Lin, Y.-H.; Hsieh, M.-Y.; Su, F.-C.; Wang, S.-H. Assessment of the kinetic trajectory of the median nerve in the wrist by high-frequency ultrasound. *Sensors* **2014**, *14*, 7738–7752. [CrossRef] [PubMed]
13. Krix, M.; Weber, M.A.; Krakowski-Roosen, H.; Huttner, H.B.; Delorme, S.; Kauczor, H.U.; Hildebrandt, W. Assessment of skeletal muscle perfusion using contrast-enhanced ultrasonography. *J. Med. Ultrasound* **2005**, *24*, 431–441. [CrossRef] [PubMed]

14. Lin, Y.H.; Huang, C.C.; Wang, S.H. Quantitative assessments of burn degree by high-frequency ultrasonic backscattering and statistical model. *Phys. Med. Biol.* **2011**, *56*, 757–773. [CrossRef]
15. Maruvada, S.; Shung, K.K.; Wang, S.-H. High-frequency backscatter and attenuation measurements of porcine erythrocyte suspensions between 30–90 MHz. *Ultrasound Med. Biol.* **2002**, *28*, 1081–1088. [CrossRef]
16. Shung, K.K. *Diagnostic Ultrasound: Imaging and Blood Flow Measurements*; CRC Press: Boca Raton, FL, USA, 2006.
17. Cosgrove, D. Ultrasound contrast agents: An overview. *Eur. J. Radiol.* **2006**, *60*, 324–330. [CrossRef]
18. Sboros, V. Response of contrast agents to ultrasound. *Adv. Drug. Deliv. Rev.* **2008**, *60*, 1117–1136. [CrossRef]
19. Huang, C.C.; Lin, Y.H.; Wang, S.H. The Effect of Kinetic Properties on Statistical Variations of Ultrasound Signals Backscattered from Flowing Blood. *Jpn. J. Appl. Phys.* **2009**, *48*, 027002. [CrossRef]
20. Kaul, S. Myocardial contrast echocardiography: A 25-year retrospective. *Circulation* **2008**, *118*, 291–308. [CrossRef]
21. Siracusano, S.; Bertolotto, M.; Ciciliato, S.; Valentino, M.; Liguori, G.; Visalli, F. The current role of contrast-enhanced ultrasound (CEUS) imaging in the evaluation of renal pathology. *World J. Urol.* **2011**, *29*, 633. [CrossRef]
22. Duerschmied, D.; Olson, L.; Olschewski, M.; Rossknecht, A.; Freund, G.; Bode, C.; Hehrlein, C. Contrast ultrasound perfusion imaging of lower extremities in peripheral arterial disease: A novel diagnostic method. *Eur. Heart J.* **2006**, *27*, 310–315. [CrossRef] [PubMed]
23. Duerschmied, D.; Maletzki, P.; Freund, G.; Olschewski, M.; Seufert, J.; Bode, C.; Hehrlein, C. Analysis of muscle microcirculation in advanced diabetes mellitus by contrast enhanced ultrasound. *Diabetes Res. Clin. Pract.* **2008**, *81*, 88–92. [CrossRef] [PubMed]
24. Thomas, K.N.; Cotter, J.D.; Lucas, S.J.; Hill, B.G.; van Rij, A.M. Reliability of contrast-enhanced ultrasound for the assessment of muscle perfusion in health and peripheral arterial disease. *Ultrasound Med. Biol.* **2015**, *41*, 26–34. [CrossRef] [PubMed]
25. Hotfiel, T.; Heiss, R.; Swoboda, B.; Kellermann, M.; Gelse, K.; Grim, C.; Strobel, D.; Wildner, D. Contrast-enhanced ultrasound as a new investigative tool in diagnostic imaging of muscle injuries—A pilot study evaluating conventional ultrasound, CEUS, and findings in MRI. *Clin. J. Sport Med.* **2018**, *28*, 332–338. [CrossRef] [PubMed]
26. Krix, M.; Krakowski-Roosen, H.; Kauczor, H.U.; Delorme, S.; Weber, M.A. Real-time contrast-enhanced ultrasound for the assessment of perfusion dynamics in skeletal muscle. *Ultrasound Med. Biol.* **2009**, *35*, 1587–1595. [CrossRef]
27. Song, Y.; Li, Y.; Wang, P.J.; Gao, Y. Contrast-enhanced ultrasonography of skeletal muscles for type 2 diabetes mellitus patients with microvascular complications. *Int. J. Clin. Exp. Med.* **2014**, *7*, 573–579.
28. Forsberg, F.; Merton, D.; Liu, J.; Needleman, L.; Goldberg, B. Clinical applications of ultrasound contrast agents. *Ultrasonics* **1998**, *36*, 695–701. [CrossRef]
29. Tao, Q.; Wang, Y.; Fish, P.; Wang, W.; Cardoso, J. The wall signal removal in Doppler ultrasound systems based on recursive PCA. *Ultrasound Med. Biol.* **2004**, *30*, 369–379. [CrossRef]
30. Tsui, P.-H.; Wang, S.-H. The effect of transducer characteristics on the estimation of Nakagami parameter as a function of scatterer concentration. *Ultrasound Med. Biol.* **2004**, *30*, 1345–1353. [CrossRef]
31. Tsui, P.H.; Chang, C.C. Imaging local scatterer concentrations by the Nakagami statistical model. *Ultrasound Med. Biol.* **2007**, *33*, 608–619. [CrossRef]
32. Tsui, P.H.; Yeh, C.K.; Chang, C.C. Microvascular Flow Estimation by Microbubble-Assisted Nakagami Imaging. *Ultrasound Med. Biol.* **2009**, *35*, 653–671. [CrossRef]
33. Gu, X.; Wei, M.; Zong, Y.; Jiang, H.; Wan, M. Flow quantification with nakagami parametric imaging for suppressing contrast microbubbles attenuation. *Ultrasound Med. Biol.* **2013**, *39*, 660–669. [CrossRef] [PubMed]
34. Shankar, P.M. Ultrasonic tissue characterization using a generalized Nakagami model. *IEEE Trans. Ultrason. Ferroelectr. Freq. Control* **2001**, *48*, 1716–1720. [CrossRef] [PubMed]
35. Tsui, P.-H.; Ma, H.-Y.; Zhou, Z.; Ho, M.-C.; Lee, Y.-H. Window-modulated compounding Nakagami imaging for ultrasound tissue characterization. *Ultrasonics* **2014**, *54*, 1448–1459. [CrossRef]
36. Krix, M.; Kiessling, F.; Farhan, N.; Schmidt, K.; Hoffend, J.; Delorme, S. A multivessel model describing replenishment kinetics of ultrasound contrast agent for quantification of tissue perfusion. *Ultrasound Med. Biol.* **2003**, *29*, 1421–1430. [CrossRef]

37. Everitt, B.S.; Skrondal, A. *The Cambridge Dictionary of Statistics*, 4th ed.; Cambridge University Press: Cambridge, UK, 2010.
38. Tsui, P.-H.; Wang, S.-H.; Huang, C.-C.; Chiu, C.-Y. Quantitative Analysis of Noise Influence on the Detection of Scatterer Concentration by Nakagami Parameter. *J. Med. Biol. Eng.* **2005**, *25*, 45–51.
39. Sassaroli, E.; Hynynen, K. Resonance frequency of microbubbles in small blood vessels: A numerical study. *Phys. Med. Biol.* **2005**, *50*, 5293–5305. [CrossRef]
40. Dave, J.K.; Halldorsdottir, V.G.; Eisenbrey, J.R.; Liu, J.B.; McDonald, M.E.; Dickie, K.; Leung, C.; Forsberg, F. Noninvasive Estimation of Dynamic Pressures In Vitro and In Vivo Using the Subharmonic Response From Microbubbles. *IEEE Trans. Ultrason. Ferroelectr. Freq. Control* **2011**, *58*, 2056–2066. [CrossRef]
41. Chen, Q.; Zagzebski, J.; Wilson, T.; Stiles, T. Pressure-dependent attenuation in ultrasound contrast agents. *Ultrasound Med. Biol.* **2002**, *28*, 1041–1051. [CrossRef]
42. Sboros, V.; MacDonald, C.A.; Pye, S.D.; Moran, C.M.; Gomatam, J.; McDicken, W.N. The dependence of ultrasound contrast agents backscatter on acoustic pressure: Theory versus experiment. *Ultrasonics* **2002**, *40*, 579–583. [CrossRef]
43. Ho, M.-C.; Lin, J.-J.; Shu, Y.-C.; Chen, C.-N.; Chang, K.-J.; Chang, C.-C.; Tsui, P.-H. Using ultrasound Nakagami imaging to assess liver fibrosis in rats. *Ultrasonics* **2012**, *52*, 215–222. [CrossRef] [PubMed]

Article

Vibroarthrographic Signal Spectral Features in 5-Class Knee Joint Classification

Adam Łysiak [1,*], Anna Froń [1], Dawid Bączkowicz [2] and Mirosław Szmajda [1]

[1] Faculty of Electrical Engineering, Automatic Control and Informatics, Opole University of Technology, 45-758 Opole, Poland; a.fron@doktorant.po.edu.pl (A.F.); m.szmajda@po.edu.pl (M.S.)

[2] Faculty of Physical Education and Physiotherapy, Opole University of Technology, 45-758 Opole, Poland; d.baczkowicz@po.edu.pl

* Correspondence: a.lysiak@doktorant.po.edu.pl

Received: 31 July 2020; Accepted: 1 September 2020; Published: 3 September 2020

Abstract: Vibroarthrography (VAG) is a non-invasive and potentially widely available method supporting the joint diagnosis process. This research was conducted using VAG signals classified to five different condition classes: three stages of chondromalacia patellae, osteoarthritis, and control group (healthy knee joint). Ten new spectral features were proposed, distinguishing not only neighboring classes, but every class combination. Additionally, Frequency Range Maps were proposed as the frequency feature extraction visualization method. The results were compared to state-of-the-art frequency features using the Bhattacharyya coefficient and the set of ten different classification algorithms. All methods evaluating proposed features indicated the superiority of the new features compared to the state-of-the-art. In terms of Bhattacharyya coefficient, newly proposed features proved to be over 25% better, and the classification accuracy was on average 9% better.

Keywords: vibroarthography; VAG; knee joint; spectral features; frequency analysis; non-invasive examination

---

1. Introduction

The knee is one of the most loaded joints within the human body, highly susceptible to injuries and an increased risk of early degeneration of the articular surfaces. Classical radiography is a basic diagnostic tool for imaging knee injuries. In advanced degeneration of this joint, X-ray examination correlates with arthroscopy evaluation which is used as a "gold standard". However, the lower sensitivity and specificity of X-rays are limitations for diagnosis of early stages of chondral disorders, e.g., chondromalacia. On the other hand, the availability of modern imaging methods such as magnetic resonance imaging is limited due to high expense. One of the experimental methods developed for sensitive assessment of articular function is vibroarthrography (VAG) [1–3]. This relatively cheap and potentially widely available non-invasive method is based on the analysis of high frequency vibroacoustic emission, which is a natural phenomenon acquired from the relative motion of articular surfaces of the synovial joint. Although the VAG method is still under development, it reveals high accuracy, sensitivity, and specificity. Previously, it has been used to differentiate disorders of the patellofemoral joint, due to the specific, disorder-related character of the VAG signal pattern [4–6].

Exemplary VAG signals, generated by knee joints with different conditions were presented in Figure 1. Conditions included in the Figure and in the rest of this paper are as follows: control group (healthy knee joint, abbreviated as ctrl), three stages of chondromalacia patellae (cmp1, cmp2, and cmp3, consecutively), and osteoarthritis (oa). As it can be seen, a lot of information about the specific condition could potentially be embedded in the frequency spectrum of the signal.

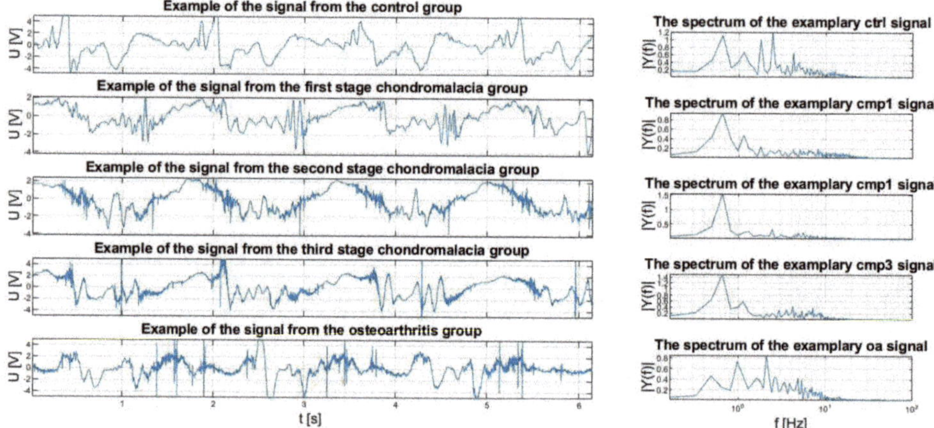

Figure 1. Exemplary signals from each condition group. The differences between signals results from the differences in condition of articular surfaces of the synovial joints.

Although frequency analysis is present in literature devoted to vibroarthrography, the state-of-the-art spectral features are designed to distinguish all classes at once [1–6], i.e., to utilize a single value to differentiate between specific conditions. The objective of this study was to find better frequency ranges, focusing on the distinction between each particular class pairs (i.e., ctrl-cmp1, ctrl-cmp2, ctrl-cmp3, ... , cmp3-oa).

This results in ten spectral features, instead of one. There are ten spectral features, as there are two classes in a pair and five distinct classes, i.e., 5 choose 2 combinations. Specific combinations were presented in Tables 1–3. That way, a single numeric value was changed to a 10-element feature vector, distinguishing conditions more precisely, ensuring that each class is described by the feature set as unambiguously as possible. That approach prevents the potential problem of some values of the feature being typical for more than one condition, making it impossible to state a precise diagnosis.

To amplify the higher frequencies in a way and enable better distinction between classes with potentially useful information hidden in wider frequency spectrum, analysis of the derivative of the signal was conducted.

2. Standard Descriptors

Wu in [7] specified three main methods of the knee joint VAG signal analysis: the spatiotemporal, time-frequency, and statistical analysis methods. Using the spatiotemporal analysis method Wu recommends to use adaptive segmentation [8,9] to avoid redundancy in given segments, and emphasizes its usefulness for calculating sets of features and signal classification. The statistical analysis is often used to determine distribution measures and statistical parameters [10,11], giving the possibility to display data in the form of a tabular summary [12] and graphic forms such as the histogram [6] or the box plot [13–15], and to analyze variations using various types of tests (t-test [16], Kruskal–Wallis test [13], one-way ANOVA test [4], Wilcoxon rank-sum test [11]). The time-frequency analysis and the frequency analysis supported by using statistical analysis are commonly used in the field of signal processing. The frequently used VAG signal analysis is the short-time Fourier transform (STFT) extended by the visual representation of a spectrogram. Łysiak et al. [14] used STFT and spectrograms, on whose specific analysis the new three descriptors have been proposed. Dołęgowski et al. [15] proposed the incremental decomposition of voltage in time and the spectrogram as the methods to identify the knee joint disease stage and compared the statistical parameters of normalized energy values of the band 50–250 Hz (P1) and 250–450 Hz (P2). Befrui et al. [2] in their measurement system used the two accelerometers, the piezoelectric disk and the potentiometer (four channels) in their measurement system. The signals were acquired at 16 kHz sampling frequency and extension and flexion cycles were extracted by using semi-automatic segmentation. The power spectra were calculated, frequency features were normalized and averaged, and the classification by a linear support vector machine (SVM) using the knee-specific feature vectors was performed.

Wavelet transformation is another tool which gives the possibility to analyze the change of signal frequency in the function of time. Mascarenhas et al. [17], for the analysis of VAG signals, proposed in their paper the tunable Q wavelet transform (TQWT) in comparison to the traditional wavelet packet decomposition (WPD). To overcome the imbalance set problem they used the synthetic minority oversampling technique (SMOTE) and afterwards they compared the performance of the random forest classifier (RF) they used to the soft margin support vector machine (SVM).

Nalband et al. [18] proposed for the VAG signals analysis a CAD system using time-frequency analysis and nonstationary signal processing techniques. Their methods were the Wavelet packet decomposition algorithm (WPD), the smoothed pseudo Wigner–Ville distribution (SPWVD) as a nonstationary time-frequency analysis, and a modified version of Hilbert–Huang transform (HHT) where instead of empirical mode decomposition (EMD) for computing intrinsic mode functions (IMF) they proposed complete ensemble empirical mode decomposition with adaptive noise (CEEMDAN). The Least square support vector machine (LS-SVM) was used.

Bączkowicz et al. [1,4] in their research presented the frequency characteristics of VAG signals by STFT. The spectra were obtained by computing Discrete Fourier Transform (DFT) of the segments (150 samples/segment), the Hanning window, and 100 samples overlap/segment. They analyzed spectral activity by summing spectral power in two different bands. The first parameter P1 concerned the range of 50–250 Hz, and the second P2 the range of 250–450 Hz. In [1] Bączkowicz et al. obtained two additional parameters by computing power of spectral density using Fast Fourier Transform (FFT): F470 at 470 Hz, and F780 at 780 Hz. They performed 2-class classification (normal/abnormal) and 5-class classification (healthy knee, the first to the third stages of chondromalacia, osteoarthritis), used the genetic search algorithm to select the best features of VAG signals, and applied four different algorithms to classify the selected features, wherein one signal feature was distinguishing all classes at once.

3. Enhanced Descriptors

As can be seen in Figure 1, VAG signals of different conditions consist of different frequency distortions, which could be potentially used as the features to classify them. The question arises of which frequency ranges would differentiate those classes in the best possible manner. Another question concerns the method of measuring the quality of obtained features.

3.1. *Quality Measure of the Feature*

Boxplot, as a visualization tool, gives very intuitive concept of the quality of a feature. It shows the variation of some numerical data, enabling the visual comparison. Particularly, it shows the median of a sample (indicated by the red line), interquartile range (indicated by the blue box), variability outside the interquartile range (indicated by the whiskers) and outliers (indicated by the red crosses). The question arises, though, how to quantitatively compare two features visualized on the boxplot? Lots of different measures are available in literature [19]. One of the most straightforward ones is the Bhattacharyya coefficient.

In a brief preliminary research, the Bhattacharyya coefficient was compared to several different coefficients (some existing in literature, like DBM/OVS [20–23] or Jaccard index [24], some newly defined ones). This comparison was conducted in the following way:

- The best frequency ranges were generated by every coefficient.
- Obtained frequency ranges were used to train 10 different classifiers (two decision trees, LDA, naïve Bayes, SVM, two knn classifiers, two random forests and a neural network).

- The largest mean classification accuracies were compared.
- The Bhattacharyya coefficient proved to be the best coefficient in this application.

It is defined as [25]:

$$B_{coef}(p,q) = \int \sqrt{p(x) \cdot q(x)} dx, \tag{1}$$

where p and q are probability distributions of two current classes and x is a domain of current feature values. The probability densities were obtained using kernel density estimation with a window:

$$h = \left(\frac{4}{3n}\right)^{1/5} \sigma, \tag{2}$$

as defined in [26].

The Bhattacharyya coefficient indicates the overlap between two statistical samples. It ranges from 0 when distributions are completely separated to 1, when distributions overlap entirely. Therefore, in our study, the coefficient is minimized. The coefficient is equal to 0.401 for the g example from Figure 7 and 0.876 for the j.

3.2. Optimal Frequency Range

Finding frequency ranges in a strict brute-force manner would be too computationally expensive, since the DFT of the analyzed signal is composed of too many samples. Consequently, the search was done in three iterations, starting sparsely in the whole frequency domain, gradually narrowing the searched range and increasing precision. To visualize the results, a kind of map was generated, called in the rest of this paper the frequency range map (FRM). An exemplary map is shown in Figure 2.

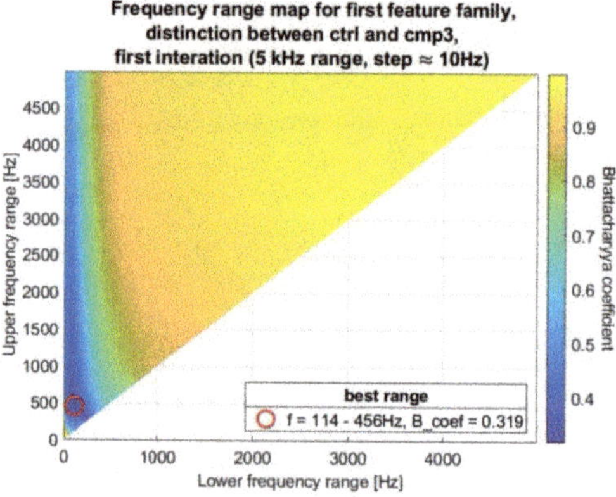

Figure 2. Example of the frequency range map (FRM).

In the FRM, the abscissa x-axis is the lower frequency range and the ordinate is the upper frequency range. As a result, in the first iteration, only the upper-left half of the map consists of possible results. Applicate, or color, indicates quality of feature, defined as the Bhattacharyya coefficient. Similar maps were defined and used in [2], but with lower resolution and for slightly different features. The z-axis, or feature evaluation coefficient used in [2] was the area under the ROC curve, not the Bhattacharyya coefficient. Consequently, the coefficient in [2] was maximized, not minimized.

On each map, the best frequency range was emphasized by a red circle. The exemplary FRM in Figure 2 shows that the best range for this (first) iteration is 114–456 Hz, in which neighboring consecutive iteration will be conducted.

3.3. Definition of the Features

Four types or "families" of features were defined. The first family is the sum of some frequency range of the discrete Fourier transform of the VAG signal:

$$d_1 = \delta f \sum_{i=f_1}^{f_2} \mathcal{DFT}(s_{VAG})(f_i), \text{ where} \quad (3)$$

d_1 is the first feature family,
δf is the normalization factor, equal to the $\frac{1}{f_2 - f_1}$. It ensures that the value of the feature is affected only by the shape of the spectrum and not by its size,
f_1 is the lower frequency range,
f_2 is the upper frequency range,
f_2 is the upper frequency range,
s_{VAG} is the VAG signal,
DFT is the Discrete Fourier Transform operator,
f_i is the i-th frequency amplitude.

The visualization of the first feature family is provided in the (b) plot of Figure 3.

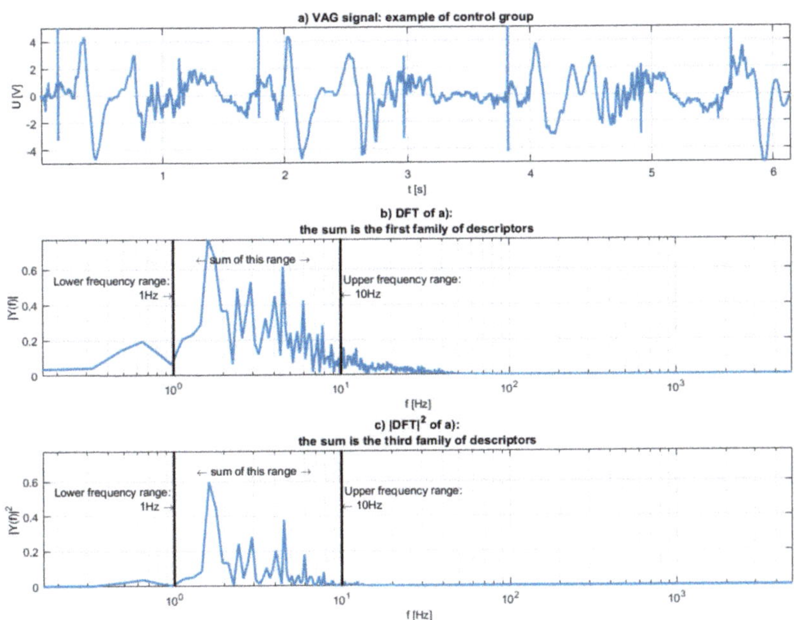

Figure 3. Visualization of the first and the third family of features: (**a**) the analyzed VAG signal, (**b**) the spectrum of the (**a**,**c**) the squared spectrum.

The second family of features was defined similarly, but instead of the signal, the derivative of the signal is used to obtain DFT. One of the fundamental traits of the Fourier transform is that the transform of the derivative is the transform of the original signal multiplied by the frequency:

$$\mathcal{F}(f')(\xi) = 2\pi i \xi \cdot \mathcal{F}(f)(\xi) \quad (4)$$

where $\mathcal{F}$ is the Fourier transform. The derivative can be then interpreted as high-pass filter. The motivation for using the derivative was the certainty that potentially useful information, which could be embedded in wider frequency spectrum, will not be omitted. The second feature family was then defined as

$$d_2 = \delta f \sum_{i=f_1}^{f_2} \mathcal{DFT}\{s'_{VAG}\}(f_i) = \delta f \sum_{i=f_1}^{f_2} (f_i \cdot 2\pi i) \cdot \mathcal{DFT}\{s_{VAG}\}(f_i) \tag{5}$$

The derivative and this family of features were plotted on a and b of Figure 4 respectively.

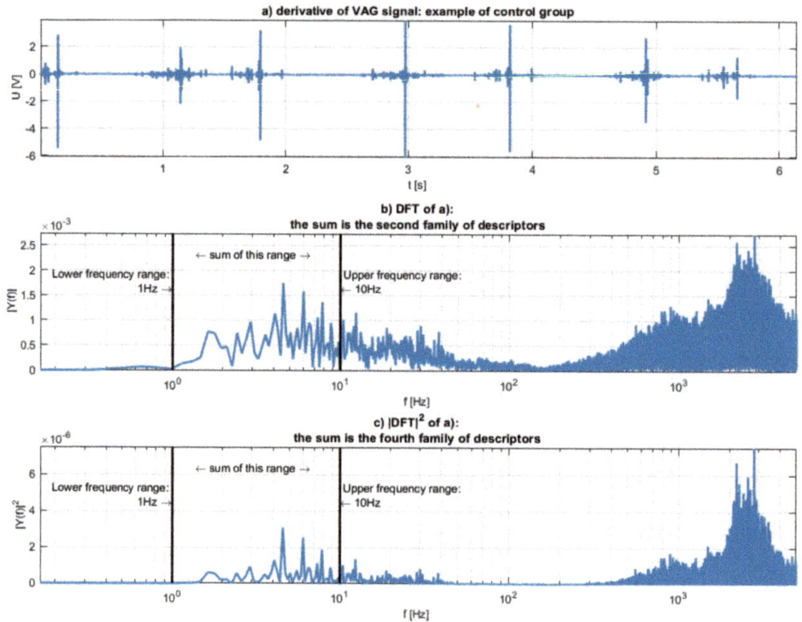

Figure 4. Visualization of the second and the fourth family of features: (**a**) the derivative of the analyzed VAG signal, (**b**) the spectrum of the (**a**,**c**) the squared spectrum.

The third (Equation (6)) and the fourth (Equation (7)) families of features were defined like the first and the second, but instead of summing the DFT amplitudes, the energy was summed (so the square of the values of the previous feature families):

$$d_3 = \delta f \sum_{i=f_1}^{f_2} \left|\mathcal{DFT}(s_{VAG})(f_i)\right|^2, \tag{6}$$

$$d_4 = \delta f \sum_{i=f_1}^{f_2} \left|\left(f_i \cdot 4\pi^2 i\right) \cdot \mathcal{DFT}\{s_{VAG}\}(f_i)\right|^2. \tag{7}$$

The illustrations of those families were shown in Figures 3c and 4c respectively.

Interpretation of the derivative of the signal as a high pass filter can clearly be seen in Figures 3 and 4. Higher frequencies are much more potent in Figure 4, while lower frequencies seem relatively unaltered. Energies, or the c plots of both Figures appear "sharper". It is to be expected, as the energy is the square of the spectrum.

The calculation of the feature values was followed by the evaluation by Bhattacharyya coefficient. The point with coordinates: x = lower frequency range, y = upper frequency range, z = the Bhattacharyya coefficient of the feature generated by x-y frequency range, is then plotted on FRM. Then, the next feature, composed of slightly different frequency range is evaluated, plotted, etc.

4. Research Methodology

For the purposes of this research, VAG signals were acquired from groups of patients qualified by specialists as having first stage chondromalacia (cmp1), second stage chondromalacia (cmp2), third stage chondromalacia (cmp3), and osteoarthritis (oa) and from the control group, healthy knee (ctrl). All subjects underwent routine medical interviews, physical examination, and imaging via MRI. Patients with stages I–III of chondromalacia patellae were classified in accordance with the Outerbridge classification [27]. In turn osteoarthritis (OA) patients were diagnosed with mild to moderate knee OA (grade II and III according to the Kellgren-Lawrence grading system [28]) with a disease duration of more than 2 years. All imaging examinations were analyzed by single radiologist, who was blinded to patients' symptoms. To prevent any signal artifacts from deteriorations other than chondral lesions, individuals with a history of knee fracture, knee surgery, history of meniscal tears, significant knee instability, or patellar maltracking were not enrolled in the study. Moreover, due to the methodology of the VAG assessment, individuals with a restricted knee joint range of motion (required 0° to 100°), significantly weakened muscles, and substantially swollen knees in the affected lower limb were excluded from the study. The VAG examination was performed in a sitting position and consisted of exactly four full cycles of alternating extension and flexion of the knee joint (90°–0°–90°) lasting 6 s. Mounted 1 cm above the apex of patella, the acceleration sensor was recording vibration and acoustic processes occurring during the knee movements. The acceleration sensor attachment was shown in Figure 5.

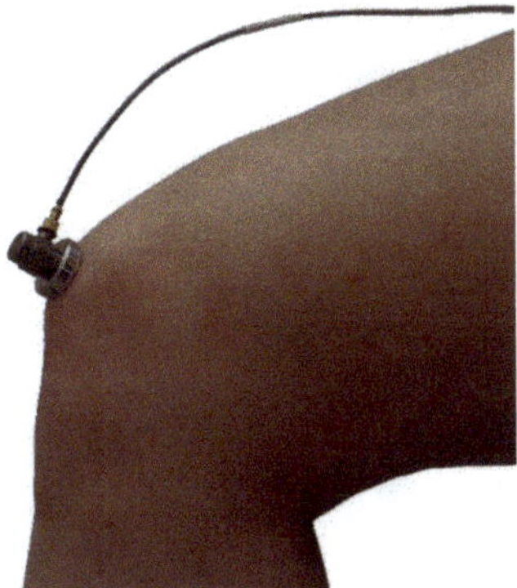

Figure 5. The accelerometer attachment: 1 cm above the apex of patella.

After acquiring the signals, their spectra were obtained, and four feature families were proposed. For each family, a set of features containing different frequency ranges was generated, and every feature from this set was evaluated by the Bhattacharyya coefficient. The best features constructed final 10-element vector. This vector was compared to the state-of-the-art feature vector [1], by using both of those vectors as an input to some classification algorithms. The accuracies of trained algorithms determined which feature vector is superior.

4.1. Acquisition of the VAG Signal

The signals analyzed in this paper were obtained from knee joints of patients diagnosed by physicians into five groups: control group (containing 66 signals), I stage chondromalacia patellae (26 signals), II stage chondromalacia patellae 30 signals), III stage chondromalacia patellae (36 signals), and osteoarthritis (26 signals). The diagnoses were based on the X-ray.

4.2. The Process of Defining New Frequency Features

For every sample of 184 VAG signals, the frequency analysis was conducted, resulting in four spectral vectors:

- the DFT of the VAG signal, from which the first feature family is obtained, using Equation (3),
- the DFT of the derivative of the VAG signal, from which the second feature family is obtained, using Equation (5),
- squared DFT of the VAG signal, from which the third feature family is obtained, using Equation (6),
- squared DFT of the derivative of the VAG signal, from which the fourth feature family is obtained, using Equation (7).

To generate FRMs, Bhattacharyya coefficient was calculated for the sums of desired frequency ranges. As already mentioned, evaluating all possible frequency ranges would be too computationally expensive, therefore the search was done in three iterations. The search started from the full frequency range (0–5 kHz) with quite big frequency step and gradually narrowing the range and reducing the step. The last iteration was done with highest precision (the smallest frequency step) in quite narrow frequency range. Specifically, three iterations were conducted as follows:

1. the first one in range 0–5 kHz (since signal was sampled with the frequency of 10 kHz), with about 10 Hz step. Features were then defined as sums, defined by Equations (3) and (5)–(7), in subsequent ranges: 0–10 Hz, 0–20 Hz, ... , 0–5000 Hz, 10–20 Hz, 10–30 Hz, ... , 4980–4990 Hz, 4980–5000 Hz, 4990–5000 Hz. Every range was evaluated by Bhattacharyya coefficient and plotted as a point on the FRM.
2. the second iteration was conducted in range ±800 Hz from the best range obtained in previous iteration with about 2.5 Hz step,
3. the last iteration was conducted in range ±80 Hz from the best range obtained in previous iteration with about 0.16 Hz step.

Three exemplary maps were shown on Figure 6, illustrating three consecutive iterations.

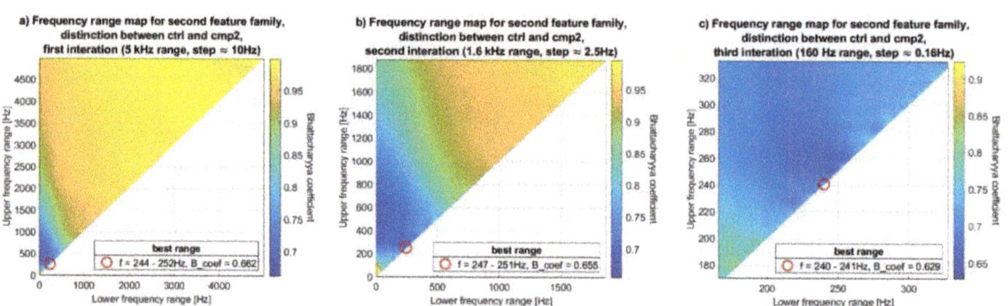

Figure 6. Three examples of the FRM; all of them were generated for the second family of features (i.e., the sum of the DFT of the derivative of the VAG signal). (**a**–**c**) are the consecutive iterations.

The FRMs were separately generated for every class combination (details in Tables 1–3), for every feature family, so 120 maps in total, 40 maps for final iteration. Generating separate maps for each class combination may seem redundant. Only distinction between neighboring classes appears to be sufficient. Different approach, i.e., temporary ensemble all classes but one, generating maps only for distinguishing between, for example, the second stage chondromalacia condition and the rest, generating characteristic features for each class also seems enough. As it certainly would be less computationally expensive, it probably would be less robust. Neighboring classes are not completely separate and independent, as the first stage chondromalacia precedes the second stage etc. Consequently, it would be quite difficult to find the frequency range typical for particular class only. Minimizing Bhattacharyya coefficient for every class pair ensures two things:

- that the found frequency ranges for each classes are as optimal as possible, without sacrificing the distinction between more different conditions (maybe one frequency range would be appropriate for the distinction between the first and the second stage chondromalacia, but not as effective with distinguishing between first stage chondromalacia and the osteoarthritis; the question would arise, which distinction should be dominant, and which should be sacrificed),
- that the obtained frequency ranges provide as unambiguous distinction as possible; the situation can be imagined in which two classes neighboring one another (for example cmp1 and cmp3 neighboring cmp2) can be distinguished from the one with very similar frequency ranges. Then the statement that the particular frequency range is typical for, e.g., cmp2 would be true, but the contrary would not unambiguously point out cmp1 or cmp3.

Analysis of the final 40 maps led to definition of 10 frequency ranges in different feature families, indicated by bolded font in the Table 2. Obtained features were compared to standard frequency descriptors defined in [1].

4.3. The Verification of Defined Features as a Classifier Input

The comparison between newly defined features and the state-of-the-art was done in two ways. First, the Bhattacharyya coefficients of features were compared. However, as mentioned before, a 4-element (P1, P2, F470, F780) spectral feature vector was proposed in [1]. In this study, 10 features were proposed, making it hard to directly compare the quality of obtained features.

To compare obtained features, both newly defined and state-of-the-art were used to train a couple of classifier algorithms. Utilized algorithms were listed in Table 4. To find which feature vector is superior, the final classification accuracies were compared.

The classification accuracy of every classifier strongly depends on the division of the data into the teaching, testing, and validation sets. To surpass that potential bias, 1024 classifiers of each type were constructed with random division of the data into mentioned sets.

5. Results and Discussion

The results obtained for every combination of classes, for each feature family, were presented in Tables 1 and 2. The use of the derivative was quite fruitful for some class pairs, even though intuition behind it, i.e., the informativeness of higher frequency spectrum, was not fully correct. For example, the distinctions between the first and the second stage chondromalacia or the third stage chondromalacia and osteoarthritis were better with the use of the derivative, but the range of the frequencies was very narrow. Generally, the most optimal results obtained for the primitive VAG signal and its derivative are comparable.

Boxplots of the best results for each combination of classes were shown in Figure 7.

The new features were defined with the emphasis on distinction between each particular class pair. Then, it can be seen, that some of the pairs are well distinguished, especially when the knee joint conditions strongly differ (for example, the control group and the third stage chondromalacia or the first stage chondromalacia and the osteoarthritis). The problem persists, though, with the distinction between neighboring pairs, especially the third stage chondromalacia and the osteoarthritis. This is to be expected to some extent, as the chondromalacia condition commonly occurs with the osteoarthritis.

To better evaluate the quality of the obtained features, they were compared to the features defined in [1] (used signal database was the same in this research and in [1]). Firstly, the Bhattacharyya coefficient was generated for the frequency features from [1], then some classification algorithms were used to quickly check potential classification accuracies, using both the new features and the features defined in [1].

The comparison between features defined in [1] and features defined in this paper was presented in Table 3. The boxplots of the features from [1] were given in Figure 8.

None of the newly obtained frequency ranges was close to the range of P1 or P2 from [1]. The frequencies around 250 Hz, the interface of P1 and P2, occur in the Table 2 relatively often, however the ranges are significantly narrower. The new feature closest to F470 from [1] is feature e from Table 2, so the sum of the DFT of the derivative of the signal for the frequencies around 400 Hz. Nonetheless, the difference in frequencies is quite big. The F780 feature from [1] does not overlap with any features obtained in conducted research.

The new features are superior, in terms of Bhattacharyya coefficient, for every class combination. Improvement of distinction is quite significant for some pairs (like the pair cmp1–cmp3, for which the coefficient dropped by more than 40%), and for others is smaller (line the cmp3–oa pair, for which the coefficient dropped by around 5%).

To test the features as a diagnostic tool, ten different types of classifiers were constructed for the newly defined features and another ten classifiers for the features defined in [1]. The comparison between their accuracies is presented in Table 4.

Table 1. Research results: the Bhattacharyya coefficients, for best frequency range, for each feature family, for each class combination. Corresponding frequency ranges are presented in Table 2.

Boxplot Letter (Figure 7)	Class Combination	The Bhattacharyya Coefficient for Feature Family:			
		1 (DFT of the Signal)	2 (DFT of the Derivative)	3 (Square of the DFT of the Signal)	4 (Square of the DFT of the Derivative)
a	ctrl, cmp1	0.863	0.844	0.842	0.872
b	ctrl, cmp2	0.634	0.629	0.659	0.662
c	ctrl, cmp3	0.318	0.316	0.453	0.464
d	ctrl, oa	0.308	0.367	0.458	0.462
e	cmp1, cmp2	0.726	0.717	0.769	0.768
f	cmp1, cmp3	0.446	0.442	0.560	0.577
g	cmp1, oa	0.401	0.486	0.587	0.603
h	cmp2, cmp3	0.637	0.667	0.694	0.693
i	cmp2, oa	0.594	0.696	0.741	0.744
j	cmp3, oa	0.900	0.919	0.897	0.876

Table 2. Research results: the best frequency range for each feature family, for each class combination. Corresponding Bhattacharyya coefficient values are presented in Table 1.

Boxplot Letter (Figure 7)	Class Combination	The Frequency Range (Hz) for Feature Family:			
		1 (DFT of the Signal)	2 (DFT of the Derivative)	3 (Square of the DFT of the Signal)	4 (Square of the DFT of the Derivative)
a	ctrl, cmp1	235.51–235.51	279.95–279.95	235.51–235.51	226.56–226.56
b	ctrl, cmp2	240.23–240.56	240.23–240.56	331.22–331.38	331.22–331.38
c	ctrl, cmp3	111.17–452.96	103.19–359.21	109.70–428.39	26.20–303.71
d	ctrl, oa	15.79–1110.68	0.00–554.85	47.69–5000.00	0.00–649.25
e	cmp1, cmp2	398.11–398.93	405.76–405.76	258.3–258.63	239.10–239.10
f	cmp1, cmp3	78.78–465.82	15.14–417.97	78.61–428.39	9.11–290.36
g	cmp1, oa	8.46–849.61	0.81–470.70	42.64–5000.00	0.81–690.92
h	cmp2, cmp3	94.40–394.86	91.80–287.43	88.05–392.42	79.92–263.83
i	cmp2, oa	8.79–955.73	0.00–513.02	71.45–5000.00	233.40–233.89
j	cmp3, oa	0.00–4384.44	0.98–166.18	16.28–193.20	157.88–157.88

Table 3. The Bhattacharyya coefficients of frequency features defined in [1] compared to present study. The improvement is calculated as $\frac{(B_{coef_{OLD}} - B_{coef_{NEW}})}{B_{coef_{OLD}}} \cdot 100$.

Class Combination	Bhattacharyya Coefficient					
	P1 (50–250 Hz)	P2 (250–450 Hz)	F470 (470 Hz)	F780 (780 Hz)	The Best Feature from Table 1	Improvement (%)
ctrl, cmp1	0.964	0.960	0.951	0.963	0.842	11.46
ctrl, cmp2	0.906	0.795	0.924	0.943	0.659	17.11
ctrl, cmp3	0.549	0.668	0.837	0.947	0.316	42.44
ctrl, oa	0.582	0.516	0.860	0.943	0.308	40.31
cmp1, cmp2	0.942	0.874	0.936	0.904	0.717	17.96
cmp1, cmp3	0.607	0.782	0.882	0.955	0.442	43.48
cmp1, oa	0.670	0.635	0.901	0.948	0.401	36.85
cmp2, cmp3	0.809	0.881	0.952	0.965	0.637	21.26
cmp2, oa	0.809	0.809	0.950	0.951	0.594	26.58
cmp3, oa	0.924	0.962	0.985	0.980	0.876	5.19

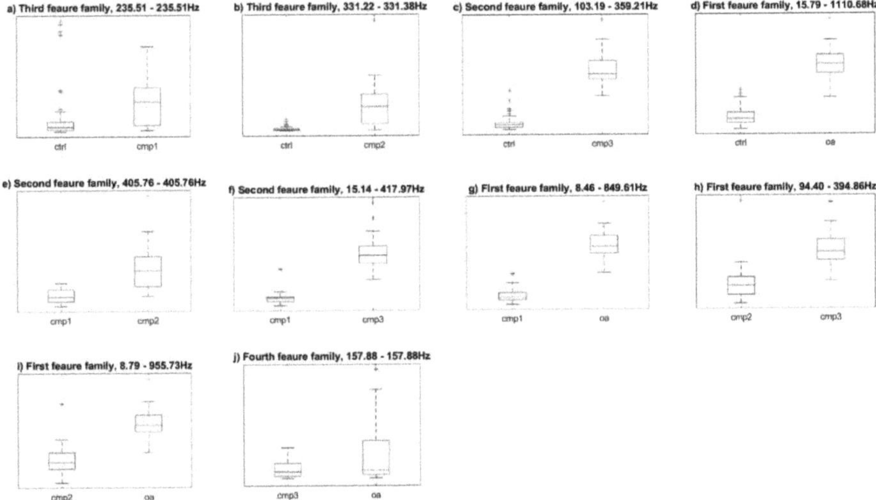

Figure 7. Boxplots of the best results for each class combination. The class distinctions are following: (**a**) ctrl-cmp1, (**b**) ctrl-cmp2, (**c**) ctrl-cmp3, (**d**) ctrl-oa, (**e**) cmp1-cmp2, (**f**) cmp1-cmp3, (**g**) cmp1-oa, (**h**) cmp2-cmp3, (**i**) cmp2-oa, (**j**) cmp3-oa. All of the letters correspond with Tables 1 and 2.

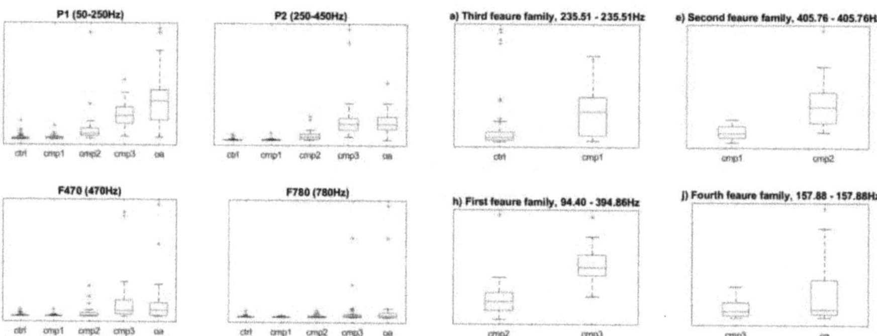

Figure 8. Boxplots of the features defined in [1] compared to boxplots of the new features defined for neighboring classes. The letters in titles of the new features correspond with Tables 1 and 2.

Table 4. Comparison between classification accuracies for the new features and the features defined in [1]. The improvement is calculated as $\frac{(classification\_accuracy_{OLD} - classification\_accuracy_{NEW})}{classification\_accuracy_{OLD}} \cdot 100$

No.	Classifier	Classification Accuracy			Additional info. about Classifier
		The New Features	The Features Defined in [1]	Improvement (%)	
1	decision tree	0.62	0.59	5.1	max. 10 splits
2	decision tree	0.60	0.59	1.7	max. 5 splits
3	discriminant analysis	0.63	0.53	18.9	
4	naïve Bayes	0.64	0.48	33.3	
5	support vector machine	0.67	0.63	6.3	linear kernel
6	k nearest neighbors	0.64	0.61	4.9	$k = 20$, euclidean distance
7	k nearest neighbors	0.63	0.62	1.6	$k = 5$, euclidean distance
8	decision forest	0.62	0.57	8.8	bagging
9	decision forest	0.63	0.60	5.0	boosting, max 10 splits
10	neural network	0.63	0.60	5.0	10 hidden neurons, tansig function
	mean	0.63	0.58	9.1	
	max	0.67	0.63	33.3	

All accuracies presented in the Table 4 are actually the mean values of 1024 classifiers of a given type.

As expected, classifiers constructed on new feature set proved to be more accurate. Additional visual comparison was given with the use of boxplot on the Figure 9. All the data came from the Table 4.

Although the new features are more useful for classifying different knee joint conditions, obtaining them is more computationally expensive. Additionally, they are probably somewhat correlated to each other, which could be undesirable while using them as an input for some classification algorithms [29]. A principal component analysis could be a useful step before constructing classifier, but it would increase computational cost even further.

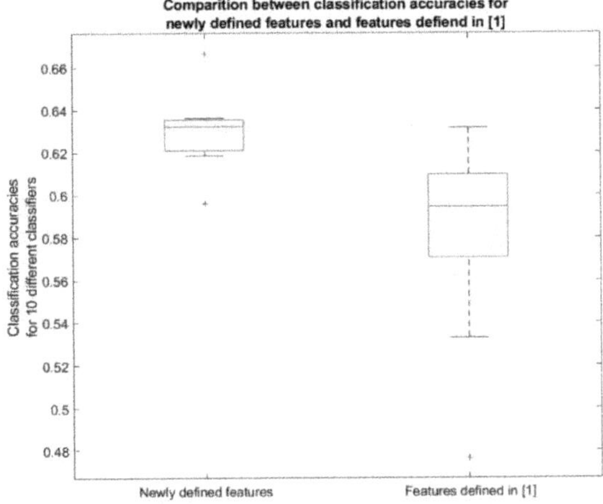

Figure 9. Visual representation of the data given in Table 4.

6. Conclusions

Conducted research resulted in the definition of ten new spectral features, emphasizing differences between every possible knee condition class pair. This ensured that the found frequency ranges were as optimal as possible, without the risk of sacrificing potentially valuable information about the differences between some class pairs. Proposed features were used to train ten different classification algorithms proving their superiority as signal descriptors. In comparison to [1], in terms of Bhattacharyya coefficient, newly proposed features proved to be over 25% better. The classification accuracy has increased on average by 9%.

Frequency Range Maps were proposed as a spectral feature visualization tool, utilizing Bhattacharyya coefficient as a feature quality measure. Visual analysis of those maps may help better understand the nature of the signal. The maps are not limited to VAG signals and can be used in different fields of research.

Proposed features may constitute the reference point in future studies, utilizing proposed frequency ranges as a method of measuring effect of some factor on quality of knee joint itself.

Author Contributions: Conceptualization, A.Ł. and M.S.; methodology, A.Ł. and M.S.; software, A.Ł.; validation, D.B. and M.S.; formal analysis, A.Ł. and A.F.; investigation, A.Ł.; resources, A.Ł., A.F., D.B. and M.S.; data curation, A.Ł. and D.B.; writing—original draft preparation, A.Ł., A.F. and D.B.; writing—review and editing, A.Ł., A.F. and M.S.; visualization, A.Ł.; supervision, M.S.; project administration, A.Ł. All authors have read and agreed to the published version of the manuscript.

Funding: This research received no external funding.

Conflicts of Interest: The authors declare no conflict of interest.

References

1. Kręcisz, K.; Bączkowicz, D. Analysis and multiclass classification of pathological knee joints using vibroarthrographic signals. *Comput. Methods Prog. Biomed.* **2018**, *154*, 37–44. [CrossRef] [PubMed]
2. Befrui, N.; Elsner, J.; Flesser, A.; Huvanandana, J.; Jarrousse, O.; Le, T.N.; Müller, M.; Schulze, W.H.W.; Taing, S.; Weidert, S. Vibroarthrography for early detection of knee osteoarthritis using normalized frequency features. *Med. Biol. Eng. Comput.* **2018**, *56*, 1499–1514. [CrossRef] [PubMed]
3. Bączkowicz, D.; Kręcisz, K. Vibroarthrography in the evaluation of musculoskeletal system a pilot study. *Ortop. Traumatol. Rehabil.* **2013**, *15*, 407–416. [CrossRef] [PubMed]
4. Bączkowicz, D.; Majorczyk, E. Joint motion quality in vibroacoustic signal analysis for patients with patellofemoral joint disorders. *BMC Musculoskelet. Disord.* **2014**, *15*, 426–433. [CrossRef]

5. Bączkowicz, D.; Majorczyk, E.; Kręcisz, K. Age-related impairment of quality of joint motion in vibroarthrographic signal analysis. *BioMed Res. Int.* **2015**, *2015*, 1–7. [CrossRef]
6. Bączkowicz, D.; Majorczyk, E. Joint motion quality in chondromalacia progression assessed by vibroacoustic signal analysis. *PM R* **2016**, *8*, 1065–1071. [CrossRef]
7. Wu, Y. *Knee Joint Vibrographic Signal Processing and Analysis*; Springer: London, UK, 2015.
8. Krishnan, S.; Rangayyan, R.M.; Bell, G.D.; Frank, C.B.; Ladly, K.O. Adaptive filtering, modelling, and classification of knee joint vibroarthrographic signals for non-invasive diagnosis of articular cartilage pathology. *Med. Biol. Eng. Comput.* **1997**, *35*, 677–684. [CrossRef]
9. Moussavi, Z.M.K.; Rangayyan, R.M.; Bell, G.D.; Frank, C.B.; Ladly, K.O.; Zhang, Y.T. Screening of vibroarthrographic signals via adaptive segmentation and linear prediction modeling. *IEEE Trans. Biomed. Eng.* **1996**, *43*, 15–23. [CrossRef]
10. Rangayyan, R.M.; Wu, Y.F. Screening of knee-joint vibroarthrographic signals using statistical parameters and radial basis functions. *Med. Biol. Eng. Comput.* **2008**, *46*, 223–232. [CrossRef]
11. Wu, Y.; Chen, P.; Luo, X.; Huang, H.; Liao, L.; Yao, Y.; Wu, M.; Rangayyan, R.M. Quantification of knee vibroarthrographic signalirregularity associated with patellofemoral jointcartilage pathology based on entropy and envelopeamplitude measures. *Comput. Methods Programs Biomed.* **2016**, *130*, 1–12. [CrossRef]
12. Andersen, R.E.; Arendt-Nielsen, L.; Madeleine, P. Knee joint vibroarthrography of asymptomatic subjects during loaded flexion-extension movements. *Med. Biol. Eng. Comput.* **2018**, *56*, 2301–2312. [CrossRef] [PubMed]
13. Nalband, S.; Prince, A.A.; Agrawal, A. Entropy-based feature extraction and classification of vibroarthographic signal using complete ensemble empirical mode decomposition with adaptive noise. *IET Sci. Meas. Technol.* **2018**, *12*, 350–359. [CrossRef]
14. Łysiak, A.; Froń, A.; Bączkowicz, D.; Szmajda, M. The new descriptor in processing of vibroacoustic signal of knee joint. *IFAC PapersOnLine* **2019**, *52*, 335–340. [CrossRef]
15. Dołegowski, M.; Szmajda, M. Use of incremental decomposition and spectrogram in vibroacoustic signal analysis in knee joint disease examination. *Przegląd Elektrotech.* **2018**, *7/2018*, 162–166. [CrossRef]
16. Rangayyan, R.M.; Wu, Y. Analysis of vibroarthrographic signals with features related to signal variability and radial-basis functions. *Ann. Biomed. Eng.* **2009**, *37*, 156–163. [CrossRef]
17. Mascarenhas, E.; Nalband, S.; Fredo, A.R.J.; Prince, A. Analysis and Classification of Vibroarthrographic Signals using Tuneable 'Q' Wavelet Transform. In Proceedings of the 2020 7th International Conference on Signal Processing and Integrated Networks, Noida, India, 27–28 February 2020. [CrossRef]
18. Nalband, S.; Valliappan, C.A.; Prince, A.A.; Agrawal, A. Time-frequency based feature extraction for the analysis of vibroarthographic signals. *Comput. Electr. Eng.* **2018**, *69*, 720–731. [CrossRef]
19. Guyon, I.; Gunn, S.; Nikravesh, M.; Zadeh, L.A. (Eds.) Feature Extractoin. In *Foundations and Aplications*, 1st ed.; Springer: Heidelberg, Germany, 2006.
20. Wild, C.J.; Pfannkuch, M.; Regan, M.; Horton, N.J. Towards more accessible conceptions of statistical inference: Conceptions of Statistical Inference. *J. R. Stat. Soc. A* **2011**, *174*, 247–295. [CrossRef]
21. Nachkebia, N.; Alexander, M.; House, W.; North, H. The Simple Theory of Informal Rules. *Math. Teach. Res. J. Online* **2013**, *6*, 83–99.
22. Rao, J.S.; Liu, H. Discordancy Partitioning for Validating Potentially Inconsistent Pharmacogenomic Studies. *Sci. Rep.* **2017**, *7*, 15169. [CrossRef] [PubMed]
23. Pramono, R.X.A.; Imtiaz, S.A.; Rodriguez-Villegas, E. Evaluation of features for classification of wheezes and normal respiratory sounds. *PLoS ONE* **2019**, *14*, e0213659. [CrossRef] [PubMed]
24. Jaccard, P. Distribution comparée de la flore alpine dans quelques régions des Alpes occidentales et orientales. *Bull. Soc. Vaud. Sci. Nat.* **1901**, *37*, 241–272.
25. Kailath, T. The Divergence and Bhattacharyya Distance Measures in Signal Selection. *IEEE Trans. Commun.* **1967**, *15*, 52–60. [CrossRef]
26. Bowman, A.W.; Azzalini, A. The Kernel Approach with S-Plus Illustrations. In *Applied Smoothing Techniques for Data Analysis*, 1st ed.; Oxford University Press: New York, NY, USA, 1997.
27. Perera, J.R.; Gikas, P.D.; Bentley, G. The present state of treatments for articular cartilage defects in the knee. *Ann. R. Coll. Surg. Engl.* **2012**, *94*, 381–387. [CrossRef] [PubMed]

28. Culvenor, A.G.; Engen, C.N.; Øiestad, B.E.; Engebretsen, L.; Risberg, M.A. Defining the presence of radiographic knee osteoarthritis: A comparison between the Kellgren and Lawrence system and OARSI atlas criteria. *Knee Surg. Sports Traumatol. Arthrosc.* **2015**, *23*, 3532–3539. [CrossRef] [PubMed]
29. Wang, W.; Carreira-Perpiñán, M.Á. The role of dimensionality reduction in linear classification. *arXiv* **2014**, arXiv:1405.6444. Available online: http://arxiv.org/abs/1405.6444 (accessed on 2 July 2020).

Letter

Quality Assessment during Incubation Using Image Processing

Sheng-Yu Tsai [1], Cheng-Han Li [1], Chien-Chung Jeng [2] and Ching-Wei Cheng [3,*]

1 Department of Bio-Industrial Mechatronics Engineering, National Chung Hsing University, Taichung 402, Taiwan; d105040001@mail.nchu.edu.tw (S.-Y.T.); d108040001@mail.nchu.edu.tw (C.-H.L.)
2 Department of Physics, National Chung Hsing University, Taichung 402, Taiwan; ccjeng@phys.nchu.edu.tw
3 College of Intelligence, National Taichung University of Science and Technology, Taichung 404, Taiwan
* Correspondence: cwcheng@nutc.edu.tw; Tel.: +886-4-2219-5795

Received: 21 September 2020; Accepted: 20 October 2020; Published: 21 October 2020

Abstract: The fertilized egg is an indispensable production platform for making egg-based vaccines. This study was divided into two parts. In the first part, image processing was employed to analyze the absorption spectrum of fertilized eggs; the results show that the 580-nm band had the most significant change. In the second part, a 590-nm-wavelength LED was selected as the light source for the developed detection device. Using this device, sample images (in RGB color space) of the eggs were obtained every day during the experiment. After calculating the grayscale value of the red layer, the receiver operating characteristic curve was used to analyze the daily data to obtain the area under the curve. Subsequently, the best daily grayscale value for classifying unfertilized eggs and dead-in-shell eggs was obtained. Finally, an industrial prototype of the device designed and fabricated in this study was operated and verified. The results show that the accuracy for detecting unfertilized eggs was up to 98% on the seventh day, with the sensitivity and Youden's index being 82% and 0.813, respectively. On the ninth day, both accuracy and sensitivity reached 100%, and Youden's index reached a value of 1, showing good classification ability. Considering the industrial operating conditions, this method was demonstrated to be commercially applicable because, when used to detect unfertilized eggs and dead-in-shell eggs on the ninth day, it could achieve accuracy and sensitivity of 100% at the speed of five eggs per second.

Keywords: incubation; spectrum; image processing; ROC curve

---

1. Introduction

Eggs—rich in protein, minerals, and vitamins, and an important source of nutrients for humans—play a vital role in the human daily diet. The importance of fertilized eggs is increasing because fertilized eggs are currently used to manufacture most vaccines, wherein fertilized eggs are usually screened and delivered to vaccine manufacturers by the 10th day [1–3]. A fertilized egg takes about 21 days to hatch, with the chick embryo completely formed after the 18th day. Fertilized eggs are generally screened via light illumination. With the light penetrating the shell, the embryo development is judged. The eggs are preliminarily classified as fertilized and unfertilized; the fertilized eggs can be further divided into normal and dead-in-shell embryos. Artificial screening mainly aims to remove unfertilized and dead-in-shell eggs. Eggs are usually subjected to three light-based inspections during the hatching process. The first inspection is conducted on the sixth day, when spiderweb-like blood vessels develop along the eggshell membrane inside the normally fertilized eggs. In contrast, the interior of an unfertilized egg is devoid of blood vessels and appears bright when held against light, while only a black blood clot-like partial embryo can be identified in a dead-in-shell egg. The second inspection is conducted on the 10th day; dead-in-shell eggs with only tiny blood vessels, if found, are removed to avoid the malodorous amine gas produced by the fermentation of eggs due to long-term decomposition of dead embryos, which can cause the eggs to burst and generate pollution. To eliminate embryos that might die after the 10th day, the third inspection is conducted on the 18th day, when the eggs are about to be placed in the hatching room; subsequently, these eggs are removed to avoid resource wastage. Moreover, non-removal of dead embryos can pollute the hatching environment and reduce the hatching rate [4].

Recently, there have been numerous nondestructive-detection-based studies on monitoring the hatching process of eggs, including machine vision, percussion-vibration methods [5], optical detection method, and dielectric-characteristic measurement method. To develop a nondestructive detection technique capable of distinguishing fertilized eggs from eggs not suitable for vaccine production, LEDs can be used where visible light and near-infrared (VNIR) spectroscopy techniques are employed in conjunction with a LED light source [3]. In [6], the authors used machine vision to identify whether an egg was fertilized. In their study, a LED lamp was used to penetrate the eggshell at a close range, with a CCD camera used to film the embryo growth inside the egg. Using an image-processing method, relevant characteristics, such as the growth of blood vessels inside the eggs, were obtained as the basis for identification. The identification accuracy rates corresponding to the 1st–5th days of incubation were 47.13%, 81.41%, 93.08%, 97.73%, and 98.25%, respectively. The optical detection method has a wide range of applications, including the detection of the sugar content, pH value, and water content [7–9] of agricultural products; VNIR spectroscopes are usually employed as a nondestructive light source to detect the protein quality and freshness [10,11] of egg products. Whether an egg is fertilized [12,13] is determined by the light absorption ratios of different egg compositions, such as eggshell, egg white, blood vessels, and erythrocytes developed during the embryo growth inside the egg as they exhibit different absorption spectral bands. In the optical detection method, similar to machine vision, one side of the egg is illuminated with strong light, and a CCD camera placed on the other side to detect the absorption spectral wavelength of the egg to determine the fertilization and development status. Using a 50-W tungsten halogen lamp to illuminate the eggshell, Liu and Ngadi collected spectral images in the wavelength range of 900–1700 nm using a NIR hyperspectral imaging system, with the signals filtered to determine the egg status [4]. Using a similar method to collect data on spectral wavelengths between 400 and 1000 nm for identification, Smith et al. achieved an identification accuracy rate of 71%, 63%, 65%, and 83% on the 0th–3rd day of incubation, respectively [14]. Collecting data on wavelengths between 400 and 1000 nm, Zhu et al. extracted 155 spectral characteristic variables from the 520-nm waveband and used a support vector machine (SVM) to classify and model the image, spectrum, and image-spectrum fusion information to identify fertilized eggs, unfertilized eggs, and dead embryos, with the identification accuracy rate being 84%, 90%, and 93%, respectively. They concluded that the image-spectrum fusion information-based accuracy was higher than the single characteristic identification-based accuracy [15]. In the dielectric-characteristic measurement method, a high-frequency wave was input through a parallel plate without destroying the eggshell, where characteristic values, such as the dielectric constant and loss factor of the egg at different frequencies, were continuously measured and imported into an artificial neural network and an SVM classifier to classify the samples [16].

In this study, an image-processing method was used to analyze egg spectra. In this method, a 590-nm-wavelength LED was selected as the light source to obtain sample data, where sample colors were layered and converted into grayscale images using an imaging device, and a receiver operating characteristic (ROC) curve was employed to analyze the daily data to obtain the area under the curve (AUC) as the basis for determining the optimal screening threshold. Finally, the actual operation and verification were conducted using the detection device developed in this study.

2. Materials and Methods

2.1. Sample and Experimental Equipment

The eggs used in the experiment were of the Lohmann variety and were cold-stored and had not started hatching (0-day age). They were randomly selected at the moment of purchase. Two main experiments were conducted in this study. The first experiment aimed to seek the most suitable light source for image processing, with ten fertilized eggs and four unfertilized eggs used. In the second experiment, 150 eggs were used to establish the threshold, with the other 150 eggs used for verification. The environmental parameters of the incubation equipment were set in accordance with the literature, with temperature and humidity set to 100 °C and 70% [5,17], respectively.

In Experiment 1, a 50-W halogen lamp with an illuminance of 2300 lux was used as the light source. After passing through the egg, the light passed through an equilateral dispersive prism and was received by a CCD imaging sensor (ICX274, Sony, size type: 1/1.8, 1600 × 1200 pixels). The spectral wavelength of the received sample is shown in Figure 1. In [18], the authors used a spectrometer and a 475–810-nm mercury neon lamp to calibrate the wavelength. This setup was used to determine the freshness of brown-shelled and white-shelled eggs. After obtaining the light wavelength penetration response spectra of brown-shelled and white-shelled eggs, the spectral data were processed using standard normal variate (SNV). Finally, multiple regression analysis (MLR) was used to classify and judge the freshness of white and red eggs. Experiment 2 of this study aimed to develop the detection device. According to the results of the Experiment 1, a 5-W LED lamp with a wavelength of 590 nm was used as the light source. The images were taken with a color camera manufactured by ACTi Corporation (model no. E23) with a resolution of 1920 × 1080 (2 million pixels). This development device has 50 LEDs inside. When the whole plate of eggs (150 eggs) passes, the camera at the bottom takes pictures, analyzes the data in real time, and displays the identification results of the whole plate.

Figure 1. Spectrum experimental setup.

2.2. Experimental Apparatus and Measuring Methods

2.2.1. Experiment 1

In this experiment, ten fertilized eggs and four unfertilized eggs were hatched simultaneously, with the spectra of the eggs on the 1st–9th day of incubation measured to analyze the daily trend.

With wavelengths between 320 and 1100 nm, a halogen lamp was used in this experiment to seek the absorption spectra of the fertilized eggs. The captured images were analyzed using the software developed by Andor Technology Ltd. to understand the changing trend of the absorption spectra. This experiment mainly aimed to identify a low-watt lamp with a single wavelength most suitable for the eggs as the light source of the detection device. Low-power LEDs were used as high-power LEDs can cause heat dissipation and the light passes through the egg with too much power, seriously overexposing and affecting the surrounding eggs; this causes serious problems in sampling the entire screen. The industrial prototype of the device designed and fabricated in this study would require 50 LEDs.

2.2.2. Experiment 2

In this experiment, 150 unselected eggs were incubated simultaneously, with the eggs filmed daily from the inception of incubation. Following the hatching, the unfertilized, fertilized, and dead-in-shell eggs were classified, and daily ROC curves were established to determine the screening threshold value. Finally, the accuracy of the detection equipment was verified using the other 150 unselected eggs. Figure 2a shows a three-dimensional view of the detection device drawn using SolidWorks. The main function of Section 1 of the device is to capture images and detect the 150 eggs, as exhibited in Figure 2b. Following the detection, the eggs are transferred to Section 2 through the conveyor belt, where the colors are projected on eggs by the projection device, as shown in Figure 2c, with unqualified eggs removed manually.

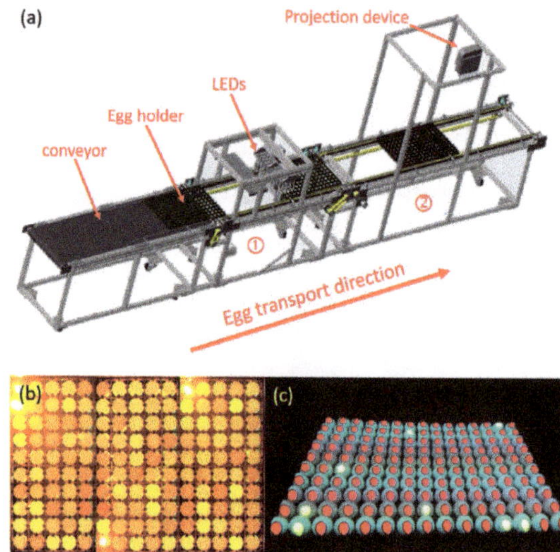

Figure 2. (**a**) Detection device (**1**) the image-capturing zone; and (**2**) the projection marking zone); (**b**) schematic of image capturing; and (**c**) schematic of projection marking.

2.2.3. Digital Image Processing

A digital image comprises array pixels, and image position pixels can be represented by a matrix. A pixel comprises three primary colors: Red (R), Green (G), and Blue (B). The colored image can be converted to a grayscale image by converting the RGB color space into the YIQ color counterpart, where Y indicates luminance, representing the brightness of the light; I indicates the channel of in-phase; and Q indicates the quadrature phase, representing the color details. The color image can be converted into the brightness of the three primary colors, individually, using the formula $Y(m,n) = 0.299 \cdot R(m,n) + 0.587 \cdot G(m,n) + 0.114 \cdot B(m,n)$, with the brightness ranging from 0 to

255, where 0 represents full darkness, 255 indicates full brightness, and the brightness value is the grayscale value [19]. In this study, images of the bottom of the eggs were taken, and each egg was positioned separately. The individual values could be obtained from the grayscale value from the red layer; these individual values were used as the basis for the selection of fertilized, unfertilized, and dead-in-shell eggs.

2.2.4. Statistical Analysis

The ROC curve (statistical analysis) is a receiver operating characteristic curve [20] or a relative operating characteristic curve. It is a coordinate graphical analysis tool used for selecting the best signal detection model as well as setting the optimal threshold value in the same model. Without being influenced by costs or effectiveness, the ROC analysis gives objective and neutral advice [21] to help users make a decision. In 1997, Hanson indicated that the AUC [22,23] could be used to describe the accuracy of the risk assessment scale. Therefore, in this study, the RGB-layered images of egg samples (numbered) of each day were used to calculate the average grayscale value of each egg's red layer. After the sample eggs were hatched, the unhatched eggs were sorted. The samples were divided into fertilized, unfertilized, and dead-in-shell eggs, and ROC classification was performed with the values of unfertilized and fertilized eggs to obtain the AUC value of unfertilized eggs each day. Next, the values of dead-in-shell and fertilized eggs were used. After ROC classification, the AUC value on each day for dead-in-shell eggs was obtained. Finally, the number of days for which AUC is maximum was determined and used to evaluate the accuracy of the classification results.

The ROC curve analyzes a binary classification model, that is, a model with only two types of output results. To evaluate the accuracy and sensitivity of the detection device, the detection device-based and actual classification results were divided into four categories using a confusion matrix, as shown in Table 1. The unfertilized eggs were taken as an example for explanation. The unfertilized eggs confirmed by the detection device were found to be unfertilized; in this setting, unfertilized eggs are considered as True Positive (TP). The unfertilized eggs detected to be not unfertilized proved to be not unfertilized; in this setting, unfertilized eggs are considered as True Negative (TN). The unfertilized eggs detected to be unfertilized proved to be not unfertilized; in this setting, unfertilized eggs are considered as False Positive (FP). The unfertilized eggs detected to be not unfertilized were found to be unfertilized; in this setting, unfertilized eggs are considered as False Negative (FN) [23].

Table 1. Prediction and allocation table and related formulas.

			Real State	
			True	**False**
Predict		**True**	*True Positive (TP)*	*False Positive (FP)*
		False	*False Negative (FN)*	*True Negative (TN)*
$Accuracy = \frac{TP+TN}{TP+TN+FP+FN}$	$Precision = \frac{TP}{TP+FP}$	$Sensitivity = \frac{TP}{TP+FN}$	$Specificity = \frac{TN}{TN+FP}$	$Negative\ Prediction = \frac{TN}{TN+FN}$

The optimal discrimination threshold is an important indicator used to evaluate the ROC curve. The ROC curve is drawn using "sensitivity" as the y-axis and "1 − specificity" as the x-axis [18]. The AUC has values ranging from 0 to 1. A larger AUC value indicates a higher accuracy. In real life, the random-guess AUC value for a dichotomy problem is not less than 0.5 [23]. In this study, ROC curves were used to determine the maximum AUC, i.e., the judgment threshold value.

Youden's index uses the comprehensive performance of sensitivity and specificity to determine the optimal discrimination threshold and calculates the value of "$Sensitivity + Specificity - 1$". The calculated values range from 0 to 1; the closer the value is to 1, the better is the overall performance of sensitivity and specificity, where $Sensitivity = TP/(TP + FN)$ is defined as the probability of a correct prediction in a group with true results and $Specificity = TN/(FP + TN)$ is defined as the probability of correct prediction in groups with false results [24,25].

3. Results and Discussion

3.1. Experiment 1

This experiment investigated the spectra of 14 eggs—ten eggs fertilized and the other four unfertilized. A halogen lamp was used to illuminate the incubated samples, and an image-capturing device was employed to record the data on Days 1–9. Then, Andor software was employed to analyze and obtain the data on spectral change on a daily basis, as shown in Figure 3. The average daily change trend of fertilized eggs is shown in Figure 3a, clearly indicating that, in the wavelength range below 680 nm, the values decrease as the number of incubation days increases, while Figure 3b exhibits the average daily change trend of unfertilized eggs, suggesting that the values show no significant change in the absorption spectra as the number of incubation days increases. Therefore, the values of unfertilized eggs, regarded as control samples, were compared with those of the fertilized eggs, as shown in Figure 3c. Dividing the grayscale values of the daily mean value of unfertilized eggs by the grayscale values of the daily mean value of fertilized eggs indicates that the peak of the ratio of absorption spectra of fertilized eggs is close to 580 nm. This band falls in the wavelength range of orange to yellow light; accordingly, a 590-nm-wavelength LED lamp was used in the detection device as it has a wavelength closer to 580 nm.

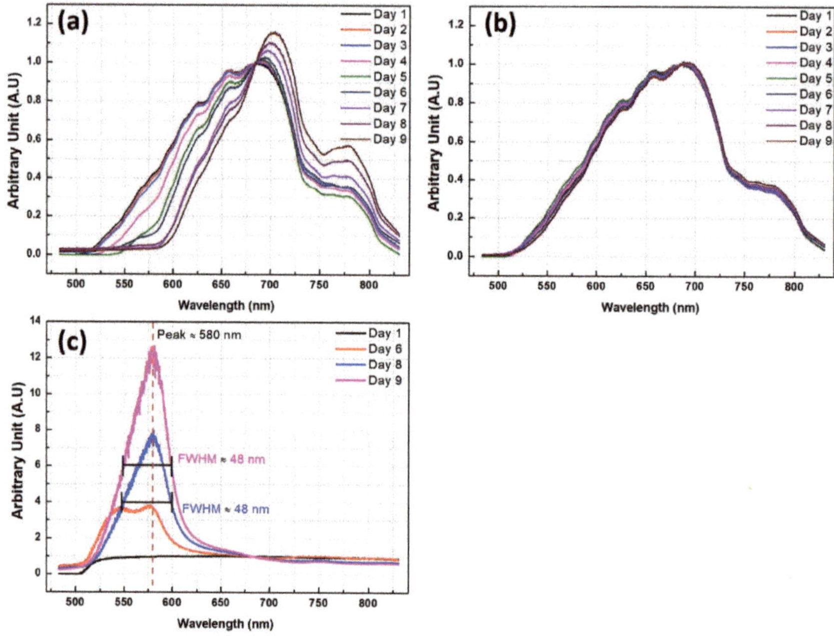

Figure 3. (**a**) Average values of fertilized egg spectral data on Days 1–9; (**b**) average values of unfertilized egg spectral data on Days 1–9; and (**c**) average values of unfertilized eggs divided by average values of fertilized eggs.

3.2. Experiment 2

In this experiment, the classification basis for the detection device was established using 150 eggs—18 unfertilized eggs, 12 early dead-in-shell eggs, and 120 successfully hatched eggs. The red layer in the images of the sample eggs on the 1st–15th days of incubation, captured using the detection device, were converted into grayscale counterparts through image processing. The change in grayscale values was recorded daily, with the average values shown in Figure 4, which indicates that the grayscale values decrease during the incubation process due to the continuous growth of embryos. Finally, the ROC curves were used to analyze the individual binary classification of unfertilized and dead-in-shell eggs from normally fertilized eggs, as shown in Table 2. The grayscale values of the unfertilized and dead-in-shell eggs can be clearly distinguished from those of the normally fertilized eggs from Day 7 onward. The AUC value of both reached 0.99 on Day 9, indicating that the method has an outstanding discrimination capability. In 2015, Kimura et al. used LED lamps with 585- and 635-nm wavelengths as the light sources to screen eggs on the 12th day of incubation. In their experiment, the detection accuracy rate was 92.9% under a single light source and 75% [3] when 36 light sources were used. In contrast, unfertilized eggs and dead-in-shell eggs could be identified with an overall accuracy rate of 100% on the ninth day of incubation, when using the light source developed in this study.

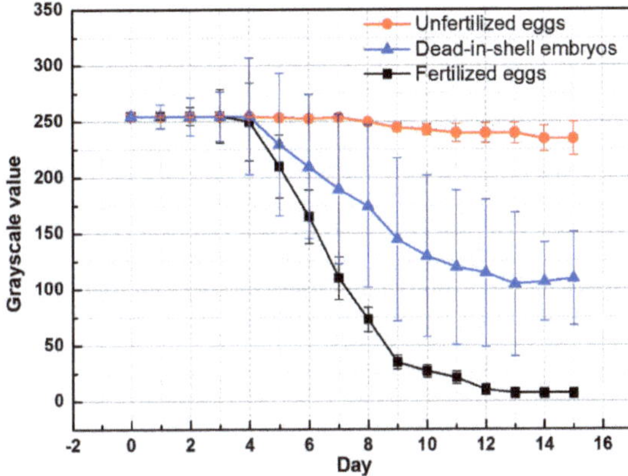

Figure 4. Change in grayscale values of fertilized eggs, unfertilized eggs, and dead-in-shell eggs.

Table 2. ROC analysis results.

	ROC	Day 7	Day 8	Day 9	Day 10
Unfertilized eggs	Grayscale values	190.5	188.0	182.5	181.0
	AUC	0.97	0.99	0.99	0.99
Dead-in-shell embryos	Grayscale values	122.0	107.5	74.0	65.0
	AUC	0.83	0.95	0.99	0.99

With the ROC analysis results imported into the detection device, the accuracy was verified using 150 eggs. The numbers of unfertilized and dead-in-shell eggs determined by the detection device were individually and statistically analyzed. In this experiment, 123 eggs were successfully hatched, of which 11 were unfertilized, 7 were early dead-in-shell, and the remaining 9 were confirmed by dissection to be late dead-in-shell (the chicks failed to move out of the shell). This study aimed to detect unfertilized and dead-in-shell eggs at an early stage. The late-stage dead-in-shell eggs fell in the category of normal eggs, as there were nearly mature chick embryos in the shells; that is, the embryos developed normally during the first half of the incubation period and the chicks failed to break shells just due to individual factors. According to the results in Figure 4, it is obvious that a threshold (grayscale value) can be used to classify unfertilized eggs from fertilized and dead-in-shell eggs (150 in total). Table 3 shows the results of the identification and classification of unfertilized eggs. After removing 11 unfertilized eggs, another threshold can be used to classify dead-in-shell eggs from fertilized eggs (139 in total). Table 4 shows the results of the identification and classification of dead-in-shell eggs.

Table 3. Using detection device to identify and classify unfertilized eggs on the 7th–10th day.

Day 7		Real State TRUE	FALSE	Grayscale Values: 190.5 Total	Accuracy (%)	AUC: 0.97 Precision (%)
Predict	TRUE	9	1	10	98.0	90.0
	FALSE	2	138	140	**Sensitivity (%)**	**Specificity (%)**
Total		11	139	150	81.8	99.3
Day 8		**Real State TRUE**	**FALSE**	**Grayscale Values: 188.0 Total**	**Accuracy (%)**	**AUC: 0.99 Precision (%)**
Predict	TRUE	10	0	10	99.3	100.0
	FALSE	1	139	140	**Sensitivity (%)**	**Specificity (%)**
Total		11	139	150	90.9	100.0
Day 9		**Real State TRUE**	**FALSE**	**Grayscale Values: 182.5 Total**	**Accuracy (%)**	**AUC: 0.99 Precision (%)**
Predict	TRUE	11	0	11	100.0	100.0
	FALSE	0	139	139	**Sensitivity (%)**	**Specificity (%)**
Total		11	139	150	100.0	100.0
Day 10		**Real State TRUE**	**FALSE**	**Grayscale Values: 181.0 Total**	**Accuracy (%)**	**AUC: 0.99 Precision (%)**
Predict	TRUE	11	0	11	100.0	100.0
	FALSE	0	139	139	**Sensitivity (%)**	**Specificity (%)**
Total		11	139	150	100.0	100.0

Table 4. Using detection device to identify and classify dead-in-shell eggs on the 7th–10th day.

Day 7		Real State TRUE	FALSE	Grayscale Values: 122.0 Total	Accuracy (%)	AUC: 0.83 Precision (%)
Predict	TRUE	4	8	12	92.1	33.3
	FALSE	3	124	127	**Sensitivity (%)**	**Specificity (%)**
Total		7	132	139	57.1	93.9
Day 8		**Real State TRUE**	**FALSE**	Grayscale Values: 107.5 Total	Accuracy (%)	AUC: 0.95 Precision (%)
Predict	TRUE	5	1	6	97.8	83.3
	FALSE	2	131	133	**Sensitivity (%)**	**Specificity (%)**
Total		7	132	139	71.4	99.2
Day 9		**Real State TRUE**	**FALSE**	**Grayscale Values: 74.0 Total**	**Accuracy (%)**	**AUC: 0.99 Precision (%)**
Predict	TRUE	7	0	7	100.0	100.0
	FALSE	0	132	132	**Sensitivity (%)**	**Specificity (%)**
Total		7	132	139	100.0	100.0
Day 10		**Real State TRUE**	**FALSE**	**Grayscale Values: 65.0 Total**	**Accuracy (%)**	**AUC: 0.99 Precision (%)**
Predict	TRUE	6	0	6	99.3	100.0
	FALSE	1	132	133	**Sensitivity (%)**	**Specificity (%)**
Total		7	132	139	85.7	100.0

In Figure 4, the grayscale values of the fertilized eggs began to drop on the fourth day, increasing the difference in the values between fertilized and unfertilized eggs. Table 3 also indicates that the accuracy of detecting unfertilized eggs on the seventh day was up to 98%, with a sensitivity of 82% and a Youden's index of 0.813, and that both the accuracy and sensitivity reached 100%, with Youden's index reaching 1 on the ninth day, suggesting that the judgment accuracy increases day-by-day. Table 4 exhibits the results of detecting dead-in-shell eggs. Note that the sensitivity reached the peak value on Day 9. At this moment, both accuracy and sensitivity reached 100%, with Youden's index reaching 1.

The embryos in dead-in-shell eggs stopped growing due to sudden death. However, the difference in grayscale values between dead-in-shell eggs and normally fertilized eggs on the 5th–7th days was not significant. Generally, the difference was not significant until Day 8, when the accuracy, sensitivity, and Youden's index reached 97.8%, 71.4%, and 0.693, respectively. The detection result was worse than that on the ninth day, which was mainly accounted for by the fact that each egg developed at its own speed, causing the numbers and formation rates of blood vessels and hemoglobin molecules to be different. On the 10th day, the sensitivity dropped to 85.7%, which was mainly due to the misjudgment caused by the erythrocyte protein deterioration resulting from dead embryos in the dead-in-shell eggs [12,13], which changed the absorption spectrum of the substance. Therefore, according to this study's results, the two grayscale values on the ninth day can be used as the threshold to effectively classify unfertilized and dead-in-shell eggs.

4. Conclusions

In this study, a spectral wavelength of 580 nm proved to be favorable for the detection of egg embryos, inspiring us to use a 5-W LED lamp with a wavelength of 590 nm as the detection light source, as the LED lamp has a wavelength closer to 580 nm. Using this device, we extracted data on 150 eggs, with ROC curves and AUC values used as a reference to obtain the daily optimal discrimination threshold values suitable for detecting unfertilized and dead-in-shell eggs. The detection device developed in this study was verified using another 150 eggs and was found to be capable of identifying the unfertilized eggs on the seventh day with an accuracy of 98% and a sensitivity of 82% at a screening threshold of 190.5, while dead-in-shell eggs could be identified on the ninth day with an accuracy of 100% and a sensitivity of 100% at a screening threshold of 74. At present, most vaccines are made from eggs, which are generally screened and delivered by the hatchery personnel to vaccine suppliers by the 10th day. Therefore, if 182.5 and 74 are used as the screening thresholds, unfertilized and dead-in-shell eggs, which are not suitable for vaccine culture, could be removed on the ninth day. With the developed screening thresholds introduced into the detection device, 150 eggs could be detected at a time—the eggs were detected at a speed of 30 s per tray, or five eggs per second, at an accuracy of 100%. The detected eggs were then marked with projected colors, which enabled the users to effectively screen eggs. This research provides a fast and accurate detection method. If commercial detection equipment is developed in the future, the target sample can be effectively detected using the research results' screening threshold.

5. Patents

We successfully obtained the Taiwanese invention patent "Method for judging the hatching shape of poultry eggs by using image processing", patent number TWI644616B.

Author Contributions: S.-Y.T., C.-C.J. and C.-W.C. conceived and designed the experiments; S.-Y.T. performed the experiments; S.-Y.T., C.-W.C. and C.-H.L. analyzed the data; S.-Y.T. and C.-H.L. wrote the paper; and S.-Y.T., C.-H.L. and C.-W.C. reviewed and edited the paper. All authors have read and agreed to the published version of the manuscript.

Funding: This research was funded by Ministry of Science and Technology, Taiwan, grant number 106-2622-B-005-007-CC1.

Acknowledgments: The authors thank the Ministry of Science and Technology of Taiwan for its financial support.

Conflicts of Interest: The authors declare no conflict of interest.

References

1. Rajão, D.S.; Pérez, D.R. Universal Vaccines and Vaccine Platforms to Protect against Influenza Viruses in Humans and Agriculture. *Front. Microbiol.* **2018**, *9*, 123. [CrossRef] [PubMed]
2. DeMarcus, L.; Shoubaki, L.; Federinko, S. Comparing influenza vaccine effectiveness between cell-derived and egg-derived vaccines, 2017–2018 influenza season. *Vaccine* **2019**, *37*, 4015–4021. [CrossRef]

3. Kimura, K.; Nakano, K.; Ohashi, S.; Takizawa, K.; Nakano, T. LED measurement for development of a non-destructive detector of unsuitable chicken eggs in influenza vaccine production. *Biosyst. Eng.* **2015**, *134*, 68–73. [CrossRef]
4. Liu, L.; Ngadi, M.O. Detecting Fertility and Early Embryo Development of Chicken Eggs Using Near-Infrared Hyperspectral Imaging. *Food Bioprocess. Technol.* **2013**, *6*, 2503–2513. [CrossRef]
5. Coucke, P.M.; Room, G.M.; Decuypere, E.M.; de Baerdemaeker, J.G. Monitoring embryo development in chicken eggs using acoustic resonance analysis. *Biotechnol. Prog.* **1997**, *13*, 474–478. [CrossRef] [PubMed]
6. Hashemzadeh, M.; Farajzadeh, N. A Machine Vision System for Detecting Fertile Eggs in the Incubation Industry. *Int. J. Comput. Intell. Syst.* **2016**, *9*, 850–862. [CrossRef]
7. Guan, X.; Liu, J.; Huang, K.; Kuang, J.; Liu, D. Evaluation of moisture content in processed apple chips using NIRS and wavelength selection techniques. *Infrared Phys. Technol.* **2019**, *98*, 305–310. [CrossRef]
8. Li, J.; Huang, W.; Zhao, C.; Zhang, B. A comparative study for the quantitative determination of soluble solids content, pH and firmness of pears by Vis/NIR spectroscopy. *J. Food Eng.* **2013**, *116*, 324–332. [CrossRef]
9. Ying, Y.; Liu, Y. Nondestructive measurement of internal quality in pear using genetic algorithms and FT-NIR spectroscopy. *J. Food Eng.* **2008**, *84*, 206–213. [CrossRef]
10. Giunchi, A.; Berardinelli, A.; Ragni, L.; Fabbri, A.; Silaghi, F.A. Non-destructive freshness assessment of shell eggs using FT-NIR spectroscopy. *J. Food Eng.* **2008**, *89*, 142–148. [CrossRef]
11. Abdel-Nour, N.; Ngadi, M.; Prasher, S.; Karimi, Y. Prediction of Egg Freshness and Albumen Quality Using Visible/Near Infrared Spectroscopy. *Food Bioprocess. Technol.* **2011**, *4*, 731–736. [CrossRef]
12. Bamelis, F.R.; Tona, K.; De Baerdemaeker, J.G.; Decuypere, E.M. Detection of early embryonic development in chicken eggs using visible light transmission. *Br. Poultry Sci.* **2002**, *43*, 204–212. [CrossRef] [PubMed]
13. Dong, J.; Dong, X.; Li, Y.; Peng, Y.; Chao, K.; Gao, C.; Tang, X. Identification of unfertilized duck eggs before hatching using visible/near infrared transmittance spectroscopy. *Comput. Electron. Agric.* **2019**, *157*, 471–478. [CrossRef]
14. Smith, D.P. Fertility and Embryo Development of Broiler Hatching Eggs Evaluated with a Hyperspectral Imaging and Predictive Modeling System. *Int. J. Poultry Sci.* **2008**, *7*, 1001.
15. Zhu, Z.H.; Liu, T.; Xie, D.J.; Wang, Q.H.; Ma, M.H. Nondestructive detection of infertile hatching eggs based on spectral and imaging information. *Int. J. Agric. Biol. Eng.* **2015**, *8*, 69–76.
16. Ghaderi, M.; Banakar, A.; Masoudi, A.A. Using dielectric properties and intelligent methods in separating of hatching eggs during incubation. *Measurement* **2018**, *114*, 191–194. [CrossRef]
17. Kemps, B.J.; De Ketelaere, B.; Bamelis, F.R.; Decuypere, E.M.; De Baerdemaeker, J.G. Vibration analysis on incubating eggs and its relation to embryonic development. *Biotechnol. Prog.* **2003**, *19*, 1022–1025. [CrossRef] [PubMed]
18. Cheng, C.-W.; Jung, S.-Y.; Lai, C.-C.; Tsai, S.-Y.; Jeng, C.-C. Transmission spectral analysis models for the assessment of white-shell eggs and brown-shell eggs freshness. *J. Supercomput.* **2020**, *76*, 1680–1694. [CrossRef]
19. Li, T.; Jean, J. *Digital Signal Processing*; Academic Press: Cambridge, MA, USA, 2013; p. 896. [CrossRef]
20. Husted, J.A.; Cook, R.J.; Farewell, V.T.; Gladman, D.D. Methods for assessing responsiveness: A critical review and recommendations. *J. Clin. Epidemiol.* **2000**, *53*, 459–468. [CrossRef]
21. Metz, C.E. Basic principles of ROC analysis. *Semin Nucl. Med.* **1978**, *8*, 283–298. [CrossRef]
22. Hanson, R.K. *The Development of a Brief Actuarial Risk Scale for Sexual Offense Recidivism*; Department of the Solicitor General of Canada: Ottawa, ON, Canada, 1997.
23. Fawcett, T. An introduction to ROC analysis. *Pattern Recognit. Lett.* **2006**, *27*, 861–874. [CrossRef]
24. Youden, W.J. Index for rating diagnostic tests. *Cancer* **1950**, *3*, 32–35. [CrossRef]
25. Schisterman, E.F.; Perkins, N.J.; Liu, A.; Bondell, H. Optimal cut-point and its corresponding Youden Index to discriminate individuals using pooled blood samples. *Epidemiology* **2005**, *16*, 73–81. [CrossRef] [PubMed]

Publisher's Note: MDPI stays neutral with regard to jurisdictional claims in published maps and institutional affiliations.

MDPI
St. Alban-Anlage 66
4052 Basel
Switzerland
Tel. +41 61 683 77 34
Fax +41 61 302 89 18
www.mdpi.com

Sensors Editorial Office
E-mail: sensors@mdpi.com
www.mdpi.com/journal/sensors

www.ingramcontent.com/pod-product-compliance
Lightning Source LLC
LaVergne TN
LVHW071305170726
843515LV00006BA/1492
* 9 7 8 3 0 3 6 5 2 6 7 9 9 *